THE NILE RIVER BASIN

Water, Agriculture, Governance and Livelihoods

The Nile is the world's longest river and sustains the livelihoods of millions of people across ten countries in Africa. It provides fresh water not only for domestic and industrial use, but also for irrigated agriculture, hydropower dams and the vast fisheries resource of the lakes of Central Africa. This book covers the whole Nile Basin and is based on the results of three major research projects supported by the Challenge Program on Water and Food (CPWF). It provides unique and up-to-date insights on agriculture, water resources, governance, poverty, productivity, upstream–downstream linkages, innovations, future plans and their implications.

Specifically, the book elaborates the history, and the major current and future challenges and opportunities, of the Nile River Basin. It analyses the basin characteristics using statistical data and modern tools such as remote sensing and geographic information systems. Population distribution, poverty and vulnerability linked to production systems and water access are assessed at the international basin scale, and the hydrology of the region is also analysed. The book provides in-depth scientific model adaptation results for hydrology, sediments, benefit sharing, and payment for environmental services based on detailed scientific and experimental work of the Blue Nile Basin. Production systems as they relate to crops, livestock, fisheries and wetlands are analysed for the whole Blue and White Nile Basin, including their constraints. Policy, institutional and technological interventions that increase productivity of agriculture and use of water are also assessed. Water demand modelling, scenario analysis and trade-offs that inform future plans and opportunities are included to provide a unique, comprehensive coverage of the subject.

Seleshi Bekele Awulachew was, at the time of writing, Acting Director in Africa for the International Water Management Institute (IWMI), Addis Ababa, Ethiopia. He is now Senior Water Resources and Climate Specialist at the African Climate Policy Center (ACPC), United Nations Economic Commission for Africa (UNECA), Addis Ababa, Ethiopia.

Vladimir Smakhtin is Theme Leader – Water Availability and Access at IWMI, Colombo, Sri Lanka.

David Molden was, at the time of writing, Deputy Director General – Research at IWMI, Colombo, Sri Lanka. He is now Director General of the International Centre for Integrated Mountain Development (ICIMOD), Kathmandu, Nepal.

Don Peden is a Consultant at the International Livestock Research Institute (ILRI), Addis Ababa, Ethiopia.

THE NILE RIVER BASIN

Water, Agriculture, Governance and Livelihoods

Edited by Seleshi Bekele Awulachew, Vladimir Smakhtin, David Molden and Don Peden

First published in paperback 2024

First edition published 2012
by Routledge
4 Park Square, Milton Park, Abingdon, Oxon OX14 4RN

and by Routledge
605 Third Avenue, New York, NY 10158

Routledge is an imprint of the Taylor & Francis Group, an informa business

Publisher's Note
The publisher has gone to great lengths to ensure the quality of this reprint but points out that some imperfections in the original copies may be apparent.

British Library Cataloguing in Publication Data
A catalogue record for this book is available from the British Library

Library of Congress Cataloging-in-Publication Data
The Nile River basin : water, agriculture, governance and livelihoods / edited by Seleshi Bekele Awulachew ... [et al.].
p. cm.
Includes bibliographical references and index.
1. Watershed management–Nile River Watershed. 2. Water resources development–Nile River Watershed. 3. Water-supply–Nile River Watershed–Management. 4. Nile River Watershed–Economic conditions. 5. Agriculture–Nile River Watershed. 6. Nile River Watershed–Environmental conditions. 7. Nile River Watershed–History–20th century. I. Awulachew, Seleshi Bekele.
TC519.N6N56 2012
333.91620962–dc23
2012006183

ISBN: 978–1–84971–283–5 (hbk)
ISBN: 978–1–03–292150–1 (pbk)
ISBN: 978–0–203–12849–7 (ebk)

DOI: 10.4324/9780203128497

Typeset in Bembo
by FiSH Books, Enfield

CONTENTS

LIST OF FIGURES AND TABLES

Figures

Tables

ACKNOWLEDGEMENTS

This book is based primarily on results of several projects supported by the CGIAR Challenge Program on Water and Food (CPWF) and implemented by the International Water Management Institute (IWMI), International Livestock Research Institute (ILRI) – the CGIAR research centres – together with various partners during the period 2004–2010. We greatly acknowledge the support provided by the CPWF.

We also greatly acknowledge the support of various institutions that partnered in the projects. We particularly thank the Nile Basin Initiative (NBI), the NBI Subsidiary Action Program of Eastern Nile Technical Regional Organization (ENTRO), the World Fish Center, Cornell University (USA), Addis Ababa University, Omdurman Islamic University UNESCO-Chair on Water Resources (Sudan), Agricultural Research Corporation (Sudan), Makarrare University (Uganda), Bahir Dar University (Ethiopia), the Ethiopian Institute of Agricultural Research and Ethiopian Electricity Power Corporation.

The authors acknowledge the help and insights received from the NBI shared vision programme and its subsidiary action project management. Many national systems such as Egypt's Ministry of Water Resources and Irrigation, Nile Water Sector (Egypt), National Water Research Center (Egypt), South Sudan's Ministry of Water Resources, Makarere University (Uganda), Ministry of Water Resources (Uganda), Ministry of Water Resources – Department of Hydrology (Ethiopia), National Meteorological Service Agency (Ethiopia), Amhara Regional Agricultural Research Institute (ARARI), FAO Nile Project (Uganda), and a number of individuals participated in the various conferences and meetings during the deliberations of the research results, and many secretaries, drivers and farmers helped us plan and implement our field trips and programmed meetings.

Valuable data and insights were provided by Wim Bastiaanssen (WaterWatch, Netherlands) and Mac Kirby and Mohammed Mainuddin (both of CSIRO, Australia). Karen Conniff, Pavithra Amunugama and Upamali Surangika (IWMI, Colombo) helped coordinate finalization and submission of the Book. Sumith Fernando (IWMI, Colombo) took up several last-minute requests for graphics. And Kingsley Kurukulasuriya edited the entire book. We sincerely acknowledge all these valuable contributions.

Seleshi B. Awulachew, Vladimir Smakhtin, David Molden, Don Peden

CONTRIBUTORS

Enyew Adgo is assistant professor at Bahir Dar University, Bahir Dar, Ethiopia.

Abdalla A. Ahmed is professor and director of the UNESCO Chair in Water Resources (UNESCO-CWR), Khartoum, Sudan.

Tadesse Alemayehu is an independent consultant based in Addis Ababa, Ethiopia.

Tilahun Amede is a systems agronomist at the International Livestock Research Institute (ILRI), Addis Ababa, Ethiopia and International Water Management Institute (IWMI), Addis Ababa, Ethiopia (joint appointment).

Seleshi Bekele Awulachew was, at the time of writing, acting director in Africa for the International Water Management Institute (IWIMI), Addis Ababa, Ethiopia. He is now senior water resources and climate specialist at the African Climate Policy Center (ACPC), United Nations Economic Commission for Africa (UNECA), Addis Ababa, Ethiopia.

Tenalem Ayenew is professor of hydrogeology at Addis Ababa University, Addis Ababa, Ethiopia.

Kamaleddin E. Bashar is associate professor and a hydrologist and water resources specialist for UNESCO Chair in Water Resources (UNESCO-CWR), Khartoum, Sudan.

Ana Elisa Cascão is programme manager of capacity building at Stockholm International Water Institute (SIWI), Stockholm, Sweden

Karen Conniff was at the time of writing, an independent consultant working with the International Water Management Institute (IWMI), Colombo, Sri Lanka. She is now a consultant in Kathmandu, Nepal.

Solomon S. Demisse is a water resources systems specialist at the International Water Management Institiute (IWMI), Addis Ababa, Ethiopia.

Zachary M. Easton is an assistant professor at the Department of Biological Systems Engineering, Virginia Polytechnic Institute and State University, Blacksburg, USA.

Teklu Erkossa is an irrigation and agricultural engineer at the International Water Management Institute (IWMI), Addis Ababa, Ethiopia.

Hamid Faki works at the Agricultural Research Corporation, Sudan.

Solomon Gebreselassie is a research officer at the International Potato Center (CIP), Addis Ababa, Ethiopia.

Saliha Alemayehu Habte works at Dresden University of Technology, Dresden, Germany.

Fitsum Hagos is a researcher at the International Water Management Institute (IWMI), Addis Ababa, Ethiopia.

Amare Haileslassie is a post-doctoral scientist at the International Livestock Research Institute (ILRI), Hyderabad, India.

Mario Herrero is team leader at the International Livestock Research Institute (ILRI), Nairobi, Kenya.

Mohamed Elhassan Ibrahim is a consultant hydrogeologist based in Sudan.

Robyn Johnston is senior researcher and water resources planner at the International Water Management Institute (IWMI), Colombo, Sri Lanka.

Poolad Karimi is a research officer at the International Water Management Institute (IWMI), Colombo, Sri Lanka.

James Kinyangi is CCAFS regional programme leader at the International Livestock Research Institute (ILRI), Nairobi, Kenya.

Charlotte MacAlister is a hydrologist for the International Water Management Institute (IWMI), Addis Ababa, Ethiopia.

Everisto Mapedza is a researcher and social and institutional scientist at the International Water Management Institute (IWMI), Pretoria, South Africa.

Matthew P. McCartney is a principal hydrologist at the International Water Management Institute (IWMI), Addis Ababa, Ethiopia.

Mohamed Abdel Meguid is a researcher at the Channel Maintenance Research Institute, Kalyubia, Egypt.

David Molden was, at the time of writing, deputy director general of the IWMI, Colombo, Sri Lanka. He is now director general of the International Centre for Integrated Mountain Development (ICIMOD), Kathmandu, Nepal.

Denis Mpairwe is a senior lecturer at Makerere University in Kamapala, Uganda.

Aditi Mukherji is a senior researcher with the International Water Management Institute (IWMI) in New Delhi, India.

An Notenbaert is a spatial analyst at the International Livestock Research Institute (ILRI), Nairobi, Kenya.

Tom Ouna is a planner at the International Livestock Research Institute (ILRI), Nairobi, Kenya.

Paul Pavelic is a senior researcher in Geohydrology for International Water Management Institute (IWMI), Hyderabad, India.

Don Peden is a consultant at the International Livestock Research Institute (ILRI), Addis Ababa, Ethiopia.

Lisa-Maria Rebelo is a researcher in remote sensing and GIS at the International Water Management Institute (IWMI), Addis Ababa, Ethiopia.

Yihenew G. Selassie is an associate professor at the Department of Civil Engineering, Addis Ababa University, Ethiopia.

Yilma Seleshi is head of the Department of Civil Engineering at Addis Ababa University, Addis Ababa, Ethiopia.

Vladimir Smakhtin is the theme leader – water availability and access at the International Water Management Institute (IWMI), Colombo, Sri Lanka.

Tammo S. Steenhuis is a professor at the Department of Biological and Environmental Engineering, Cornell University, Ithaca, USA.

Tesfaye Tafesse is a researcher at the Council for the Development of Social Science Research in Africa (CODESRIA), Dakar, Senegal.

Seifu A. Tilahun is a research assistant at the Department of Biological and Environmental Engineering, Cornell University, Ithaca, USA.

Callist Tindimugaya is commissioner for water resources regulation at the Ministry of Water and Environment, Uganda.

Aster Tsige is human resources coordinator for the International Livestock Research Institute (ILRI), Addis Ababa, Ethiopia.

Paulo van Breugel is an agricultural researcher at the International Livestock Research Institute (ILRI), Addis Ababa, Ethiopia.

Aster D. Yilma was at the time of writing, an expert in GIS, IT and databases for International Water Management Institute (IWIMI), Addis Ababa, Ethiopia. She is now geographic information systems officer, ICT, Science and Technology for Development (ISTD), United Nations Economic Commission for Africa (UNECA) Addis Ababa, Ethiopia.

Birhanu Zemadim is a post-doctoral fellow at Hydrology International Water Management Institute (IWMI), Addis Ababa, Ethiopia.

ABBREVIATIONS

Aa	arid
AARI	Amhara Agricultural Research Institute
AGNPS	Agricultural Non-Point Source Pollution
AHD	Aswan High Dam
AMC	antecedent moisture condition
AR^4	artesian conditions
AWC	available water content
AWM	Agricultural Water Management
BNB	Blue Nile Basin
BoARD	Bureau of Agriculture and Rural Development
BoWRD	Bureau of Water Resources Development
reseedC-I	confidence interval
CFA	Cooperative Framework Agreement
CFW	cash for work
CIDA	Canadian International Development Agency
CN	curve number
CPWF	Challenge Program on Water and Food
CRA	cooperative regional assessment
CTI	compound topographic index
CV	coefficient of variation
CWP	crop water productivity
CWR	UNESCO Chair in Water Resources (UNESCO-CWR)
DEM	digital elevation model
DRC	Democratic Republic of Congo
Ds	dense-soil
Ds	dry-subhumid
EGS	Ethiopian Geological Survey
EGY	Egypt
EIA	environmental impact assessment
EIAR	Ethiopian Institute of Agricultural Research
ELR	Equatorial Lakes Region

ENGDA	Ethiopian National Groundwater Database
ENSAP	Eastern Nile Subsidiary Action Program
ENSAPT	Eastern Nile Subsidiary Action Program Technical
EnSe	environmentally sensitive
ENTRO	Eastern Nile Technical Regional Office
EPA	Ethiopian Environmental Protection Authority
EPE	Environmental Policy of Ethiopia
EPLAUA	Environmental Protection Land Administration and Land Use Authority
ET/Eta	evapotranspiration
ETB	Ethiopian birr
ETH	Ethiopia
FAO	Food and Agriculture Organization of the United Nations
FCC	false colour composite
FFL	institutionalized flow and linkage
FFW	food for work
FM	fencing plus manure
FMRi	fencing plus manure incorporated into the soil plus reseeding
FMRs	fencing plus manure left on soil surface plus reseeding
FO	fencing exclosures only
FR	fencing plus reseeding
GDP	gross domestic product
GEF	Global Environmental Facility
GIS	geographic information system
GOSS	Government of South Sudan
GPS	geographic positioning system
GRACE	Gravity Recovery and Climate Experiment
GVP	gross value of production
GW-MATE	Groundwater Management Advisory Team
Ha	hyper-arid
HCENR	Higher Council for Environment and Natural Resources
Hh	humid
HI	Poverty Headcount Index
HRUs	hydrologic response units
HYP	related to hyper-arid climatic regions
IC	irrigation cooperatives
ICCON	International Consortium for Cooperation on the Nile
IFL	indirect flow and linkage
ISP	Institutional Strengthening Project
ITCZ	Inter-tropical Convergence Zone
IWRM	integrated water resources management
IWSM	Integrated Watershed Management Policy
JMP	joint multi-purpose
KNN	Kohonen neural network
Ls	light-soil
LG	livestock-dominated grazing areas
LGA	arid and semi-arid grazing areas
LGH	humid grazing lands
LGP	length of growing period

LSI	large-scale irrigation
LULA	Land Use and Land Administration Policy
LWN	Lower White Nile
LWP	livestock water productivity
MAP	mean annual precipitation
MAR	managed aquifer recharge
masl	metres above sea level
Mha	million hectares
MI	rain-fed mixed crop-livestock systems
MIWR	Ministry of Irrigation and Water Resources
MoA	Ministry of Agriculture
MoAF	Ministry of Agriculture and Forests
MoARD	Ministry of Agriculture and Rural Development
MoARF	Ministry of Animal Resources and Fisheries
MoWR	Ministry of Water Resources
MR	irrigated mixed crop-livestock farming
MRA	MR related to arid and semi-arid climatic regions
MRH	MR related to humid climatic regions
MRT	MR related to temperate climatic regions
Ms	medium-soil
MUSLE	Modified Universal Soil Loss Equation
MW	megawatts
MWLE	Ministry of Water, Lands and Environment
MWRI	Ministry of Water Resources and Irrigation
MWTP	mean willingness to pay
mya	million years ago
NBC	Nile Basin Commission
NBI	Nile Basin Initiative
NBI	Nile River Basin
NBTF	Nile Basin Trust Fund
NDVI	Normalized Differenced Vegetation Index
NELSA	Nile Equatorial Lakes Subsidiary Action Program
NELSAP	Nile Equatorial Lakes Subsidiary Action Program
NELSAP-CU	Nile Equatorial Lakes Subsidiary Action Program – Coordination Unit
NELTAC	Nile Equatorial Lakes Technical Advisory Committee
NFL	no flow and linkage
NFMP	National Fluorosis Mitigation Project
NGIS	National Groundwater Information System
Nile-COM	Nile Council of Ministers
Nile-SEC	Nile Secretariat
Nile-TAC	Nile Technical Advisory Committee
NRB	Nile River Basin
NRBAP	Nile River Basin Action Plan
NRCS	Natural Resource Conservation Service
NSAS	Nubian Sandstone Aquifer System
NSE	Nash-Sutcliffe Efficiency
O&M	operation and maintenance
P	precipitation

P-E	precipitation–evapotranspiration
PCA	principal components analysis
PD	person-days
PES	payment for environmental services
PEST	parameter estimation
PET	potential evapotranspiration
PoE	Panel of Experts
PPA	participatory poverty assessments
PPP	purchasing power parity
PSNP	Productive Safety Net Program
RBO	River Basin Organization
RUE	rainwater use efficiency
S	theoretical storage capacity
S^3	storativity
Sa	semi-arid
SAP	Subsidiary Action Program
SBD	soil bulk density
SCE	shuffled complex evolution
SCRP	Soil Conservation Reserve Program
SGVP	standardized gross value of production
SOM	Self-Organizing Map
SPAM	spatial allocation model
SSI	small-scale irrigation
SUD	Sudan
SVP	Shared Vision Program
SWAT	Soil and Water Assessment Tool
SWAT-WB	SWAT–Water Balance
SWC	soil water content
T	transpiration
T^2	transmissivity
Ta	actual transpiration
TDS	total dissolved solids
TI	topographic index
TLU	tropical livestock unit
Tp	potential transpiration
TVETS	technical and vocational education and trainings
UGA	Uganda
UNDP	United Nations Development Programme
UNESCO	United Nations Educational, Scientific and Cultural Organization
USBR	United States Bureau of Reclamation
USLE	universal soil loss equation
USLE_K	soil erodibility factor of USLE
VSA	variable source areas
WaSiM	Water balance Simulation Model
WEAP	Water Evaluation And Planning model
WEPP	Water Erosion Prediction Project
Wh	wet-humid
WP	water productivity

WRMP	Water Resources Management Policy/Regulation/Guideline
WSG	Watershed Management Guideline
WTP	willingness to pay
WUA	water user association

1
Introduction

Seleshi B. Awulachew, Vladimir Smakhtin, David Molden and Don Peden

The Nile Basin covers about 10 per cent of the African land mass and hosts nearly 20 per cent of the African population, mainly dependent on crop and livestock-keeping agriculture for their livelihoods. It experiences widespread and varying degrees of poverty, food shortages, land degradation and water scarcity.

Access to water underpins human prosperity in the Nile riparian countries, which prioritize water development for agriculture, domestic consumption, power and industry. Competition for water among people and nations creates a climate of conflict that undermines human prosperity and ecosystem functions. People of the Nile require new approaches to water development and use that can sustainably reduce poverty and improve food security and human well-being in the basin. Agriculture plays an important role in the economies of all Nile Basin countries. Yet the role and potential of water for agriculture are not well understood throughout the basin, and in some parts of it massive investments in agricultural water development have not achieved the desired levels of food security and poverty reduction. This book aims to suggest promising options for future water management in the Nile Basin to help guide policy-makers, investors, and further research.

To begin with, we briefly reviewed the long, complex and eventful history of the Nile. Understanding the historical trajectory of the basin is a point of departure for developing water management solutions. The purpose of the historical review was to highlight how the Nile has been used for agriculture (crops, livestock and fish) and for economic benefits of the millions of people who live along the river. The Nile has intrigued poets and historians from the time of the Pharaohs. However, planning and development of the Nile waters were revolutionized in the twentieth century, commencing from the colonial era. In the modern period, the Nile water use increased and agriculture expanded – with environmental and human consequences and hydro-political disputes between the riparian countries.

As a further background analysis and documentation, we developed various maps of the basin displaying its current characteristics related to poverty, production systems and related information. To establish links between poverty, on the one hand, and rural agricultural production systems and water access, on the other, we used food security, poverty level and poverty inequality indicators. The poverty maps in different parts of the basin show distinct characteristics and a strong correlation to the agricultural systems and managed water access. Poverty level within the Nile Basin ranges from 17 per cent in Egypt to over 50 per cent in five of the

Nile Basin countries. The mapping also shows poverty hot spots and highly vulnerable production systems in the basin.

We further attempted to map hydronomic (water management) zones. Such zoning is instrumental in identifying and prioritizing the water management issues and opportunities in different parts of a river basin. Classifying the river basin into water management zones facilitates the development of management strategies and informed decision-making during planning and operation. Our mapping helped identify seven major zones, and eighteen detailed zones. The major ones include irrigated, mixed rain-fed, environmentally sensitive, desert, arid, semi-arid and humid zones. The detailed zones are derived from the main 'water-based' ones by adding biophysical factors that include soils, topography and climate. These sets of maps are a new addition to the Nile information and knowledge. One major finding is that the water source zone covers only 15 per cent of the area that generates most of the Nile flow.

To add value, we took a new approach when considering water resources and their management in the Nile Basin. Most previous studies considered the thin strip of the Nile River that traverses 6000 km across the riparian countries. First, we considered rain as the ultimate water resource, and then we placed high importance on evapotranspiration (ET) from landscapes as an indicator of the main water use. Second, we differentiated water access (the ease of obtaining water) from water availability (the water found in nature). Most studies focus primarily on the river water itself, without recognizing that it is access and not availability that makes the difference to people. Third, we considered a range of agricultural water management practices from soil water conservation to large-scale irrigation. Within this range we considered agriculture, including fish, livestock and crops, along with other ecosystem services that provide livelihoods. Finally, we recognized that policies and institutions are the ultimate driving force between access and productivity, and that policies and actions outside of the river, such as trade or livestock management practices, influence the river itself.

The central hypothesis of the research is that poverty is related to water access for agriculture. A second point is that poverty is related to the productivity of Nile waters, whether rain or river water is the source. And, third, we contend that poverty is related to the capability of people to cope with risks inherent in water management for agriculture such as drought. Our research provided evidence that these factors are strongly at play within the Nile.

How much water is used in the Nile, and where does the water go in broad hydrological water balance terms? A water accounting exercise used land cover, rainfall analysis and a satellite-derived map of evaporation to understand the water balance and water use patterns. It was found that the total rainfall in the basin in 2007 averages 2000 $km^3\ yr^{-1}$. The most commonly used number for water availability is based on the river into Lake Nasser, Egypt, which is about 84.5 $km^3\ yr^{-1}$. Irrigation is significant for Egypt and Sudan, and much less so for other countries, and accounts for 50–60 km^3 of water use (<3% of total rainfall). In contrast, the ET from rain-fed crops is around 200 $km^3\ yr^{-1}$. Most of the remaining rainfall is depleted as ET from other landscapes that include pastoral lands. A contentious and unclear number in the water accounts is the amount that flows to the sea, where estimates range from 2 to 30 $km^3\ yr^{-1}$.

Water productivity analysis was done for crops, livestock and aquaculture within the river. We took advantage of the ET, production system and crop yield maps to produce a comprehensive crop water productivity map within the basin. In all cases, except for Egypt, water productivity and productivity values are low. The range for crop water productivity was between US$0.01–0.20, showing a major scope for improvement in most areas. Yields are on the order of 1 ton per ha (t ha^{-1}) for grain crops outside of Egypt. In the case for low yields, improving yields is a major means for improving water productivity. A little more water supplied for crop ET, combined with fertilizers, seeds and good management, will result in increased water productivity. This is not the case in Egypt, where production can increase, but without an additional water increase.

We have also examined in detail the hydrological processes in selected parts of the basin. We used models such as the revised Soil and Water Assessment Tool (SWAT) to simulate water balance components in the Ethiopian Highlands, taking into account the specifics of the region such as steep topography and degraded watersheds. We analysed the Nile Basin sediment loss and degradation, using Blue Nile as an example (which is also the source of main sediment load in the entire basin).

The question of how much more large-scale irrigation is possible in the Nile has been examined using the Water Evaluation And Planning model for the entire basin and plans of governments for irrigation and hydropower development. While there is little existing irrigation upstream in the Nile, there are large ambitions for more irrigation. Our findings showed that more large-scale irrigation is possible, but not close to the extent planned. It also showed that coordinated planning is absolutely necessary to expand irrigated land and manage the entire river. Part of this planning is clear data-sharing, as a major uncertainty in our present analysis was the existing flow pattern. In spite of the limits on the scope for irrigation expansion, there is definitely scope to improve water productivity on irrigated lands. Analysis in the Gezira scheme suggested that overall production was far below desired levels, ET was much less than it could be, and all this was influenced by changes in policies that changed water management practices and productivity. However, increases in production in the Gezira are likely to reduce downstream flows and overall water availability in the basin.

Given that rain-fed and pastoral systems serve most areas and host more poor people, and that there are limits on the scope for large-scale irrigation, the largest investment opportunity is to focus on rain-fed areas. Here water management practices such as small-scale irrigation have high potential. In particular, adoption of land and water management practices in rain-fed areas that convert more evaporation to transpiration can greatly increase production of both food and natural vegetation without placing additional demand on river waters. Livestock are particularly important in these areas. Improving water productivity for livestock will require good water productivity of feed sources, practices to enhance feed conversion, better marketing opportunities, better vegetation and soil cover, as well as strategic placement of watering sites. Good water for livestock practices will take pressure off the mainstream river. A good example was recorded in Nakasangola, Uganda, where improved pasture plus water harvesting meant that cattle and people did not have to migrate to the Nile's Lake Kyoga, where overcrowding and disease are rampant. Because animal products such as meat and milk command high market prices, economic water productivity tends to be slightly higher for livestock than for crops.

There is a large scope to improve fisheries. Lake Victoria and Lake Nasser are significant sources of fish, but Lake Victoria's fisheries are threatened by land management practices surrounding the lake and water management practices associated with hydropower releases. The Sudd and other wetlands have huge untapped potential. Over 90 per cent of aquaculture is done in Egypt, and there certainly are opportunities elsewhere. Markets are at present a key constraint to improved fish production.

There is ample water in the Nile wetlands. While there are plans to drain parts of these wetlands, the present situation about how people use the wetlands and their future potential is poorly understood. There are 14 Ramsar wetland sites across the Nile, all of which not only support fisheries, livestock and other forms of agriculture, but are threatened by poor agricultural practices. Looking to the future, wetlands management could either lead to prosperity, or be a flashpoint for conflict. Our special studies in the Sudd confirmed that there is potential for more agriculture within these areas, but it also confirmed the need for a much better understanding in order to do agriculture sustainably.

Ultimately, water governance will facilitate sustainable and productive development of Nile waters. The Nile Basin Initiative (NBI) was formed with the realization of the need for cooperation among the Nile countries. The NBI has made significant progress in this regard. An important finding of this study was that too little attention was, and is, given to water and agriculture within the NBI, especially in rain-fed areas. There needs to be better consideration for fisheries and livestock practices. There are numerous other institutions involved in water and agriculture. Overall, there is a dire need for improved human and institutional capacity to implement programmes for the benefit of the rural poor.

In summary, key messages of the book are:

- Agriculture is the mainstay of the economy of the countries and source of livelihood of the majority of the basin people. It is crucial to provide sufficient attentions and investment in agriculture to reduce poverty.
- Agricultural Water Management (AWM) is crucial for economic growth, food security and poverty reduction. AWM needs to be better integrated in the NBI programmes.
- Rainwater is Nile Water, so start from rain in the analysis. Water availability for food production can be enhanced through conversion of some 'non-beneficial' water to managed land and water use.
- There is some (but limited) scope for large-scale irrigation expansion. There is ample scope to improve productivity in irrigation systems south of Lake Nasser. Further addition in large-scale irrigation needs to come through improved cooperation and integrated management of the water resources.
- Water access, rainwater management, livestock, productivity gains, fisheries and small-scale irrigation are important, and need more attention.
- Consider rainwater options by looking beyond the river to improve productivity and significant gains in livelihood. Productivity potential within the landscape is high and can be greatly improved.
- All-inclusive sustainable cooperation, such as a comprehensive agreement and the Nile Commission, can contribute to the agriculture, socio-economic development and regional integration in the Nile Basin.
- The Nile Basin is wide and complex, and it varies in poverty, productivity, vulnerability, water access and socio-economic conditions. It is essential to make further in-depth research and local analysis for further understanding of issues and systems, and to design appropriate measures.
- Further research should also target analysis related to rainwater management interventions, impacts, upstream–downstream relationships, trade-off analysis, economic modelling and new innovations.
- It is necessary to improve human and institutional capacity to make this happen, from community to national to regional scale.

2

Nile water and agriculture

Past, present and future

Karen Conniff, David Molden, Don Peden and Seleshi B. Awulachew

Key messages

- Agriculture has been a dominant feature of Nile Basin countries for centuries. Irrigated agricultural expansion over the last hundred years, often driven by foreign powers, has caused significant change in the use of the Nile water, and continues to be a major influence on the decisions around the Nile River use today.
- Use of Nile River water is a cause for transboundary cooperation and conflict. More than ever, the Nile Basin countries feel the pressure of expanding population requirements for food production and energy to develop their economies. However, historical treaties and practices continue to significantly shape directions of future Nile water use.
- Power development is changing the Nile River. Many dams are planned and several are under construction. The dam projects will have direct consequences for local populations and governments as they negotiate for water resources, land and power.

Introduction

This chapter highlights the use of the Nile River in the past and the present, and its future possibilities for both agriculture (crops, livestock and fish) and the economic benefit of the millions of people who live along the Nile. This brief glance at the geographical, historical and current developments of Nile water includes the socio-political, environmental and human consequences of these developments, and the direction towards which future changes in the Nile Basin might lead. Ultimately, the benefits of the Nile River need to be shared among the ten basin countries, with populations totalling approximately 180 million, of whom half are below the poverty line (Bastiaanssen and Perry, 2009).

Geographical Nile

A short introduction to the physical Nile will help to visualize the situation and understand the dynamics of historical and current power struggles. Figure 2.1 is used by the Nile Basin Initiative (NBI) and the Nile Equatorial Lakes Subsidiary Action Programme (NELSAP), and shows the areas and countries drained by the Nile River.

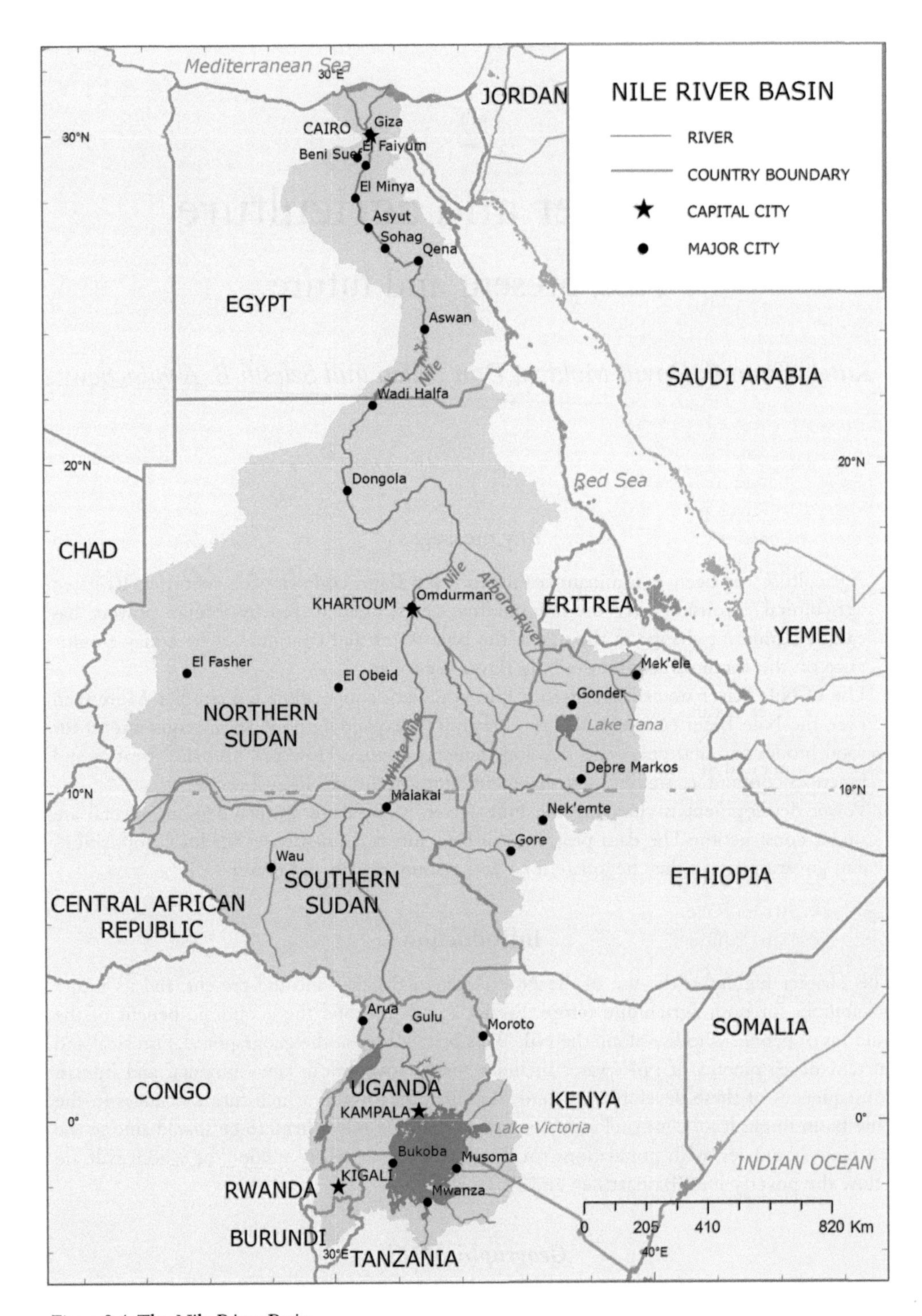

Figure 2.1 The Nile River Basin

Source: World Bank, 1998

The Nile River we know is quite different from the deep Eonile formed during the late Miocene period, 25 to 5.3 million years ago (mya), when the Mediterranean Sea dried up (Warren, 2006). The Cenozoic period of the Blue Nile was one of upheavals, plate movements and volcanic eruptions that occurred more than 30 mya, and this is what defines the hydrological differences between the Blue and White Niles (Talbot and Williams, 2009). The meeting of the Blue and White Niles is explained by two theories. Said (1981) believes that Egypt supplied most of the water to the early Nile, and the Nile we know now was formed within one of several basins more than 120,000 years ago – fairly recent in geological years. The other theory is of a Tertiary period river when the Ethiopian rivers flowed to the Mediterranean via the Egyptian Nile (Williams and Williams, 1980). Sedimentation studies and the discovery of an intercontinental rift system by Salama (1997) supports the Tertiary period Nile that formed a series of closed basins that connected during wet periods 120,000 years ago; the filling of the basins connected the Egyptian, Sudanese and Ethiopian Nile basins. The oldest part of the Nile drainage is associated with the Sudd, believed to have formed 65 mya. What we see of the Nile now is also in a state of change as the landscape is excavated to construct large dams to re-divert water and change the river's physiography.

The Nile River passes through several distinct climatic zones, is fed from different river sources, and creates vast wetlands, high surface evaporation and a huge amount of energy that is tapped for hydropower. Seventeen river basins feed into Lake Victoria, where John Speke identified the source of the Nile in 1862, with the greatest contribution from the Kagera River (Howell *et al.*, 1988).

The Nile is a river with many names. Exiting Lake Victoria, it is the Victoria Nile or White Nile; then, as it flows through Lakes Kyoga and Albert, it is called the Albert Nile; arriving in Sudan, the river is called Bahr el Jebel, or Mountain Nile; where it winds through the Sudd and flows into Lake No it is called the Bahr al Abyad, or White Nile, because of the white clay particles suspended in the water. Near Malakal the Ethiopian Sobat River joins the White Nile. Originating in Ethiopia, the Blue Nile born from volcanic divisions in the landscape that brought up the Ethiopian plateau has carved impressive gorges and brings silt-laden Blue Nile waters coiling around and collecting from many tributaries before flowing into the White Nile confluence at Omdurman Sudan, where the name becomes Nile or Main Nile. Further downstream, the Atbara River joins the Nile. The current Nile is supplied mainly from the Blue Nile called Abay in Ethiopia and fed by 18 tributaries. Contributions to the Main Nile from Ethiopian rivers are 86 per cent (composed of 59% from the Blue Nile, 13% from the Atbara and 14% from Sobat), all flowing into Egypt (Sutcliff and Parks, 1999). With this as background the history of the Nile will unfold.

Early Nile history

Before the common era

Ancient rock carvings in Egypt depicting cattle show that they have had a special importance within the Nile Valley cultures for thousands of years (Grimal, 1988). Pastoral production systems using wild cereals have been documented from the mid-Holocene period with evidence dating back to 10,000 BCE when the Nile Valley and the Sahara were one ecosystem. Early hunter, gatherer settlements have been documented in Nubian areas of Sudan dating to 9000 years back (Barich, 1998). Cave drawings from Ethiopia depicting sheep, goats and cattle date back to 3000 to 2000 BCE (Gozalbez and Cebrian, 2002). People at that time were in areas of unstable climatic conditions; they began herding to replace fishing as a primary source of protein. Herding became a strategy aimed at reducing the effects of climate variations (Barich, 1998). Bantu-speaking people spread across the eastern and central areas of Africa over 4000 years ago. They were excellent pastoralists and farmers.

From 1000 BCE, agricultural patterns were established that remain characteristic to the present time. Ethiopian cultural and agrarian history is determined in part by its geography. The high plateau formed by volcanic uplifts is split by the African rift valley (Henze, 2000), where the Northern fringe of the rift constitutes the deepest and hottest land surface on Earth, 126 m below sea level (Wood and Guth, 2009). The origin of human beings had occurred in the split in the rift valley highlands; from here began the domestication of several economically important crops, such as coffee, teff, khat, ensete, sorghum and finger millet (Gozalbez and Cebrian, 2002). Vavilov (1940), a Russian plant palaeontologist studying the origins of ensete and teff, said the Ethiopian Highland region was one of the more distinctive centres of crop origin and diversification on the planet. In the 1920s, Vavilov found hundreds of endemic varieties of ancient wheat in an isolated area of Ethiopia.

Ethiopia, considered to be the cradle of humankind, was the site where anthropologists found early hominid remains from 3.18 mya (known as 'Lucy'; Johanson and Edey, 1990). From 8000 BCE fishers and gatherers settled along lakes and rivers. The earliest reference to Ethiopia was recorded in ancient Egypt in 3000 BCE, related to Punt or Yam, where the early Egyptians were trading for myrrh. Ethiopia was a kingdom for most of its early history, tracing its roots back to the second century BCE. Ethiopia's link with the Middle East is from Yemen on the Red Sea, thought to be the source of Yemeni migrants speaking a Semitic language related to Amharic, and bringing along animals and several grain crops (Diamond, 1997). Ethiopia has connections to the Mediterranean where its religious and cultural ties to the ancient cultures of Greece and Rome have played a role in its history.

In Egypt irrigated agriculture and control over Nile water have been continuous for more than 5000 years (Postel, 1999). Traces of ancient irrigation systems are also found in Nubia or North Sudan, where people grew emmer, barley and einkorn (a primitive type of wheat). The people who settled along the Nile River in the Pharaonic kingdoms were cultivating wheat, barley, flax and various vegetables. They raised fowl, cattle, sheep and goats, and fished. Irrigation was also practised in Sudan and Ethiopia thousands of years ago, but to a lesser extent than in Egypt. There are few, if any, ancient irrigation records from the upstream riparian countries of Burundi, Democratic Republic of Congo (DRC), Kenya, Rwanda, Tanzania and Uganda, but most have sufficient rainfall and still rely more on rain-fed than on irrigated agriculture.

There are many theories about the spread of farming in the Nile Valley, but the consensus is that people moved to the upper Nile River when the fertile plains of the Sahara began to dry up, forming deserts. The Nile River was too large to control in ancient times but irrigation came naturally with the annual Nile floodwaters, particularly in Egypt where early Egyptians practised basin irrigation (Cowen, 2007). Most of what is known about ancient irrigation practices was from the Pharaonic civilization in 4000 BCE, recorded in hieroglyphics where an ancient king was depicted cutting a ditch with a hoe to let water flow into the fields (Postel, 1999). The early Egyptians were linked to the Nile; they worshipped it, based their calendar on it, drank from it and lived in harmony with the alternating cycles of flow. It was the control of the yearly floods that allowed the Egyptians to irrigate using a system of dikes and basins. When the floodwaters receded water was lifted with a device called a *shadoof* to get it to where it was needed. Irrigation was so successful that Egypt was referred to as the breadbasket for the Roman Empire (Postel, 1999).

Egypt's dynastic periods were characterized by periods of advancement and stagnation. The Egyptians' attachment to the Nile River, land and ability to irrigate led to wealth and a strong government, but this was followed by periods of stagnation in economy and population: 'It's not clear whether strong central government resulted in effective irrigation and good crop production, or whether strong central government broke down after climatic changes resulted in unstable agricultural production' (Cowen, 2007). Flooding periods came and went just as the ruling powers changed over and over again; in a period of decline the Assyrians took over (673–663 BCE), followed by the Persians (525 BCE), who were conquered by Alexander the Great in 332 BCE. Alexander the Great's death in 323 BCE signalled the start of the Ptolemaic dynasty; Cleopatra VII, last of the Ptolemaic rulers, took over and was subsequently defeated in 31 BCE, and the Roman Empire took over until the Arab conquest (El Khadi, 1998). Years of occupation of Egypt only added to and improved the irrigation systems.

Invaders and conquerors: the common era

From 300 to 800 CE, Bantu pastoralism helped shape the economy in the wet equatorial regions of Kenya, Tanzania and Uganda (Oliver and Fage, 1962). The trade and import of crops from the Far East further changed the influence of agriculture on the people in this region. From the settlements around the equatorial lakes many of these people grew and developed a trade in many crops, especially bananas. Cultivation of bananas and other crops was very successful in the wet equatorial region, and trade from cities in Kenya, Tanzania and Uganda developed rapidly.

Slaves were sought, used and exported from Uganda, Kenya, Tanzania and Uganda from the fifth century. Between the seventh and the fifteenth centuries, Arab slave traders introduced both Islam and slave trade to many regions in Africa, where they controlled the slave trade (Ehret, 2002). Arabic and Portuguese traders shaped the economy in East Africa, bringing in goods from China and India and trading them for ivory, gold and slaves.

Muslim armies also tried to enter Ethiopia in the 1540s, but unsuccessfully. Only later (in the 1850s) did Ethiopia begin to open up and interact more with foreign powers (Henze, 2000). Ruled for centuries by kings from the Solomonic Dynasty and isolated by its Coptic faith in a sea of Islam, Ethiopia was spared from conquests by foreign powers until the Italians caught up in the conquest for Africa invaded Ethiopia during 1936–1941 (Pankhurst, 1997; Henze, 2000).

For centuries, both Egypt and Sudan have viewed the Nile as their main lifeline, because they lacked other main freshwater sources. However, the dominant user of Nile water historically has been Egypt, which maintained its highly successful irrigation systems throughout the conquest periods – beginning with the Arab conquest in 641, followed by Turkish Mamelukes in 1250, until the Ottomans took over in 1517. The Arabs realized how important the Nile was to the success of their control over Egypt (El Khadi, 1998). It was the Arabs that made improvements in the irrigation practices with new types of water-lifting devices, building embankments and canals, and monitoring the Nile flow with about 20 Nilometers (devices that allowed them to measure river levels, compare flow over years and predict the oncoming floods). While the Mamelukes were warriors with periods of fighting, they were also builders, as evidenced by several beautiful mosques in Cairo. Their main agrarian successes were in land tenure and property rights, which also had an effect on land productivity (El Khadi, 1998).

The Ottomans took over in 1517 and were defeated in 1805. The Ottomans did not change irrigation much, but they did keep detailed records. The Ottomans were also very attentive to the Nilometer because it determined the health of the country, predicting floods, which meant how much tax the Ottomans would put on the Egyptian farmers. The French, under Napoleon, attacked Cairo in 1798 and defeated the Mamelukes at the battle of the pyramids;

during the attack they partially destroyed the Nilometer at Roda. In a costly expedition to Egypt in 1798, Napoleon wrote:

> There is no country in the world where the government controls more closely, by means of the Nile, the life of the people. Under a good administration the Nile gains on the desert, under a bad one the desert gains on the Nile.
>
> *(Moorehead, 1983)*

To obtain peace with the Egyptians again, they rebuilt the Nilometer. Mohamed Ali Pasha finally drove the Mamelukes and the Ottomans from Egypt.

When Mohamed Ali Pasha took control of both Egypt and Northern Sudan in 1805 there was an active period of agricultural expansion, increased irrigation and excavation of more canals. Mohamed Ali's polices gave priority to agricultural production, and yields of cotton and other crops were boosted by year-round irrigation. Mohammed Ali established the first agricultural school in 1829. This school closed and reopened several times in several locations; education has been a high priority for Egyptians (IDRC and MoA, 1983). Later, Mohammed Ali, with help from French engineers, began the construction of two barrages at the Damietta and Rosetta branches to control water going into the delta (El Khadi, 1998).

Colonial past and control of the Nile

Beginning in the early 1700s and continuing to the late 1800s, Europeans began to realize the importance of understanding the Nile River, where it came from, how much water there was and how to control it. Finding the source of the Nile was a necessary step needed to make treaties and 'legalize' the use of Nile waters. The English, because they had much to gain, were very central to most of the actions taken on the White Nile, mapping, measuring, clearing canals for navigation in the Sudd and allocating water. Scientific measurements of Nile flow began in the early 1900s with the installation of modern meters along the Nile (Hurst *et al.*, 1933).

Explorations by Europeans on the upstream sections of the Nile began mainly during the period 1770–1874. A Portuguese monk who founded a Catholic church at Lake Tana is believed to be the first European to note the Blue Nile source in Ethiopia in 1613 (Gozalbez and Cebrian, 2002). The length of the main Nile River, plus the physical dangers of passing through cataracts and the swamps in southern Sudan, gave explorers trouble for many years. The White Nile source caused confusion and acrimony between Richard Burton and John Speke, and it was 1862 when John Speke's claim was confirmed that the river flowed out of Lake Victoria through Rippon Falls. Grant and Speke would also follow the flow to Lakes Kyoga and Albert, and on to Bahr el Jebel. Europeans began vigorous scientific explorations, making maps and hydrological measurements. In 1937 a German scientist explorer named Dr Burkhart Waldecker traced the Kagera River to its southern-most source, with its headwaters in Burundi. In 2006, a National Geographic group of explorers have claimed to be the first to travel the length of the Nile to its true source in Rwanda's Nyungwe Forest (Lovgren, 2006). Using modern geographic information system (GIS) equipment they believe they have accurately identified the source. To ease the confusion, the National Geographic Society has in the past recognized two sources of the Nile, one in Rwanda and one in Burundi.

Several foreign powers were involved in shaping the history of the southern Nile nations. The United Kingdom, Germany and Belgium were all colonial power players in Kenya, Rwanda, Tanzania and Uganda. They brought diseases that are estimated to have killed off

40–50 per cent of the population in Burundi. The colonialists were interested in large plantations for growing sugar cane and cotton to send back to Europe.

In 1888, the Imperial British East African Company was given administrative control over all of east Africa. The Ugandan leader signed a treaty of friendship with the Germans in 1890, but in 1894 the British quickly made Kenya and Uganda protectorates. An Anglo-German agreement put Burundi, Rwanda and Tanganyika into German control. The United Kingdom's reason for wanting Kenya and Uganda was to protect their interests on the Nile in Egypt and Sudan.

Because of Egypt's total dependence on the Nile River, Egypt continued to develop more control, making embankments and creating more structures. The first structure was a diversion dam called the delta barrage, which was built across the Nile north of Cairo, where the delta begins to spread to raise the water level for upstream irrigation and for navigation on the river. The barrage was completed in 1861 after rebuilding and improving the structure. The main purpose of the dam was to improve irrigation and expand the agricultural area in the delta.

In 1890, Ethiopia was the only independent country in Africa. British influence was responsible for the allocation of Nile water, beginning in 1890 with a treaty, the Anglo-German Agreement, between Great Britain and Germany, which put the Nile under Great Britain's influence (Tvedt, 2004). The following year, Great Britain signed a protocol with Italy when they held interests in the Blue Nile region of Eritrea in which Italy pledged not to undertake any irrigation work which might significantly affect the flows of the Atbara into the Nile (Abraham, 2004). Anglo-French control over Egypt ended with outright British occupation in 1892. With these agreements Great Britain secured control of the Nile, and occupied Egypt to watch over its interests in the Suez Canal and to grow cotton for its textile mills. In 1898, the British took over Sudan and established cotton as a major export crop.

The British were fairly secure in their control over the Nile River, but there were threats from other European powers. In the middle of the nineteenth century Italy had colonized the area of Eritrea, and they wanted more of Ethiopia. Italy's show of power in Ethiopia and Eritrea was supported by Britain hoping to squash the Mahdist threat in Sudan; on the other side, the French were supporting King Menelik's opposition to the Italians to back their own interests (Ofcansky and Berry, 1991). King Menelik's weakening health and control over the country alerted Britain, France and Italy and to avoid a more serious situation in the region; the three negotiated an agreement that later became known as the Tripartite Agreement of 1906 (Keefer, 1981).

In 1902, the British formed the Nile Project Commission, with expansive development plans for projects on the Nile River. The plans included dams on the Sudan/Uganda border, the Sennar for irrigation in Sudan and one to control the summer flooding in Egypt. Egypt disliked these plans. Control over Nile water by the British was the same as control over Egypt. By 1925, a new water commission made acceptable for development plans led to the 1929 water agreement between Egypt and Sudan. Great Britain sponsored the 1929 agreement that gave 4 billion cubic metres per year (m^3 yr^{-1}) to Sudan, and the rest of the yearly flow from January to July, plus 48 billion m^3 yr^{-1} to Egypt. The most important statement in the agreement for Egypt was 'Being guaranteed that no works would be developed along the river or on any of its territory, which would threaten Egyptian interests.' According to Wolf and Newton (2007), 'The core question of historic versus sovereign water rights is complicated by the technical question of where the river ought best be controlled – upstream or down.'

A committee of international engineers in Egypt built the first true dam on the Nile; the Aswan Low Dam was completed in 1902 at the time of the first signing of treaties. The dam was raised several times, and as irrigation demands increased and floods threatened it was raised once more in 1929. Extensive measurements and studies made it clear that another dam was needed after the near overflow of the low dam in 1946. The main reason for building the Aswan High Dam (AHD) was to control the flow of water and to protect Egypt from both drought and floods. This allowed Egypt to expand irrigated agriculture and supply water to attract industries (Abu-Zeid, 1989; Abu-Zeid and El-Shibini, 1997). The British and Americans originally agreed to fund the AHD under Nasser, but the funding was suddenly halted just before construction began due to the Cold War and Arab/Israeli tensions in the region (Dougherty, 1959). The Soviet Union agreed to assist the construction and funding to help complete the project with the Egyptians.

Unhappy with plans for the AHD, Sudan demanded that the 1929 treaty be renegotiated. Great Britain was involved in all Nile water concessions until 1959 when Egypt and Sudan signed a bilateral agreement to allocate Nile water between the two countries. Egypt did not bother to consult or include upstream riparians, except Sudan in the reallocation of Nile water (Arsano, 1997; Abraham, 2002). This 1959 agreement set the maximum amount of water that could be withdrawn by these two countries with Egypt getting 55.5 billion m^3 out of a total average flow of 84 billion m^3 yr^{-1}, allowing 10 billion m^3 yr^{-1} for evaporation and 18.5 billion m^3 yr^{-1} to Sudan.

They also had a growing demand for irrigation and energy for their expanding population. A section of the 1929 treaty was integral to the 1959 agreement, which said 'Without the consent of the Egyptian government, no irrigation or hydroelectric works can be established on the tributaries of the Nile or their lakes if such works can cause a drop in water level harmful to Egypt' (Carroll, 2000). This guaranteed Egypt a set amount of Nile water, which could not be changed. This agreement included a pact to begin construction of the AHD in Egypt, and Roseires Dam and the Jonglei Canal in Sudan with benefits to be gained by Egypt and northern Sudan.

Post-colonial Nile

Many African countries gained their independence from colonial powers in the late1950s and early 1960s. The return to African rule was difficult after the colonizing powers had realigned borders and tribal ethnic groups. This was particularly true in the equatorial countries of Burundi, Rwanda and Uganda. The negotiations that took place between Great Britain and Egypt were not as important in Burundi, where they have enough rainfall; but neither did they feel obligated to acknowledge the accords that were made prior to independence. The equatorial countries (Burundi, DRC, Kenya, Rwanda, Uganda and Tanzania) agree that agreements prior to independence were no longer valid. Only the post-independence agreement in 1959 between Egypt and Sudan, which did not include any of the equatorial countries, can be disputed.

Once the agreement between Egypt and Sudan was signed in 1959, work began on the second Aswan Dam. The construction began in 1960 after moving ancient temples and a large Nubian population, and was completed in 1970. There were complications, including increased salinity and reduced fertility, but also many benefits, of which power was the most important for the development of Egypt (Abu-Zeid and El-Shibini, 1997; Abu Zeid, 1998; Biswas, 2002).

In 1964, while the negotiations were going on between Egypt and Sudan, Ethiopia had employed the United States Bureau of Reclamation (USBR) to study the hydrology of the upper Blue Nile Basin (US Dept of Interior, 1964). The study identified potential new irrigation and hydroelectric projects within Ethiopia. Preliminary designs for four large dams were prepared for both the Blue Nile and Atbara rivers that would increase power production by

5570 megawatts (MW). Block *et al.* (2007) took a renewed look at the USBR study using a model that shows little benefit/cost ratios for the use of both hydropower and irrigation for agriculture due to limitations in timely water delivery. Recent climate change studies by Block and Strzebek (2010) present a cost/benefit analysis giving several climate change scenarios that report favourable results for water conservation behind the dams in Ethiopia, but less favourable results given that success of the dams, purpose of hydropower and irrigation will depend greatly on the timing of water, climate variability and climate change. They emphasize close cooperation and economic planning that secure energy trade between neighbouring countries.

Waters of the Nile have been used for centuries for irrigation in Sudan, taking advantage of the annual Nile flood that builds up from heavy summer rains on the Ethiopian plateau. When the world market was in need of cotton it became Sudan's first commercial crop. The Sennar Dam, funded and built by the British just south of Khartoum in 1918, was to supply water to irrigate 126,000 ha of cotton in Gezira to supply British textile mills. But by the 1950s more land was needed; indeed, it had to double, and irrigation water was not enough (Wallach, 2004). Sudan needed the 1959 agreement mentioned above to increase its allotment of Nile water and to proceed with building the Roseires Dam on the Blue Nile, which was completed in 1966 to improve irrigated agriculture and to supply hydropower. Approximately 800,000 ha yr^{-1} were irrigated in the Gezira scheme by the end of the 1960s.

Until 1959, treaties were geared towards allocation of Nile water resources for irrigation. The 1929 and 1959 treaties were meant to secure irrigation water for the Gezira scheme in Sudan and Egypt, respectively. However, after 1959, and following independence of other Nile basin states, the focus of Nile agreements shifted away from water-sharing to more cooperative frameworks. As a consequence, irrigation water demand was overlooked in favour of Nile negotiations, and power supply took over (Martens, 2009).

The 1959 treaty left a legacy for potential conflict between Egypt and Sudan, on one side, and Ethiopia and the seven other riparian countries, on the other. Experts who have analysed the 1997 United Nations Watercourses Convention say it cannot resolve the legal issues concerning allocation of Nile water (Shinn, 2006). Egypt says that all Nile countries must recognize the 1959 treaty before any new agreements are implemented, including benefit-sharing proposals. This is not negotiable, according to Egypt; this claim has not been favoured by the rest of the riparian countries (Wolf and Newton, 2007). The issue becomes more complex, as several upstream riparian countries have recently criticized the 1959 treaty. Several riparian nations, especially Ethiopia, state that (i) they were not included in the 1929 and 1959 treaties and (ii) these treaties violate their right to equitable utilization as stated in the 1997 UN convention. The upstream countries with their own development issues do not feel that they need Egyptian permission to use Nile water. Attempts to unite the Nile Basin countries led to the development of the NBI.

The NBI was formed in 1999 to 'cooperatively develop the Nile and share the benefits, develop the river in a cooperative manner, share substantial socio-economic benefits and promote regional peace and security' (NBI, 2001). Funded by several donors, including the World Bank, the NBI is headed by a council of ministers of water affairs, comprising nine permanent members and one observer, Eritrea. In May 2010, Ethiopia, Kenya, Rwanda, Uganda and Tanzania signed a Cooperative Framework Agreement (CFA) to equitably share the Nile waters. Later, Burundi and the DRC also signed the CFA. Meantime, and due to lack of agreement between the different parties, a proposal emerged to rephrase Article 14b to include the ambiguous term 'water security' in order to accommodate and harmonize the differing claims of the upstream and downstream riparian countries (Cascão, 2008). Egypt refused to sign the CFA if the change in Article 14b on 'benefit sharing' was not made and has threatened to back out of the NBI if all the other countries sign the agreement. Arsano (1997) believes the NBI has been able to bring the riparian states on board for dialogue towards establishing plans for cooperative utilization and management of the water resources, and to make an effort towards establishing a legal/institutional framework.

The NBI has been instrumental in promoting information-sharing and initiating small projects, but it is still struggling to be a permanent river organization, to obtain signatories for the ratification for a new Nile Treaty as agreed by all members and to implement new large Nile Water projects (Cascão, 2008). Criticism has been aimed at a lack of coordination in development activities between the NBI and the governments of the Nile Basin countries. Basin-wide collaboration has, as Bulto (2009) states, 'hit a temporary glitch, casting doubt over the prospect of reaching a final framework agreement over the consumptive use of Nile waters'.

Recent agricultural expansion

Except for Egypt and Sudan (who have a longer history of irrigation and control of the Nile water), the other Nile Basin countries have relied more on rain-fed agricultural potential for their available water resources. Some reasons include limited financial resources, infrastructure, governance and civil wars. Increasing populations in the riparian countries mean a greater demand for food, water and energy. Poverty has led some countries to sell land to foreign buyers in hope of developing agricultural infrastructure and bringing jobs and money into areas of extreme poverty (IFPRI, 2009). Since 2008 Saudi investors have bought heavily in Egypt, Ethiopia, Kenya and Sudan (BBC, 2010; Vidal, 2010; Deng, 2011).

The Egyptian government has built over 30,300 km of channels and large canals in Egypt (El Gamal, 1999). Egypt has used desert lands to expand cultivation of horticultural crops such as fruits, nuts, vineyards and vegetables. Reclamation of desert lands allows Egypt to expand production and use drip irrigation from groundwater reserves. The use of drip irrigation and plastic greenhouses has increased attempts to save water. These are more expensive for the traditional farmers, but many investors have put these to use and have successfully supplied produce to the local markets.

Aquaculture in the Nile Delta is booming, and demonstrates high water productivity while using drainage water flows. Aquaculture in Egypt's Delta makes use of recycled water and also shows promise of providing an important source of dietary protein and income generation. Aquaculture has rapidly expanded, and yields have grown from 20,000 tonnes in the 1980s to more than 600,000 tonnes in 2009, due to run-off from sewage and fertilizer-enriched water (Oczkowski *et al.*, 2009).

Egypt's plan to develop the North Sinai Desert is a huge land reclamation project, estimated to cost nearly US$2 billion when completed. From October 1997 the Al-Salam Canal was delivering water to irrigate about 20,200 ha of the Tina Plain, which is actually outside the natural course of the Nile. The north Sinai agriculture development project is planned to eventually divert 4.45 billion m^3 yr^{-1} of Nile water to develop irrigated agriculture west and east of the Suez Canal. The 261 km-long Al-Salam Canal is the summation of both first and second phases. The Al-Salam Canal runs eastwards, taking Nile water horizontally across the Sinai. The second phase extends further east, passing under the Suez Canal to open nearly 168,000 ha of irrigated agricultural lands; both phases require a 1:1 mix of Nile water with drain water, keeping salinity and pollutants at a minimum (Mustafa *et al.*, 2007).

According to several reports, Ethiopia has about 3.7 million ha that can be developed for irrigation; about half of this is in the Nile Basin, but only 5–6 per cent has been developed so far (Awulachew *et al.*, 2007, 2009). Their current irrigated area is about 250,000 ha, with less than

20,000 ha in the Nile Basin; most of the current agricultural production is rain-fed. Ethiopia plans to expand agricultural production by an additional 3 million ha with the addition of many small- and large-scale irrigation schemes, both private and government-funded, and distributed in various parts of the country. Water for the schemes will come from several new multi-purpose hydropower projects that are completed and planned for future development.

Sudan developed two schemes in the 1960s that proved untenable and too expensive to complete (Wallach, 2004). The scheme at Rahad was intended to use water from the Roseires Dam in 1960 but lacked financial resources to develop canals. It was delayed, and finally one-tenth was operational in 1978. It is a 122,000 ha scheme irrigated from the Rahad River, using Blue Nile water; it is plagued by siltation and irrigation inefficiency, and yields have been below average (Wallach, 2004). New Halfa built to use water from the Khashim el Girba Dam was basically a resettlement scheme for Halfawis people who lost land and homes when Lake Nasser was filled. The Khashim el Girba Dam built for the Halfa scheme silted too quickly and was poorly planned (El Arifi, 1988). These two schemes were good lessons for the planners, forcing them to consider rehabilitation of existing schemes. Wallach (2004) describes the challenges of rehabilitation and modernization of irrigated agriculture in Sudan.

In the 1990s Sudan had a 2 million ha modern irrigation system developed in a fertile valley south of Khartoum between the Blue and White Niles. More than 93 per cent was government-managed. This area was originally the Gezira scheme started by the British in the 1920s. When the Nile agreement with Egypt was signed in 1959 the area was expanded to the west. Cotton is a mandatory crop, but in the 1970s cotton was partly replaced by sugar cane. Generally, yields are poor, partly due to government policies, poor canal maintenance, lack of irrigation water and inefficient use of water (Molden *et al.*, 2011).

Water, land, food, energy and development are tightly and crucially interlinked. Water is also very much linked to the potential for peace in the region. Dialogue between the riparians is necessary for the area to solve water-sharing issues. Rehabilitation of irrigation systems, improved water management including rain-fed agriculture and policy reforms will help improve existing agricultural performance.

Nile environmental challenges: wetlands, lakes and Blue Nile

The Sudd

Large wetlands forming about 6 per cent of the basin area are found in eight of the Nile Basin countries. The largest and the most important wetland to the hydraulics of the downstream Nile River is the Sudd (meaning 'blockage'), located in South Sudan (Sutcliff and Parks, 1999). The Sudd is an extensive area (average 30,000 km^2) of mainly papyrus swamp legendary for being impenetrable, high in biodiversity, having high evaporation and transpiration rates, and more recently oil and conflict.

More than 50 per cent of the water that flows into the Sudd and circulates within its ecosystems evaporates; thus, less than half of the water flowing into the Sudd actually flows out again to continue north to Khartoum (Sutcliffe and Parks, 1999). Estimates of inflow to outflow ratios vary, as do the size and evaporation rates; for example, Mohamed *et al.* (2004) estimate 38.4 billion m^3 yr^{-1}, and Sutcliff and Parks (1999) estimate 16.1 billion m^3 yr^{-1}. Reasons for different values include different means of calculation, as well as different times of measurement. Satellite images have helped improve information on the Sudd. Mohamed *et al.* (2006) have found that the swamps are larger than previously thought and the fluctuation in evaporation is difficult to estimate due to the size, and estimated the evaporation at 29 billion m^3 yr^{-1}.

From the early 1900s, British engineers began to think of creating a canal through the swamps in South Sudan (Howell *et al.*, 1988). Their concern over high rates of evaporation from the swamps prompted the development of the Jonglei Canal to channel water out of the swamps and send it north for the benefit of agricultural production in North Sudan and Egypt. In the 1980s, a large digging machine from France was brought to begin the canal. Scheduled for completion in 1985, the still incomplete canal, due to its size (80 m wide, 8 m deep and roughly 245 km long), is visible from satellite images (NASA, 1985). Work on the canal was abruptly stopped during the civil war in 1983.

The Jonglei Canal was supposed to bring about 5–7 per cent more water to irrigation schemes of both Egypt and Sudan. Egypt and Sudan have been in discussions to revive the project, but without the consent of southern Sudan this will not be possible (Allen, 2010). However, the social-environmental issues of drying wetlands, loss of traditional grazing lands, biodiversity and collapse of fisheries are reasons given to discontinue the canal (Ahmad, 2008; Lamberts, 2009). Counter-arguments state that the drying wetlands will create new grazing lands and the canal can be used for fisheries (Howell *et al.*, 1988). Furthermore, the consequences of draining the wetland on the larger regional and micro-climatic conditions are difficult to predict. Meanwhile, southern Sudan and the Sudd area have vast potential for development and improvement of agriculture with large areas of land suitable for mechanized farming (UNEP, 2007).

Oil development is an added threat to both the ecosystem and human communities. Oil drilling conducted by foreign companies brings in workers from outside of Sudan, providing very little benefit to the local communities. Conflicts arise over loss of grazing lands, loss of traditional livelihoods and increases in diseases (e.g. AIDS/HIV). Damages to the human communities, wetlands and ecosystems need monitoring and responsible management for future generations (UNEP, 2008).

The Sudd possesses huge potential for enhancing the livelihoods of local inhabitants through development of improved agriculture, pasturelands and fisheries, while supporting a rich ecosystem, and water resources. Controversy continues over the decision to finish the Jonglei Canal or abandon it. At present, it remains to be seen what the newly formed government of South Sudan will decide.

Victoria Nile

The outflow at Jinja, which was once thought to be the source of the White Nile, is a determinate point for measuring Nile flow from the lake. Nearly 80 per cent of the water entering Lake Victoria is from precipitation on the lake surface, and the remainder is from rivers, which drain the surrounding basin (Howell *et al.*, 1988). Some 85 per cent of water leaving the lake does so through direct evaporation from its surface, and the remaining 15 per cent leaves the lake near Jinja in Uganda, largely by way of the Victoria Nile. Three countries – Kenya (6%), Tanzania (51%) and Uganda (43%) – share the lake shoreline, and six countries share the basin: Burundi, DRC, Kenya, Rwanda, Tanzania and Uganda. The area around Lake Victoria has the fastest-growing population in East Africa, estimated to be more than 30 million in 2011 (LVBC, 2011).

In the 1940s Nile perch were introduced to boost fisheries production, which had already begun to fall. Nile perch depleted the endemic species and hundreds became extinct. Recently, Nile perch populations have dropped due to overfishing allowing some remaining endemics to make a slow comeback (Hughes, 1986). Lake Victoria produces a catch of over 800,000 tons of fish annually, an export industry worth US$250 million (LVFO, 2011).

Lake Victoria is important for agriculture, industry, domestic water supplies, hydropower, fisheries, travel, tourism, health and environment. It is highly sensitive to climate change and climate variability. The shallow lake is threatened by sewage, industrial and agricultural pollution, algal growth, overfishing, invasive flora and fauna, low water levels and deforestation (Kull, 2006; Johnson, 2009). Many of the biological effects directly affect the socio-economic factors of the people living on, and supported by, Lake Victoria fisheries (Onyango, 2003).

Dropping water levels have caused alarm in many of the downstream countries. Kull (2006) estimated that in the past two years, the Ugandan dams have released water at an average of almost 1250 m^3 sec^{-1}; that is 55 per cent more than the flow permitted for the relevant water levels. Diminishing water levels have acute consequences for several economic sectors dependent on the lake, such as fisheries and navigation. Variations in the water level affect shallow waters and coastal areas, which are of particular importance for numerous fish species and health of the lake. The largest projected climate change for rainfall and temperature changes for the interior of East Africa is over the Lake Victoria Basin (Conway, 2011).

Some of the greatest concerns for management are related to reducing vulnerability and poverty and improving livelihoods of the people living beside Lake Victoria. Coping with climate change effects on Lake Victoria requires a range of strategies, including proactive measures to improve the health of Lake Victoria and a reduction of the dependency on Nile perch exports (Johnson, 2009). Victoria Basin countries and organizations need to address all the factors that affect the watershed and the lake. The Lake Victoria Basin Commission (LVBC) was formed with the East African Community (EAC) to develop strategic plans for the basin countries to sustainably develop and protect the lake from further destruction (LVBC, 2011).

Blue Nile

Most of the main Nile flow can be explained by rainfall variability in both Lake Victoria and the Blue Nile Basin (Conway, 2005). The upper Blue Nile Basin is the largest section of the Nile Basin in terms of volume of discharge and second largest in terms of area in Ethiopia and is the largest tributary of the Main Nile. It comprises 17 per cent of the area of Ethiopia, where it is known as the Abay, and has a mean annual discharge of 48.5 billion m^3 (1912–1997; 1536 m^3 s^{-1}; Hughes and Hughes, 1992), with variation from less than 30 billion m^3 to more than 70 billion m^3, according to Awulachew *et al.* (2007).

This part of the sub-basin is characterized by a highly seasonal rainfall pattern, most of the rain falling in four months (June to September), with a peak in July or August. Soil erosion is a major threat in the Blue Nile Basin (Conway and Hulme, 1993). A report prepared by ENTRO (2006) estimates the total soil eroded within the Abay Basin alone is nearly 302.8 million tonnes per annum (t yr^{-1}) and erosion from cultivated land is estimated to be 101.8 million t yr^{-1} (33%). Thus about 66 per cent of soil being eroded is from non-cultivated land, (i.e. mainly from communal grazing and settlement areas; Molden *et al.*, 2011). About 45 per cent of this reaches the stream system annually causing heavy siltation of downstream reservoirs.

The Nile's ecosystems are threatened by many human activities, but by far the most damaging are agricultural practices. Good agricultural practices to control erosion, pollution and land degradation can enhance other ecosystem services. From Lake Victoria, the Sudd's and the Blue Nile's control over pollution, overgrazing, mining and erosion will help good governance, policies and organizations that could, in turn, help in regulating and monitoring the ecosystems.

Recent irrigation developments

WaterWatch (2008) estimates that the total irrigated area in the Nile Basin is 4.3 million ha, based on GIS measurements. Irrigated agriculture in the basin is dominated by Egypt, with 35 milion ha, while Sudan has 1.8 million ha and Ethiopia 0.3 million ha (CIA, 2011). Egypt and Sudan are almost completely dependent on Nile water for irrigated agriculture. Sudan's arable land is estimated to be 105 million ha, with about 18 million ha under cultivation, most of the latter being rain-fed. Sudan now irrigates about 1 per cent of its arable land.

Huge new irrigation projects in Egypt and Sudan are being planned and executed. Egypt's Al-Salam Canal, some parts of which are built under the Suez Canal, diverts Nile water and drains water from the Damietta Canal to irrigate land in northern Sinai, which will require 4.45 billion m^3 yr^{-1}. The Toshka Lakes were formed in 1997 when Egypt developed pumping stations and canals to send 'excess' Nile water into a depression in the southwest desert about 300 km west from Lake Nasser. Named the New Valley Project, if complete in 2020, it will require an additional 5 billion m^3 yr^{-1} of water, be home to 3 million people and irrigate 25,000 m^3 ha^{-1}. This expansion is to be sustained by transferred water but it could mean disaster in the future if there is no spill water available to send to the lakes.

Upstream Nile countries, dependent on rainfall, have experienced a greater frequency of droughts and cannot ignore the need to grow more food and expand production even at great costs; this requires irrigation. Ready to claim its share of Nile water, Tanzania is planning to build a pipeline 170 km long that will take water from Lake Victoria south to the Kahama irrigation project in an arid poverty-stricken area where thousands of people will benefit. Other upstream Nile countries also feeling more confident and pressed to find solutions to fight hunger and poverty will try to use their 'share' of Nile water and are planning to build multipurpose hydropower dams.

Increasing climate variability, population growth, food prices and vulnerability to food shortages mean that larger, wealthier countries need to look elsewhere to buy or produce more food. How can they produce more when all their arable land is currently cultivated or they lack the resources to produce more food?

Power on the Nile: past and future

Many of the Nile Basin countries are classified as some of the poorest in the world in terms of GDP and food security (FAO, 2010). Most people lack electrical power and the necessary means of obtaining electricity; average electrification rate is 30 per cent (per capita per annum) and this drops to 15 per cent when Egypt and DRC are excluded. This is a very low proportion, according to Economic Consulting Associates (2009). Hydropower is underexploited in most of the basin countries affecting growth and development. Deforestation for charcoal production in the Lake Victoria Basin is one example of the need to produce affordable power for people in the area. Hydropower dams and the generated power offer other benefits that include additional irrigation water, controlled releases, water storage and further social and industrial development. Figure 2.2 shows the placement of early dams on the Nile (Nicol, 2003).

Aswan High Dam (AHD), completed in 1970, is the largest man-made reservoir and produces 2100 megawatts (MW) of electricity – about half of Egypt's total power supply. It was planned to resolve both floods and droughts and irrigate about 283,000 ha. At the time, only Sudan was consulted before the AHD was built. Now other Nile countries will build more dams on the Nile (Table 2.1). Egypt feels that they should give permission for any developments on the Nile.

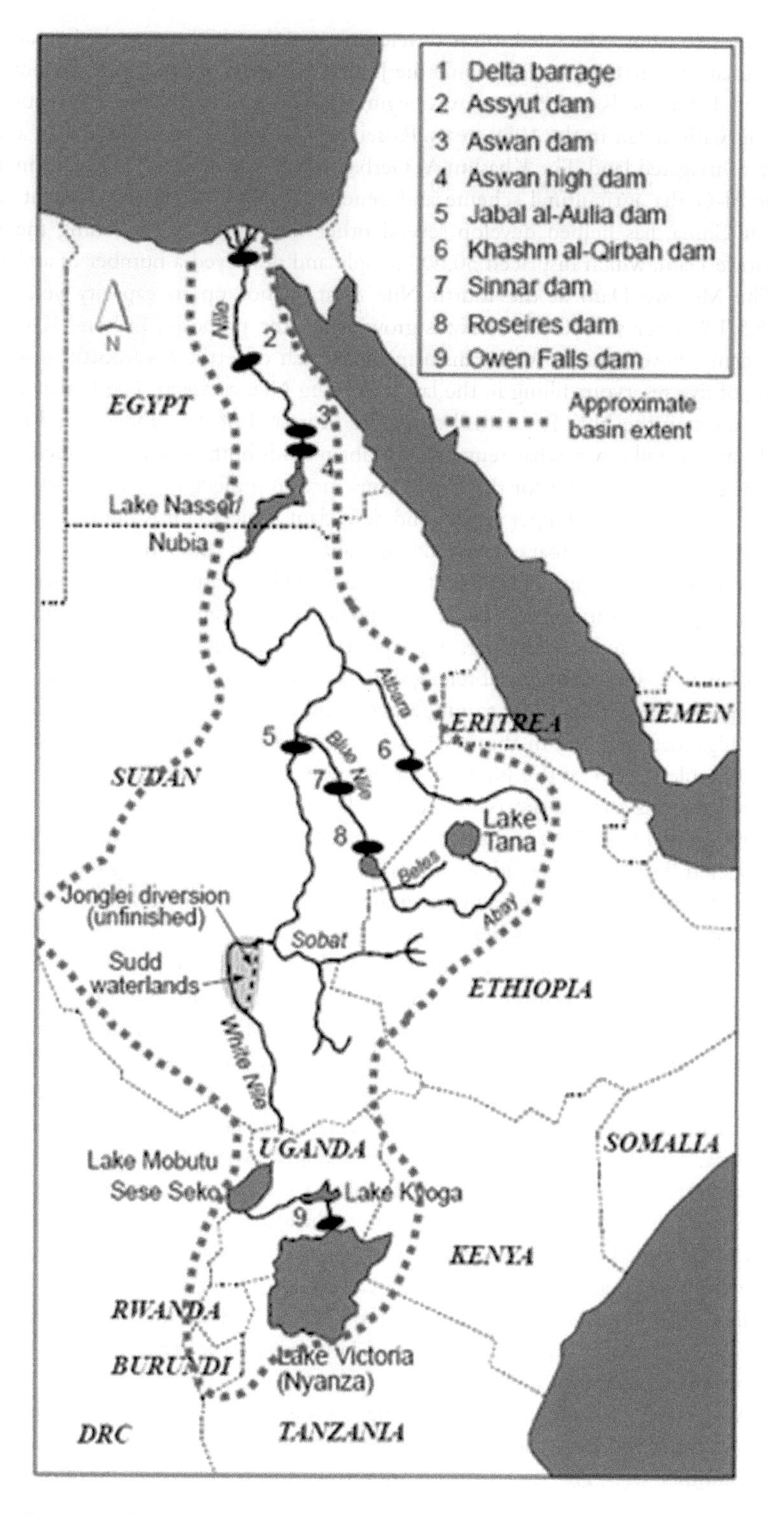

Figure 2.2 Placement of early dams on the Nile. (The map is not to scale.)

Sudan built two dams, which led to the development of the 1929 treaty: the Sennar in 1926 was built primarily to irrigate cotton, while the Jabal Awliya was built in 1936 for both power and irrigation. Later, the Roseires Dam was begun in 1950, and Egypt protested, but reached an agreement with Sudan in the 1959 treaty. Roseires Dam will be raised to add an additional 420,000 ha of irrigated land. The Khashm Al Gerba on the Atbara River was built in 1964 to irrigate the Al-Gerba agricultural scheme and generate 70 MW of power. Recent funding, mostly from China, has helped develop several other huge projects, including the US$1.2 billion Merowe Dam, which displaced 50,000 people and destroyed a number of archaeological sites. The Merowe Dam at the fourth Nile cataract, not up to capacity yet, currently generates 5.5 TWh per year. Controversy is growing on the proposed Dal and Kajbar dams. These two dams above Merowe will transform the stretch of fertile land north of Khartoum into a string of five reservoirs filling in the last remaining Nile cataracts. The Kajbar, located at the third cataract on the Nile River, will cover the heartland of the Nubians, and the Dal at the second cataract will cover what remains of Nubian lands both present and ancient.

The signing of an agreement for the largest construction project any Chinese company has taken on in Sudan is for the Upper Atbara and Setit dam project consisting of two dams, the Rumela Dam on the upper Atbara River and Burdana Dam on the Setit River, located south of the Khashm El-Qurba Dam. The project benefits include an aim to increase irrigated area and agricultural production in New Halfa area currently irrigated by the Khashm El Qurba Dam, regulate flow and reduce flooding, and support development in eastern Sudan.

South Sudan, now after a successful referendum to secede from North Sudan, will consider several hydropower projects in order to modernize southern Sudan. The Nimule Dam, on the border with Uganda, proposed in the 1970s is being considered again as South Sudan needs the Juba to Nimule stretch of the Nile for further power generation. Three large dams have been proposed for this part of the Nile (Mugrat, Dugash and Shereik dams, with projected power generation of 1140 MW) and three additional smaller run-of-the-river projects would also be considered (FAO, 2009). South Sudan might also consider reviewing the Jonglei Canal, but this is a highly political issue and a hypersensitive area in southern Sudan, such that a restart of the canal will need careful consideration (Moszynski, 2011).

Ethiopia has the capacity to become the basin's main power broker as it has huge hydropower potential in the volume of water with a steeply sloping landscape. The estimated potential from the Blue Nile (Abay) alone is about 13,000 MW. Ethiopia has at least six new dams proposed and four under construction. Ethiopia's first big dam, the Finchaa Dam, was completed in 1973 on the Finchaa River that feeds into the Blue Nile. According to Tefera and Sterk (2006), land use changes from the dam have increased soil erosion from expansion of agricultural area, displaced people and reduced grazing areas, swamps and forests.

The Tekeze Dam on the Tekeze River was completed and began operating in 2009. Located at the border with Eritrea, Tekeze was an expensive headache for the Chinese construction company that engineered and built it. Trouble with landslides destabilized the dam and delayed construction. Now the tallest arched dam in Africa (188 m), it cost the Ethiopian government US$350 million. The benefits are the ability to provide water year-round for the downstream areas, besides the generation capacity of 300 MW. The concerns are the construction cost, the sedimentation issues and loss of ecosystems.

On the Abay River, downstream of Lake Tana, the Tis-Abay I hydroelectric project began to transmit power in 1964; later, the CharaChara weir in 1997 was built to boost power supplies and in 2001 another weir, Tis-Abay II was commissioned to boost power supplies by another 20 per cent. Awulachew *et al.* (2009) analysed the possible effects of development on the water resources of Lake Tana and found that water levels would be affected. Later, McCartney *et al.*

(2010) reported that the natural environment around Lake Tana has been affected by the variability in lake levels caused by the weirs. The Tana Beles Dam will transfer water via a tunnel to the Beles catchment for hydropower and irrigation.

In April 2011, the third and newest large dam in Ethiopia, the Grand Millennium Dam, was started. Placed about 40 km from the Sudan border the dam is expected to benefit both Egypt and Sudan. The new dam, estimated to generate 5250 MW, will be completed in about 2017 (Verhoeven, 2011). Funded by the Ethiopian government, it is hoped that future power sales to its neighbours will cover the construction cost, besides helping to boost its own domestic energy supply and access.

Nalubaale Dam, previously Owen Falls at Jinjawas, the first large dam constructed in Uganda, was completed in 1954. Downstream, a new dam is under construction at Bujagali Falls. Delayed four years by many setbacks, the dam was finally started in 2010. It is projected to double power production for Uganda. When Bujagali is finished the Isimba Dam will be built further downstream at Karuma falls. Two smaller run-of-the-river dams, North Ayago, and South Ayago together will boost power by at least 500 MW. Uganda still needs more hydropower and plans to build a total of 14 hydropower dams in the future (Onyalla, 2007). Concerns about environmental issues and implementation of mitigation measures are essential elements that are needed, but often lacking in the planning of many dam projects.

Also in the power development scheme are Burundi, DRC and Rwanda; they have more rainfall and are desperate for power, but lack financial resources. The Kagera River, important to the water balance of Lake Victoria, originates in Burundi and defines borders with Rwanda, Tanzania and Uganda. Most of the Kagera flows through Rwanda. Burundi's interest in the Nile Basin is centred on the Kagera River, where development of hydropower generation is sought. Burundi, Rwanda and Tanzania are jointly constructing the multi-purpose Rusumo Dam and a power plant at Rusumo Falls where the Kagera River forms the boundary between Rwanda and Tanzania, which are on target to transmit power to Gitega in Burundi, Kigali to Kabarondo in Rwanda, and Biharamuro in Tanzania. The details of existing and planned major dams and barrages in the Nile Basin are summarized in Table 2.1. The table is compiled on the basis of multiple sources, as listed below the table.

Development goals of most Nile Basin countries are to reduce poverty, increase agricultural production and provide power for industrial growth. This is where the NBI's Shared Vision Program is set to help joint electrification projects across countries and by regions where ENSAP and NELSAP form joint investments for power transmission between countries. Region-wide transboundary electric trading has yet to be completed due to complications created in multi-country agreements.

Future Nile

What is the future of the Nile? Over the past 10 years the topic of Nile water conflict has appeared in countless articles and news briefs. The number of articles asking this question has increased with the signing of the 2010 Cooperative Framework and with a referendum in southern Sudan in January 2011, leading to a New Nile Basin country, South Sudan, in July 2011. Climate change is also looming in both current and future water developments on the Nile River (Hulme *et al.*, 2001). Pressure on water resources remains the key factor in the political and economic development of the Nile Basin countries, especially with population growth predicted to reach 600 million by 2030.

'Climate change will hit Africa worst', according to Waako *et al.* (2009), who states that climate change is now becoming a key driver in considerations over food and energy security

Table 2.1 Major dams and barrages finished, unfinished and planned in the Nile Basin

Country	*Name of dam*	*River*	*Year completed (or to be started)*	*Power (MW)*	*Storage (m^3)*	*Contractor*
1900s to 1970 post-independence						
Egypt	Assiut barrage	Main Nile	1902	Irrigation		
Egypt	Esna barrage	Main Nile	1908	Irrigation		
Egypt	Nag-Hamady barrage	Main Nile	1930	Irrigation		
Egypt	Old Aswan Dam	Main Nile	1933		450,000	
Egypt	High Aswan Dam	Main Nile	1970	2100	1,110,000	
Ethiopia	Tis-Abay	Lake Tana	1953	12		
Sudan	Sennar	Blue Nile	1925	48	0.93	
Sudan	Jebel Aulia	White Nile	1937	18		
Sudan	Khashm El Gibra	Atbara	1964	35	1.3	
Sudan	Roseires	Blue Nile	1966	60	2.386	
Uganda	Owen/Nalubaale	White Nile	1954	180	0.230	
1970 to present						
Ethiopia	Tekeze 5	Tekeze	2009–2010	300	9.2	
Sudan	Merowe	Main Nile	2009–2010	350	12	
Ethiopia	Finchaa	Finchaa	1971/2013	134	1050	
Ethiopia	CharaChara	Blue Nile	2000	84	9126	
Ethiopia	Koga	Blue Nile	2008	Irrigation	80	
Ethiopia	Tana Beles	Blue Nile	2011	460		
Kenya	Sondu Miriu	Victoria	2007	60	1.1	Japan
Uganda	Kiira/extension	White Nile	1993–2000	200		
Under construction (date gives completion date)						
Sudan	Roseires heightening	Blue Nile	2013			Multi-national
Sudan	Burdana	Setit/Atbara		135		China/Kuwait
Sudan	Rumela	Atbara		135		China/Kuwait
Sudan	Shiraik	Main Nile		300		
Ethiopia	FAN	Finchaa	2011			China/Italy
Ethiopia	Tekeze II	Tekeze	2020			
Ethiopia	Megech	Abay		Irrigation		Multi-national
Ethiopia	Ribb	Abay	2011			
Ethiopia	Grand Millennium	Blue Nile	2017	5250		China/Italy
Rwanda	Nyabarongo	Nyabarongo	2011	27.5		Australia/India
Uganda	Bujagali	White Nile	2011	250		Italy

Table 2.1 Continued

Country	*Name of dam*	*River*	*Year completed (or to be started)*	*Power (MW)*	*Storage (m^3)*	*Contractor*
Dams proposed (date gives potential start date)						
DRC	Semliki	Semliki				
Ethiopia	Jema	Jema				
Ethiopia	Karadobi	Blue Nile	2023	1600		ENSAP
Ethiopia	Border	Blue Nile	2026	1400		ENSAP
Ethiopia	Mabil	Blue Nile	2021			
Ethiopia	Beko Abo	Blue Nile		2000		ENSAP
Ethiopia	Mendaya	Blue Nile	2030	1700		ENSAP
Ethiopia	Chemoda/Yeda five dams proposed	Chemoga/ Yeda rivers	2015	278		China
Ethiopia	Baro I	Sobat				
Ethiopia	Baro II	Sobat				
Sudan	Nimule	Nile				
Sudan	Dal-1	Nile		400		
Sudan	Kajbar	Nile		300		
South Sudan	Bedden	Bahr el Jebel				Italy/NBI
South Sudan	Shukoli	Bahr el Jebel				Italy/NBI
South Sudan	Lakki	Bahr el Jebel				Italy/NBI
South Sudan	Fula	Bahr el Jebel				Italy/NBI
Uganda	Isimba	White Nile	2015	87		
Uganda	Kalagala	White Nile	2011	300		India
Uganda	Karuma	White Nile	2017	200		
Uganda	Murchison	White Nile	600			
Uganda	Ayago North	White Nile	2018	304		
Uganda	Ayago South	White Nile		234		
Uganda	15 small run-of-the river	Kagera				
Rwanda	Kikagate	Kagera	2016	10		
Rwanda	Nyabarongo	Kagera	2012	27		
Rwanda/ Tanzania/ Burundi	Rusumo I & II	Kagera	2012	60		NELSAP
Kenya	Goronga	Mara				
Kenya	Machove	Mara				
Kenya	Kilgoris	Mara				
Kenya	EwasoNgiro	Mara	2012	180		UK

Note: ENSAP = Eastern Nile Subsidiary Action Program

Sources: Ofcansky and Berry, 1991; Nicol, 2003; Scudder, 2005; Dams and Agriculture in Africa, 2007; McCartney, 2007; World Bank, 2007; UNEP, 2008; African Dams Briefing, 2010; Dams and Hydropower, 2010; Kizza *et al.*, 2010; Verhoeven, 2011; Sudan Dams Implementation Unit, undated

in the Nile. Models have been developed to prepare for climate change, but the results are inconclusive (Conway and Hulme, 1993, 1996; Strzepek and Yates 1996; Conway 2005; Kim *et al.*, 2008; Beyene *et al.*, 2009; Kizza *et al.*, 2010; Taye *et al.*, 2010). The issues that users of Nile water face are growing. Most upstream countries are going to want more water, but water is limited and the needs are growing. This creates the potential for conflict. Innovative policies and agricultural practices for the riparian countries are needed, before the situation comes to a hostile end. What steps need to be taken?

There is good news: the greatest potential increases in yields are in rain-fed areas, where many of the world's poorest rural people live and where managing water is the key to such increases (Molden and Oweis, 2007). Leaders need to allow for the creation of better water and land management practices in these areas to reduce poverty and increase productivity. Protecting ecosystems is vital to human survival and must be achieved in harmony. There are opportunities – in rain-fed, irrigated, livestock and fisheries systems – for preserving, even restoring, healthy ecosystems.

Upgrading the current irrigation systems and modernizing the technologies used in irrigation will improve production and make a sustainable irrigation sector in several Nile countries struggling to maintain systems that are no longer productive. Integration of livestock, fisheries and high-value crops will help boost farm incomes.

There is a need to plan for the future, using financial assistance to develop technical training in all countries, work on regional climate models for short- and long-term conditions, and develop methods for hydrometeorological forecasting and modelling of environmental conditions. Climate change will be a main factor in the Nile Basin water security in terms of filling dams and irrigation systems, and creating treaties. Extremes in longer droughts and heavier rains with floods are predicted over the vast areas of the Nile Basin; to cope with these extremes, UNEP and NBI joined forces in 2010 to create a project to prepare for climate change.

It is clear that more cooperation has the potential to generate more benefits equitably from Nile waters. However, the road to cooperation is not easy. But missing that road opens the door to unilateral decision-making, leading to more stress between communities and countries, possibly with disastrous impacts.

Conclusion

Yes, there are challenges, but there are solutions that can help: dialogue, trust and sharing benefits of the Nile water will help solve many of the conflicts between the Nile Basin countries. Cooperation is the key. While the role of the NBI has not yet ended, a comprehensive conclusion is more important now than ever, with water scarcity, increased development and climate change. While there have been important steps for cooperation, much more needs to be done urgently.

For future success in dealing with larger issues of poverty, food insecurity and climate change it will be necessary to conduct successful research with multinational teams working together effectively across borders in the Nile Basin. This is also important to avoid duplication of efforts, to ensure results are easily accessible and make all information commonly available to all the Nile Basin countries.

In the chapters that follow, new insights on poverty, water-related risks and vulnerability, including mapping of these, are provided for the Nile Basin. There is scope for improvement of crop, livestock and fish production in upstream countries, as will be described. Water productivity in the Nile Basin has a large variation. Use of hydronomic zoning in the Nile Basin has helped to identify various zones such as water source zones, environmentally sensitive zones and farming zones. Finally, through an all-inclusive sustainable and comprehensive agreement, with support from the Nile Commission and the NBI, contributions can be made to agriculture and socio-economic development in the Nile Basin. The past has made the Nile what it is today; it is up to the future to make the Nile provide for all who depend on it.

References

Abraham, Kinfe. (2002) The Nile Basin disequilibrium, *Perceptions, Journal of International Affairs*, Vol VII.

Abraham, Kinfe. (2004) *Nile Opportunities: Avenues Toward a Win Win Deal*, Dissemination Horn of Africa Democracy and Development International Lobby, Addis Ababa, Ethiopia.

Abu-Zeid, M. A. (1989) Environmental impacts of the High Dam, *Water Resources Development*, 5, 3, September, pp147–157.

Abu-Zeid, M. A. and El-Shibini, F. Z. (1997) Egypt's High Aswan Dam, *Water Resources Development*. 13, 2, 209–217.

African Dams Briefing (2010) *African Dams Briefing*, International Rivers, June, www.internationalrivers.org/files/AfrDamsBriefingJune2010.pdf, accessed February 2011.

Ahmad, Adil Mustafa (2008) Post-Jonglei planning in southern Sudan: combining environment with development, *Environment and Urbanization*, 20, 2, October, pp575–586.

Allan, T. (2001) *The Middle East Water Question: Hydro Politics and the Global Economy*, I. B. Taurus, London.

Allen, John (2010) *Transboundary Water Resources. The Sudd Wetlands and Jonglei Canal Project: Nile River Basin*, March, www.ce.utexas.edu/prof/mckinney/ce397/Topics/Nile/Nile_Sudd_2010.pdf, accessed 25 April 2012.

Arsano, Yacob (1997) Toward conflict prevention in the Nile Basin, paper presented at the Fifth Nile 2002 Conference held in Addis Ababa, 24–28 February.

Awulachew, S. B., Yilma, A. D., Loulseged, M., Loiskandl, W., Ayana, M. and Alamirew, T. (2007) *Water Resources and Irrigation Development in Ethiopia*, Working Paper 123, International Water Management Institute, Colombo, Sri Lanka.

Awulachew, S. B., Erkossa, T., Smakhtin, V. and Fernando, A. (2009) *Improved Water and Land Management in the Ethiopian Highlands: Its Impact on Downstream Stakeholders Dependent on the Blue Nile*, Dissemination Workshop, Addis Ababa, EthiopiaBarich, Barbara, (1998) People, water, grain: the beginning of domestication in the Nile Valley, in *Studia Archaeologica*, 98, L'Erma di Bretschneider (ed.), Rome Italy.

Bastiaanssen, Wim and Perry, Chris (2009) *Agricultural Water Use and Water Productivity in the Large-Scale Irrigation (LSI) Schemes of the Nile Basin*, WaterWatch, Wageningen, Netherlands.

Beyene, T., Lettenmaier, D. P., and Kabat, P. (2009) Hydrologic impacts of climate change on the Nile River basin: implications of the 2007 IPCC scenarios, *Climatic Change*, 100, 3–4, 433–461.

BBC (2010) Land grab fears for Ethiopian rural community, *BBC Business News*, www.bbc.co.uk/news/business-11991926, accessed April 2011.

Biswas, Asit K. (2002) Aswan Dam revisited: the benefits of a much-maligned dam. *D+C Development and Cooperation*, 6, November/December, pp25–27.

Block, Paul and Strzebek, K. (2010) Economic analysis of large scale up-stream river basin development on the Blue Nile in Ethiopia considering transient conditions of climate variability and climate change, *Journal of Water Resource Planning and Management, ASCE*, 136, 2, 156–166.

Block, P., Strzebek, Kenneth and Rajagopalan, Balaji (2007) *Integrated Management of the Blue Nile Basin in Ethiopia, Hydropower and Irrigation Modeling*, IFPRI, Washington, DC.

Bulto, T. S. (2009) Between ambivalence and necessity – occlusions on the path toward a basin-wide treaty in the Nile Basin, *Colorado Journal of International Environmental Law and Policy*, 291, p201.

Carroll, Christina M. (2000) Past and future legal framework of the Nile River Basin, *Georgetown International Environmental Law Review*, 269, 269–304.

Cascão, A. E. (2008) Changing power relations in the Nile River Basin: Unilateralism vs. cooperation? *Water Alternatives*, 2, 2, 245–265.

CIA (Central Intelligence Agency). (2011) *World Factbook Irrigated Land*, www.cia.gov/library/publications/the-world-factbook/fields/2146.html, accessed October 2011.

Conway, D. (2005) From headwater tributaries to international river: observing and adapting to climate variability and change in the Nile Basin, *Global Environmental Change*, 15, 99–114.

Conway, D. (2011) Adapting climate research for development in Africa, *Wiley Interdisciplinary Reviews: Climate Change*, 2, 3, 428–450.

Conway, D. and Hulme, M. (1993) Recent fluctuations in precipitation and runoff over the Nile sub-basins and their impact on main Nile discharge, *Climatic Change*, 25, 127–151.

Cowen, Richard (2007) Ancient Irrigation (Egypt and Iraq), chapter 17 in *Exploiting the Earth* (unpublished book), http://mygeologypage.ucdavis.edu/cowen/~GEL115/index.html, accessed August 2010.

Dams and Agriculture in Africa (2007) FAO AQUASTAT, May www.fao.org/nr/water/aquastat/damsafrica/Aquastat_Dams, accessed February 2011.

Dams and Hydropower (2010) Ministry of Water and Energy, Ethiopia, www.mowr.gov.et/index.php, accessed November 2010.

Deng, D. (2011) Land belongs to the community: demystifying the 'global land grab' in Southern Sudan, paper presented at IDS conference, April, Brighton, UK.

Diamond, Jared (1997) *Guns, Germs and Steel: The Fates of Human Societies*, W. W. Norton & Co, New York, NY.

Dougherty, James E. (1959) The Aswan decision in perspective, *Political Science Quarterly*, 74, 1, March, pp21–45.

Ehret, Christopher (2002) *The Civilizations of Africa: A History to 1800*, James Currey Publishers, London.

El Arifi, Salih A. (1988) Problems in planning extensive agricultural projects: the case of New Halfa, Sudan, *Applied Geography*, 8, 1, January, pp37–52.

El Gamal, Fathy (1999) Irrigation in Egypt and role of National Research Center, *Proceedings of the Annual Meeting of the Mediterranean Network on Collective Irrigation System*, Malta, 3–6 November, Option Méditerranéennes Series B, 31, http://www.iamb.it/par/activities/research/option_B31.pdf, accessed 24 April 2012.

El Khadi, Mostafa. (1998) *The Nile and History of Irrigation in Egypt*, Ministry of Public Works and Water Resources, ICID and MWRI, Cairo, Egypt.

ENTRO (Eastern Nile Technical Regional Office) (2006) *Eastern Nile Watershed Management Project, Cooperative Regional Assessment (CRA) for Watershed Management Transboundary Analysis Country Report*, ENTRO, Addis Ababa, Ethiopia.

Environmental Consulting Associates (2009) *The Potential of Regional Power Sector Integration: Nile Basin Initiative Transmission and Trading Sector Case Study*, Economic Consulting Associates Limited, London, UK

FAO (Food and Agriculture Organization of the United Nations) (2009) *Water Infrastructure in the Nile Basin*, FAO, Rome, Italy, www.fao.org/nr/water/faonile/WaterInfrastructure.pdf, accessed April 2011.

FAO (2010) *The State of Food Insecurity in the World*, www.fao.org/docrep/013/i1683e/i1683e.pdf, accessed November 2010.

Gozalbez Esteve, J. and Cebrian Flores, D. (2002) The Land of Ethiopia, in *Touching Ethiopia*, Shama Books, Addis Ababa, Ethiopia.

Grimal, Nicolas. (1988) *A History of Ancient Egypt*, Blackwell Publishers, Malden, MA.

Henze, Par Paul (2000) *Layers of Time: A History of Ethiopia*, C. Hurst and Co., London.

Howell, P.P., Lock, M. and Cobb, S. (1988) *The Jonglei Canal: Impact and Opportunity*, Cambridge University Press, Cambridge, UK.

Hughes, N. F. (1986) Changes in feeding biology of Nile perch, *Latesniloticus* (L) (Pisces: Centropomidae), in Lake Victoria, East Africa since its introduction in 1960, and its impact on the native fish community of the Nyanza Gulf, *Journal of Fish Biology*, 29, 541–548.

Hughes, R. H. and Hughes, J. S. (1992) *A Directory of African Wetlands*, IUCN/UNEP/WCMC, Gland/Nairobi/Cambridge.

Hulme, M., Doherty, R. M., Ngara, T., New, M. G. and Lister, D. (2001) African climate change: 1900–2100, *Climate Research*, 17, 2, 145–168.

Hurst, H. E. and Phillips, P. with Nile control staff (1933) Ten day mean and monthly mean discharges of the Nile and its tributaries, *The Nile Basin*, vol 4 and supplements 1–13, Government Press, Cairo, Egypt.

IDRC (International Development Research Center) and MoA (Ministry of Agriculture) (1983) *Allocation of Resources in Agricultural Research in Egypt*, IDRC and MoA, Cairo, Egypt.

IFPRI (International Food Policy Research Institute) (2009) *'Land Grabbing' by Foreign Investors in Developing Countries*, IFPRI publication written by Joachim von Braun, Ruth Meinzen-Dick and Ruth Suseela, www.ifpri.org/publication/land-grabbing-foreign-investors-developing-countries, accessed August 2011.

Johanson, D. and Edey, M. (1990) *Lucy, the Beginnings of Humankind*, Touchstone Books, New York, NY.

Johnson, J. L. (2009) Climate change and fishery sustainability in Lake Victoria, *African Journal of Tropical Hydrogeology and Fisheries*, 12, 31–36.

Keefer, Edward C. (1981) Great Britain, France, and the Ethiopian Tripartite Agreement of 1906, *Albion*, 13, 4, 364–380.

Kim, U., Kaluarachchi, J. J., and Smakhtin, V. U. (2008) *Climate Change Impacts on Hydrology and Water Resources of the Upper Blue Nile River Basin, Ethiopia*, Research Report 126, International Water Management Institute, Colombo, Sri Lanka.

Kizza, M., Kyaruzi, A., Isaboke, Z., Oonge, N., Rurihose, F., Ruhamya, C., Kigobe, M., Joel, N. and Nagawa, H. (2010) *Future Hydropower Scenarios under the Influence of Climate Change for the Riparian Countries of Lake Victoria Basin*, Coordinated by Dr Richard J. Kimwaga, University of Dar es Salaam, Tanzania, UNESCO-IHE, www.nbcbn.com/Project_Documents/Progress_Reports_2010/Integrated/Integrated-Future%20 Hydropower-Tanzania.pdf, accessed March 2011.

Kull, Daniel. (2006) *Connection between Recent Water Level Drop in Lake Victoria Dam Operation and Drought*, www.irn.org/programs/nile/pdf/060208vic.pdf, accessed April 2011.

Lamberts, Erwin (2009) *The Effects of Jonglei Canal Operation Scenarios on the Sudd Swamps in Southern Sudan*, Ms thesis, Twente University, The Netherlands, http://essay.utwente.nl/59163, accessed April 2011.

Lovgren, Stefan (2006) Nile explorers battled adversity, tragedy to find river source, *National Geographic News*, 19 April, http://news.nationalgeographic.com/news/2006/04/0419_060419_nile.html, accessed September 2010.

LVBC (Lake Victoria Basin Commission) (2011) *Lake Victoria Basin Digest*, Charles-Martin Jjuko (ed.), Lake Victoria Basin Commission, Kismu, Kenya, www.lvbcom.org.

LVFO (Lake Victoria Fisheries Organization) (2011) *State of Fish Stocks*, www.lvfo.org, accessed May 2011.

Martens, Anja Kristina (2009) *Impacts of Global Change on the Nile Basin: Options for Hydropolitical Reform in Egypt and Ethiopia*, Discussion Paper 01052, IFPRI, International Food Policy Research Institute, Washington, DC.

McCartney, Matthew P. (2007) *Decision Support Systems for Large Dam Planning and Operation in Africa*, IWMI Working Paper 119, International Water Management Institute, Colombo, Sri Lanka.

McCartney, M., Aleayehu, T., Shieferaw, S., Awulachew, Seleshi, B. (2010) *Evaluation of Current and Future Water Resources Development in the Lake Tana Basin, Ethiopia*, IWMI Research Report 139, International Water Management Institute, Colombo, Sri Lanka.

Mohamed, Y. A., Bastiaanssen, W. G. M. and Savenije, H. H. G. (2004) Spatial variability of evaporation and moisture storage in the swamps of the upper Nile studied by remote sensing techniques, *Journal of Hydrology*, 289, 145–164.

Mohamed, Y. A., Savenije, H. H. G., Bastiaanssen, W. G. M. and van den Hurk, B. J. J. M. (2006) New lessons on the Sudd hydrology learned from remote sensing and climate modeling, *Hydrology and Earth System Sciences*, 10, 507–518.

Molden, D. and Oweis, T. Y. (2007) Pathways for increasing agricultural water productivity, in *Comprehensive Assessment of Water Management in Agriculture and Water for Food, Water for Life*, Earthscan/International Water Management Institute, London/Colombo, pp279–310.

Molden, David, Seleshi Bekeli Awulachew, Karen Conniff, Lisa-Maria Rebelo, Yasir Mohamed, Don Peden, James Kinyangi, Paulo van Breugel, Aditi Mukherji, Ana Cascão, An Notenbaert, Solomon Seyoum Demise, Mohammed Abdel Meguid, Gamal el Naggar (2011) *Nile Basin Focal Project*, Synthesis Report, Project Number 59, Challenge Program on Water and Food and International Water Management Institute, Colombo, Sri Lanka.

Moorehead, Alan (1983) *The White Nile, and The Blue Nile*, Harper and Brothers, New York, NY.

Moszynski, Peter (2011) Southern Sudan needs to learn from the mistakes of the past, *The Guardian*, 31 January, www.guardian.co.uk/global-development/poverty-matters/2011/jan/31/south-sudan-future-development, accessed 24 April 2012.

Mustafa, H., El-Gamal, F. and Shalby, A. (2007) *Reuse of Low Quality Water in Egypt*, Water Management Research Institute, Delta Barrage, Cairo, Egypt.

NASA (National Aeronautics and Space Administration) (1985) STS51G-046-0001 Sudd Swamp, Sudan June 1985, http://eol.jsc.nasa.gov/sseop/EFS/photoinfo.pl?PHOTO=STS51G-46-1, accessed June 2011.

NBI (Nile Basin Initiative) (2001) *Strategic Action Program: Overview*, May, Nile Basin Secretariat in cooperation with the World Bank, Entebbe, Uganda.

Nicol, Alan (2003) *The Nile: Moving Beyond Cooperation, UNESCO-IHP – From Potential Conflict to Co-operation Potential,* www.unesco.org/water/wwap/pccp/case_studies.shtml, accessed March 2011.

Oczkowski, Autumn, Nixon, Scott W., Granger, Stephen, El-Sayed, Abdel-Fattah M. and McKinney, Richard A. (2009) Anthropogenic enhancement of Egypt's Mediterranean fisheries, *Proceedings of the National Academy of Sciences*, 106, 5, 1364–1367.

Ofcansky, Thomas P. and Berry, LaVerle (1991) The reign of Menelik II: 1889–1913, in *Ethiopia: A Country Study*, Library of Congress, Washington, DC.

Oliver, Roland and Fage, D. J. (1962) *A Short History of Africa*, sixth edition, Penguin African Library, London, UK.

Onyalla, Harriette (2007) *New Vision: Government Proposes 14 New Dam Sites*, http://newvision.co.ug/D/8/13/555453/Isimba%20Dam, accessed February 2011.

Onyango, P. O. (2003) Malnutrition prevalence in Lake Victoria catchment area, Tanzania: Is fish export the root cause? in *Proceedings of the LVEMP Tanzania 2001 Scientific Conference*, Ndaro, S. G. M. and Kishimba (eds), Jamana Printers, Dar es Salaam, pp128–143

Pankhurst, Richard (1997) *The Ethiopian Borderlands: Essays in Regional History from Ancient Times to the End of the 18th Century*, Red Sea Press, Inc., Asmara, Eritrea.

Postel, Sandra (1999) *Pillar of Sand: Can the Irrigation Miracle Last?* W.W. Norton Company, New York, NY.

Said, R. (1981) *The Geological Evolution of the River Nile*, Springer Verlag, New York , NY.

Salama, R. B. (1997) *Rift Basins of Sudan: African Basins, Sedimentary Basins of the World, 3*, Selley, R.C. (ed), Elsevier, Amsterdam, The Netherlands.

Scudder, Thayer (2005) The large dams dispute and the future of large dams, in *The Future of Large Dams: Dealing with Social, Environmental and Political Costs*, Earthscan, London.

Shinn, David, H. (2006) *Nile Basin Relations: Egypt, Sudan and Ethiopia*, Elliot School of International Affairs, George Washington University, Washington, DC, http://elliott.gwu.edu/news/speeches/shinn0706_nile-basin.cfm, accessed February 2011.

Strzepek, K. M. and Yates, D. N. (1996) Economic and social adaptation to climate change impacts on water resources: a case study of Egypt, *International Journal for Water Resources Discussion*, 12, 229–244.

Sudan Dams Implementation Unit (undated) *Government of Sudan*, www.diu.gov.sd/en/index.php, accessed February 2011.

Sutcliffe, J. V. and Parks, Y. P. (1999) *The Hydrology of the Nile*, IAHS Special Publication 5, IAHS Press, Wallingford, UK.

Talbot, M. R. and Williams, M. A. J. (2009) Cenozoic evolution of the Nile, in Henry J. Dumott (ed.) *The Nile*, Monographiae Biologicae vol 89, Springer, New York, NY.

Taye, M. T., Ntegeka, V., Ogiramoi, N. P. and Willems, P. (2010) Assessment of climate change impact on hydrological extremes in two source regions of the Nile River Basin, *Hydrology and Earth System Sciences Discussions*, 7, 5441–5465.

Tefera, Bezuayehu and Sterk, Geert (2006) Environmental impact of a hydropower dam in Finchaa's watershed, Ethiopia: land use changes, erosion problems, and soil and water conservation adoption, Sustainable Sloping Lands and Watershed Management Conference, Luang Prabang, Lao PDR.

Tvedt, Terje. (2004) *The River Nile in the Age of the British – Political Ecology and the Quest for Economic Power*, IB Taurus & Co, London.

UNEP (United Nations Environment Programme). (2007) *Sudan: Post-conflict Environmental Assessment*, United Nations Environment Programme, Nairobi, Kenya, www.unep.org/climatechange/adaptation/EcosystemBasedAdaptation/NileRiverBasin, accessed September 2010.

US Department of the Interior (1964) *Land and Water Resources of the Blue Nile Basin, Ethiopia, Appendixes III, IV, and V*, Bureau of Reclamation, Washington, DC.

Vavilov, N. I. (1940) *Origin and Geography of Cultivated Plants*, translated by Doris Löve, Cambridge University Press, Cambridge, UK.

Verhoeven Harry (2011) *Black Gold for Blue Gold? Sudan's Oil, Ethiopia's Water and Regional Integration*, Africa Programme Briefing Paper, Chatham House, London, www.chathamhouse.org.uk/research/africa/papers, accessed March 2011.

Vidal, John (2010) How food and water are driving a 21st century 'land grab', *Guardian*, 7 March, www.guardian.co.uk/environment/2010/mar/07/food-water-africa-land-grab, accessed May 2011.

Waako, T., Thuo, S. and Ndayizeye, A. (2009) *Impact of Climate Change on the Nile River Basin*, Nile Basin Initiative Secretariat, Entebbe, Uganda.

Wallach, Bret (2004) Irrigation in Sudan since independence, *The Geographical Review*, 74, 2, April, pp127–144.

Warren, John. (2006) *Evaporites: Sediments, Resources and Hydrocarbons*, Springer, Berlin, Germany.

WaterWatch. (2009) *The Nile Basin: Irrigation Water Management*, poster, www.waterwatch.nl/publications/posters/nile-basin-irrigation-water-management.html, accessed September 2010.

Williams, M. A. J. and Williams, F. (1980) Evolution of Nile Basin, in *The Sahara and the Nile*, Williams, M. A. J. and Faure, H. (eds), Balkema, Rotterdam, The Netherlands, pp207–224.

Wolf, Aaron T. and Newton, Joshua T. (2007) *Case Study of Transboundary Freshwater Dispute Resolution: The Nile Waters Agreement*, Oregon State University, Corvallis, OR, www.transboundarywaters.orst.edu/research/case_studies/Nile_New.htm, accessed February 2011.

Wood, James and Guth, Alex (2009) East Africa's Great Rift Valley: A Complex Rift System, Michigan Technological University, *Geology*, http://geology.com/articles/east-africa-rift.shtml, accessed September 2010.

World Bank (1998) *Map of the Nile Basin for NBI and Nile Equatorial Lakes Subsidiary Programme (NELSAP)*, http://siteresources.worldbank.org/INTAFRNILEBASINI/About%20Us/21082459/Nile_River_Basin.htm, accessed September 2011.

World Bank (2007) *Ethiopia/NBI Power Export Project: Ethiopia/Sudan Interconnector*, http://siteresources.worldbank.org/INTETHIOPIA/Resources/Part-3.pdf, accessed February 2011.

3

The Nile Basin, people, poverty and vulnerability

James Kinyangi, Don Peden, Mario Herrero, Aster Tsige, Tom Ouna and An Notenbaert

Key messages

- Access to water is a key determinant, but not the sole determinant, of poverty in the Nile riparian countries.
- Lack of access to water resources traps millions of people, particularly women and children, in poverty throughout the diverse agricultural and grazing lands of the Nile River Basin.
- Key indicators of poverty include lack of education, low national gross domestic product (GDP), stunting in children, and low levels of consumption, foreign investment and employment.
- Spatially, rural populations tend to have higher poverty levels than urban dwellers. This phenomenon is closely linked to limited access to services in relatively remote agricultural pastoral areas.
- Population growth is the primary driver of agricultural intensification, which appears to enhance vulnerability to biophysical shocks in pastoral, agro-pastoral and cultivated production systems.
- Lack of access to water imposes high labour costs associated with trekking to water sources. In rural areas, labour is the most important livelihood asset, and any constraint on labour supply (such as collecting water) greatly affects efforts to reduce poverty.

Introduction

The Nile Basin covers 3.1 million km^2, which includes 81,500 km^2 of lakes and 70,000 km^2 of swamps and wetlands. Current estimates of the basin population are about 173 million, or 57 per cent of the entire population of the 10 riparian countries: Burundi, Democratic Republic of Congo, Egypt, Eritrea, Ethiopia, Kenya, Rwanda, Sudan, Tanzania and Uganda. Most of the population is located in rural settlements with a high proportion of the population being dependent on rain-fed agriculture. The proportion of those living in rural settlements is highest in Rwanda (94%), Burundi (93%), Uganda (88%) and Ethiopia (85%), and lowest in Egypt (57%) and Sudan (67%). On average, more than half of the basin's population lives below the poverty line (US$1 a day). The World Bank (2007) estimates that per capita incomes range between US$100 and US$790, and the contribution of agriculture to GDP ranges from 17 per cent in Egypt to 55 per cent in Ethiopia. Therefore, the least-developed countries rely more on agriculture as a driver of economic development.

In terms of biophysical characteristics, precipitation is almost nil in the Sahara Desert but increases southward to about 1200–1600 mm yr^{-1} on the Ethiopian and equatorial lake plateaus (Conway, 2000). Mean annual rainfall is calculated as 615 mm with a maximum of 2060 mm. Monthly distribution of precipitation over the basin is characterized by unimodal rainfall patterns over the Ethiopian plateau and by bimodal rainfall patterns over the equatorial lake plateau (Mishra and Hata, 2006). Its distribution is highly variable within and across both countries and seasons, with extremes manifested in floods and droughts. Although rain-fed agriculture (including both livestock and crop production) is the most dominant production system, primary water use through developed infrastructure constitutes irrigation agriculture, industry, domestic supply, hydropower, navigation and fishing. The irrigated area of the basin is estimated at approximately 7.6 million ha, with the potential of increasing to 10.2 million ha (FAO, 1997). The basin supports large coverage of wetlands and marshes constituting environmental conservation areas. There are more than 100 protected areas recorded in nine countries, excluding Eritrea. In terms of water resources, the relative contribution to the mean annual Nile water at Aswan, estimated at 84.1 km^3, is approximately 57 per cent from the Blue Nile, 29 per cent from the White Nile and 14 per cent from the Atbara River. At the national level, with the exceptions of Egypt and Sudan, the Nile Basin countries suffer mostly from economic water scarcity as infrastructure and financial capacities rather than absolute water availability determine past and present access to water conditions in the sub-basins (Molden *et al.*, 2003).

Across the Nile Basin, there is little understanding of the relationships among water, food, poverty and vulnerability to biophysical and social risks in agricultural production systems. Our work aimed to establish a broad understanding of poverty and how it relates to access to water in the Nile's agricultural production systems. One hypothesis is that improving access to water and productivity can contribute to greater food security, nutrition, health status, income and resilience in income, and more diversified consumption patterns. Because data on direct measurements of vulnerability and poverty are scarce and are not available at high spatial resolutions, we identified multiple quantitative proxy indicators for linking poverty and access to water. Several candidate indicators – such as landholding, population and livestock density, access to markets, input use, child mortality/malnutrition, agricultural productivity, water supply and sanitation – have influenced poverty levels (Kristjanson *et al.*, 2005). Severity of floods and droughts is evaluated as water-related risks.

Population distribution

Figure 3.1 presents and projects country-level population distributions from 1970 to 2020 (WRI, 2007). Several countries (Ethiopia, Kenya, Sudan, Tanzania and Uganda) have high growth rates. From WRI projections, the population of the basin, which was about 160 million in 1990, is estimated to grow to 300 million by 2010 and 550 million in 2030. Figure 3.1 further shows projections of the population, tripling from 106 million in 1975 to an estimated 323 million in 2015.

Population pressure influences the magnitude of exposure to risk (Corbett, 1988). Thus, projections of population change are necessary when examining the likely scenarios of future trends in vulnerability to biophysical and social risks in agricultural systems in the basin. Unless there is urgent development of regional as well as national institutions to address vulnerability to water scarcity and access to water, livelihoods in most Nile riparian countries will probably deteriorate, and countries experiencing rapid population growth rates can expect to overstretch

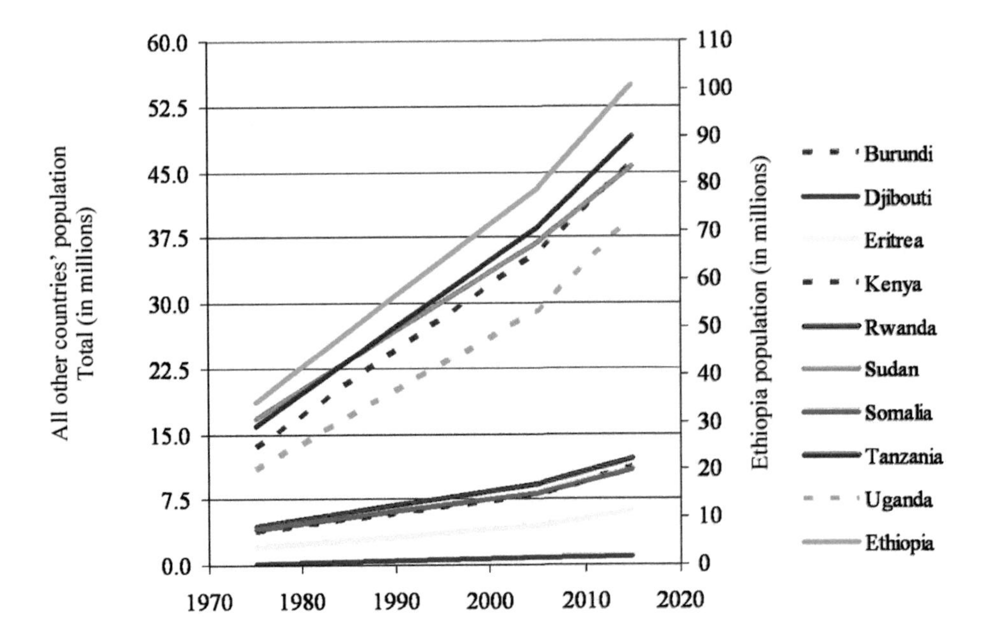

Figure 3.1 Population growth in the Nile Basin

Source: World Resources Institute, 2007

current public infrastructure. We need greater knowledge of the growth, density and distribution of human population in order to not only plan infrastructure, schools and hospitals, but also to assess future need for agricultural water development.

Demand for water resources in the basin

Among the most serious challenges facing water management in the basin are poverty and food insecurity, water shortages, land degradation and pollution from effluents (Wilson, 2007). Deforestation and cultivation of steep slopes have led to heavy soil erosion, loss of biodiversity, and sedimentation of downstream lakes and reservoirs (IWMI, 1999). The Lower Nile has also become seriously polluted by agro-chemicals, untreated sewage and industrial wastes (Mohamed *et al.*, 2005). In addition, there is poor water distribution and high loss upstream through excessive run-off, evaporation and abstraction (as much as 86% from eastern Nile basins), droughts and flooding, high rainfall variability, and high agricultural dependency accompanied by little or no technological transformation. Therefore, despite high potential, poor access to water for agriculture is a major constraint across several sub-basins. Figure 3.2 also shows that the difference in precipitation is several orders of magnitude among countries, pointing to variability in the water resources endowment.

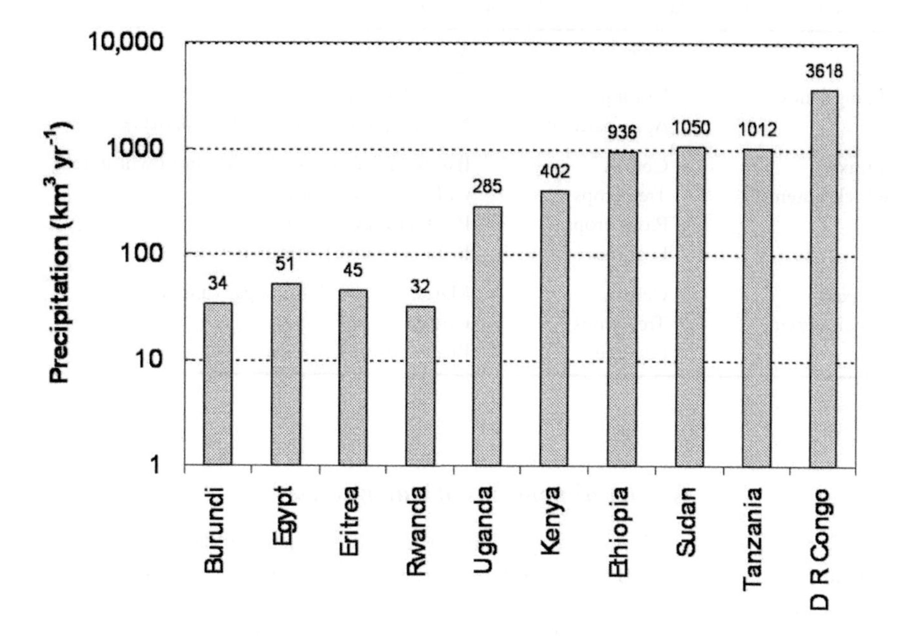

Figure 3.2 Water resources in the basin
Source: World Resources Institute, 2007

Agricultural production systems

Crop–livestock systems

The relationship between agriculture and land management has been described using a global classification system (e.g. Seré and Steinfeld, 1996; Dixon *et al.*, 2001). Four of these categories are applicable to agricultural systems in the Nile Basin. These include landless systems, livestock only/rangeland-based systems (areas with minimal cropping), mixed rain-fed systems (mostly rain-fed cropping combined with livestock) and mixed irrigated systems (a significant proportion of cropping uses irrigation and is interspersed with livestock). All but the landless systems are further disaggregated by agro-ecological potential as defined by the length of growing period (LGP; Kruska *et al.*, 2003). They comprise arid–semiarid (LGP <180 days), humid–sub-humid (LGP >180 days) and temperate tropical highland (LGP >180 days) regions.

These classifications (Table 3.1) have been widely used in a range of poverty, vulnerability and agricultural systems studies (e.g. Perry *et al.*, 2002; Thornton *et al.*, 2002, 2003, 2006; Kruska *et al.*, 2003; Fernández-Rivera *et al.*, 2004; Herrero *et al.*, 2008). We utilized the production system approach to examine variability in water, poverty and vulnerability in the Nile Basin, incorporating crop layers of rice, wheat, maize, sorghum, millet, barley, groundnut, cowpea, soybean, bean, cassava, potato, sweet potato, coffee, sugar cane, cotton, banana, cocoa and oil palm (You and Wood, 2004).

Table 3.1 Production systems classification in the Nile Basin

Broad class	*Crop group*	*Major crop types*
Rain-fed rangelands	Pastoral	Natural grasses and shrubs
	Agro-pastoral	Natural grasses, shrubs, sorghum, maize
Rain-fed mixed crop–livestock systems	Cereals	Barley, millet, maize, rice, sorghum wheat, teff
	Tree crops	Coffee, banana, cotton
	Root crops	Potato, cassava, sweet potato
	Legumes	Beans, cowpea, soybean, groundnut
Irrigated mixed crop–livestock systems	Cereals	Maize, rice, sorghum, sugar cane, wheat
	Tree crops	Cotton
	Legumes	Beans, cowpea, soybean

Gender dimensions of water access

Lack of access to water affects all poor people, but particularly women, children and the ageing (Van Koppen, 2001). In the Nile Basin, poor people are settled farther away from water sources than relatively wealthy individuals necessitating coverage of long distances to access water for livestock and domestic use. Agricultural water productivity also tends to be lower far from water sources (Peden *et al.*, 2009). In remote villages, elderly women often devote large amounts of labour in fetching water (Blackden and Wodon, 2006). Elsewhere, in Malawi for example, young mothers must choose between attending health clinics and staying at home to collect domestic water (Van Koppen *et al.*, 2007).

There is a major knowledge gap related to understanding gendered aspects of livestock and water management in the Nile Basin. Thus, we conducted a case study of the 'Cattle Corridor' in Uganda to examine gender differences related to water management and poverty. The Cattle Corridor is situated in the Victoria sub-basin, where the livelihoods of the pastoral communities are dependent on access to water and pasture in environments that are increasingly experiencing conflict for natural resources. In these communities, women play a significant role in managing household assets. Livestock herding is the dominant livelihood activity although crops are grown around the shores of Lake Kyoga in Nakasongola and throughout Kikatsi County in the Kiruhura district. Other activities such as charcoal burning, fishing, bee-keeping and local trade, help diversify rural household incomes. According to Mwebaze (2002), Uganda is divided into broad, yet distinct, farming systems depending on agro-ecological suitability resembling those shown in Table 3.1. These are pastoral, agro-pastoral with annual crops; banana–millet–cotton system and banana–coffee systems. The Nakasongola district is grouped into the banana–millet–cotton system of central Uganda, while Kiruhura is classified under pastoral and agro-pastoral with annual crops. In general, farmers in both districts cultivate similar crops such as cassava (*Manihot esculenta*), maize (*Zea mays*), sorghum (*Sorghum bicolor*), sweet potato (*Ipomea batatas*) and groundnut/peanut (*Arachis hypogea*). The exceptions are cotton, which is grown only in Nakasongola, and bananas, exclusively in Kiruhura district.

In the agro-pastoral systems of Uganda, poor people are settled away from water sources necessitating coverage of long distances to access water for livestock and domestic use. Table 3.2 provides an overview of how gender roles are disaggregated among men, women, boys, girls and hired labour in order to meet responsibilities for various water activities. In the two cattle districts of Nakasongola and Kiruhura, men and boys are primarily responsible for watering livestock while women and girls fetch water for domestic use. For both roles, hired labour is deployed at Kiruhura but less so at Nakasongola. Men, boys and girls spend much time watering animals and collecting water. Women are affected more by increased distances to water points whereas children are disrupted from attending school denying poor households opportunities to exit poverty. Providing equitable access to water for domestic use and for agriculture is essential for ensuring that investments in agricultural water development, contribute to poverty reduction.

Table 3.2 Ratings of various gender roles in water access and utilization in Uganda's Cattle Corridor

Activity/ Responsibility	*Nakasongola*					*Kiruhura*				
	Men	*Boys*	*Women*	*Girls*	*Hired labour*	*Men*	*Boys*	*Women*	*Girls*	*Hired labour*
Fetching water for domestic purposes	Low	High	High	High	Low	Very low	Very high	Low	High	High
Fetching water from wells and boreholes	Low	High	High	High	Low	Very low	Very high	Low	High	High
Watering livestock at home	High	High	High	High	Low	Medium	High	High	High	High
Watering livestock at valley tanks	High	High	Low	Low	Low	Very high	Very high	Very low	Very low	Very high
Taking livestock to the river or lake	Very high	High	Very low	Very low	High	Very high	Medium	Nil	Nil	High
De-silting wells or valley tanks	Very high	Very high	Nil	Nil	High	Medium	Low	Nil	Nil	High
Cleaning and repairing boreholes	High	Low	Nil	Nil	Very high	Low	Low	Nil	Nil	High

Note: Very high = more than 85%; high = 60–85% involved; medium = 50–60%; low = 30–50%; very low = less than 30%; nil = not involved at all

Poverty profiles of the Nile Basin

Poverty is generally thought of in terms of deprivation, either in relation to some basic minimum needs or in relation to the resources necessary to meet these minimum basic needs (ILRI, 2002; Cook and Gichuki, 2006). According to Cook *et al.* (2011), although there are varying ideas about what this basic level consists of, the three dominant approaches to poverty analysis that have featured in the development literature are the following:

- The poverty line approach, which measures the economic 'means' that households and individuals have to meet their basic needs
- The capabilities approach, which explores a broader range of means as well as ends
- Participatory poverty assessments (PPA), which explore the drivers and outcomes of poverty in more context-specific ways

Poverty line measurements equate well-being with the satisfaction individuals achieve through the consumption of various goods and services. The poverty line approach is therefore the most widely used way of establishing a threshold for the separation of poor from non-poor. Table 3.3 shows poverty line estimates in four countries across three agro-ecological regions in the mixed rain-fed production system of the Nile Basin. The range in poverty levels is large (29–70%) and the variability in the number of people living below the poverty line is a manifestation of the complex geographical as well as socioeconomic characteristics of the countries found in the basin.

Table 3.3 Poverty levels (%) in rain-fed crop-livestock production systems of selected examples of Nile riparian countries

Mixed rain-fed system	*Ethiopia*	*Uganda*	*Kenya*	*Rwanda*
Arid	56.2	42.3	62.1	60.4
Highlands	63.5	42.5	60.3	69.7
Temperate	39.2	29	50.1	64.1

Source: ILRI database (www.ilri.cgiar.org/gis)

Indicators of well-being

As indicators can be used in scientific, economic and social contexts to infer the quality of life of individuals, certain observations on the social and economic well-being of some countries can be drawn from Human Development Report Office (2007). These are highlighted below.

Education

Literacy rates indicate the level of interaction for productive economic and active social integration of members of the population older than 15 years. There are wide variations in the basin countries for this measure. The rate is significantly lower in Ethiopia (35.9%) while in the other eight countries it ranges from 59.3 to 73.6 per cent. About two-thirds of the adults over 15 years are literate, even though school enrolment in two-thirds of the countries is below 50 per cent.

Gross domestic product

Egypt has a gross domestic product that is 2 to 5 times more than that of other countries in the basin, clearly demonstrating that for its population, national investments in agricultural development continue to be a key driver of economic growth. However, some of the gains recorded in GDP growth may not be attributed to agriculture alone as Egypt has a well-developed commercial and services sector, in addition to being an oil-based economy.

Health

Except for Burundi, Ethiopia and Sudan, where 40 per cent of children under the age of 5 years are underweight, the rest of the countries show the proportion of underweight children under the age of 5 years to be 20 per cent or less indicating overall low health per capita expenditure. For this and other cultural reasons, HIV/AIDS prevalence varies from a high of 6 per cent upstream in East Africa to less than 2 per cent downstream in Egypt and the Sudan.

Consumption

With an annual change in the consumer price index of 424 per cent, it is difficult to meet basic consumption needs in the Democratic Republic of Congo as opposed to a change of less than 7 per cent in Egypt, Ethiopia and Kenya. However, part of this disparity in annual consumer price index change may be attributed to the effects of the ongoing conflict in the Congo.

Investments

Most basin countries have received low levels of direct foreign investments indicating that the economic environment is not conducive to greater trade, based on inflows of capital goods and services from foreign investments. However, this may now be changing with foreign commercial investors acquiring agricultural land in countries such as Ethiopia, Kenya, South Sudan, Sudan and Tanzania.

Employment

Apart from being the largest user of water, agriculture employs the largest proportion of available labour. It accounts for more than 80 per cent of employment in Ethiopia, Rwanda and Tanzania. Other potential employment sectors include industry and services which constitute 70 and 60 per cent of employment in Egypt and Kenya, respectively.

Gender empowerment

Taken as a measure of earned income (US$ purchasing power parity equivalents, PPP), which explains how income would be distributed among gender groups, it is lowest in Egypt (0.26) and highest in Uganda (0.6), indicating that there are significant differences in earned incomes between the genders. For this measure, Ethiopia represents an equal measure in earned income (0.48), suggesting that earned income is nearly equally distributed between the genders.

Poverty mapping

Figure 3.3 is a spatial representation and analysis of indicators of human well-being and poverty in the Nile Basin countries. The type of poverty expressed is income poverty. It is related to the ability of people to meet their income needs. This form of poverty is widespread, since many of the Nile countries have agricultural economies with rural agrarian populations. The poverty map highlights variation aggregated by national-level indicators which often hide important differences among different regions and countries in the Nile Basin. In almost all countries, these differences exist and can often be substantial. For the countries presented in Figure 3.3, recent welfare and economic well-being surveys commissioned by the World Bank reveal that poverty levels are related to rural and urban inequalities and access to services (World Bank, 2002, 2003, 2005, 2006, 2007). In Ethiopia, unique geographical disparities occur, but on average, households are 10 km away from a dry weather road and 18 km from public transport services. Therefore, it takes significantly longer to reach markets in rural Ethiopia than elsewhere. Another poverty attribute is land degradation whereby soil nutrient depletion continues at a faster rate than replenishment from mineral fertilizers. Due to population pressure, the survey found that one in five rural Ethiopian households lives on less than 0.08 ha person^{-1}, which yields, on average, only slightly more than half the daily cereal caloric needs per person, given current cereal production technologies. Gender inequalities are widespread; for example, girls are 12 per cent less likely than boys to be enrolled in school. In Uganda, the survey reported that most of the poor live in rural areas. They were characterized as subsistence farmers with limited access to infrastructure. The poor were 97 per cent rural, while the rich were classified as being more than 40 per cent urban. Inequality in Uganda continues to rise as the gap in mean income in rural and urban areas has widened, and inequality within both urban and rural areas has increased.

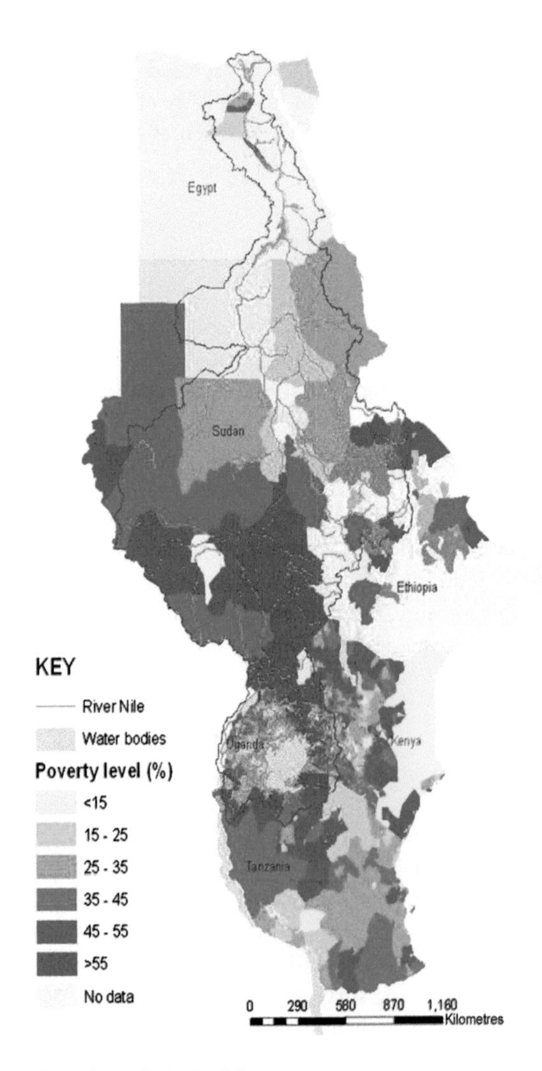

Figure 3.3 Poverty levels in the Nile Basin (%)

Source: Kinyangi *et al.*, 2009

In Kenya, the survey in 2005/2006 established that almost 47 per cent of Kenyans (17 million) were unable to meet the cost of buying the sufficient calories to meet their recommended daily requirements and minimal non-food needs. Almost one out of every five could not meet the cost of this minimal food bundle even if their entire budget was allocated to food items.

Egypt presents a slightly different situation. Due to rapid economic growth in the second half of the 1990s, average household expenditures rose, and poverty in Egypt fell compared with the early 1980s. During the survey period, less than 17 per cent of the population (10.7 million) lived below the national poverty line. In Sudan, agriculture forms the main source of livelihood and highly influences the level of poverty in the country. From the current income distribution from agriculture in the 15 states of North Sudan the overall average agricultural per capita income per day amounts to an equivalent of US$1.08 and varies from US$2.56 to US$0.61. Although there is variation among Sudanese states, these differences indicate that overall, the country has a high prevalence of poverty incidence.

In Rwanda, due to political instability, no household income and expenditure surveys have been conducted since 1994. However, using the 1996 nutritional survey (MINISANTE/UNICEF) as proxies for income it was possible to demonstrate that there is a strong correlation between high malnutrition rates in children and education. In Tanzania, the World Bank survey showed that GDP growth rates overall, and in agriculture, have increased in recent years, with an especially positive growth in 2004 when GDP overall grew by 6.7 per cent and agricultural GDP by 6.0 per cent. The survey concludes that the extent to which this growth has reduced poverty is mitigated by changes in inequality and may be affected by international and rural–urban terms of trade. In urban areas, growth had a greater impact on poverty reduction in areas where the proportion of households with incomes below the poverty line was lowest, indicating that poverty levels are sensitive to economic growth. Overall, these surveys show that the prevalence of poverty in the basin is determined by a wide set of factors, both natural and physical.

Vulnerability

Vulnerability is a very broad term, used differently in various contexts and disciplines (Turner *et al.*, 2003). Despite the multitude of meanings, most widely used definitions of vulnerability are based on the interaction of two fundamental characteristics: the frequency and magnitude of risks that a system is exposed to, and the ability of that system to withstand the impact of negative shocks (Kasperson and Kasperson, 2001).

Biophysical vulnerability

This form of exposure is linked to water-related poverty through the capacity of people and their environment to adjust to changing water capital and its related flow characteristics through the agricultural system. Capacity is determined as livelihood capital assets that modify access to water, water use, water capacity and the water environment. A biophysical vulnerability index is calculated by scoring natural asset indicators such as water and land suitability, and physical assets such as market access infrastructure (see Table 3.4).

Table 3.4 Dimensions incorporated in an index to assess biophysical vulnerability

Dimension	*Indicator*	*Dimension index*
Natural capital	Internal water resources	Dependency ratio
Physical capital (market access)	Accessibility to markets	Continuous index based on travel times to nearest urban markets
Natural capital (crop suitability)	Suitability for crop production	Suitability ranked on a score of 1–6; 1 is least suitable and 6 is most suitable

Social vulnerability

Social vulnerability is assessed through scoring social asset indicators of human conditions such as agricultural labour as well as financial assets for investing in water technologies. Table 3.5 shows that several indicators can be weighted and combined into a single index for mapping social conditions of the agricultural system.

Table 3.5 Dimensions incorporated in an index to assess social vulnerability

Dimension	*Indicator*	*Dimension index*
Agricultural dependency	Percentage of workers employed in agriculture	Agricultural dependency index or GDP as proxy
Poverty status	Human well-being	Poverty head count index (HI)

Who is vulnerable?

Vulnerable people generally have a variety of alternatives to increase their adaptability and decrease their risk in times of stress and shock (Kasperson and Kasperson, 2001). For vulnerable people, emergent changes are usually felt unequally throughout a community or region (Galvin *et al.*, 2001). In the Nile Basin, the future severity of impacts of changing water conditions on human populations will depend not only on water availability but also on the capacities of individuals and communities to respond to variability in basin water conditions.

Vulnerability mapping

We mapped several data sets that are major components of vulnerability in the three production systems. These are environmental and socio-economic resource base conditions that expose communities to vulnerability. Spatial data sets related to vulnerability or proxy indicators were used as a measure of vulnerability from earlier studies in the region (Thornton *et al.*, 2006). Risks related to three major factors (water availability and accessibility to water; biophysical resources endowment of an area; and prevailing socio-economic conditions) were mapped, analysed and combined to produce vulnerability layers which were based on the probability function.

Each of the three vulnerability layers was strictly composed of variables related to water, social and biophysical risks. Because each of these variables was measured on a different scale, it was first necessary to convert each of them into an index that ranged from 0 to 1. The indices were summed together and depending on the number of variables used, they were found to be directly proportional (i.e. the higher the index, the higher the vulnerability of a place). We adopted the calculation of the indices using the formula:

$$Vi = (Xi - Xi_{min}) / (Xi_{max} - Xi_{min})$$

where V_i is the standardized indicator *i*, X_i is the indicator before it is transformed, $X_{i'min}$ is the minimum score of the indicator *i* before it is transformed and $X_{i,max}$ is the maximum score of the indicator *i* before it is transformed.

All data were transformed into a relative score ranging from 0 to 1, which represented lowest to highest level of risk, respectively. However, the inverse applied to a number of variables mentioned below, where lowest values meant higher risk (e.g. in the dryness indicator a lower number of growing days means higher stress). Therefore, such indicators were further transformed using the formula $1 - X_i$.

The indices were then grouped together depending on the number of quality data sets available and used in each the three outputs that correspond to the agricultural production systems (5, 5 and 5 for social, water and biophysical risks, respectively).

Vulnerability in agricultural systems

The outcome from these combinations was vulnerability severity indices ranging from 0 to 6 levels. The vulnerability index represents how many risk levels a certain area is exposed to. The risks range from very high risk → high risk → moderate risk → low risk → very low risk. Table 3.6 shows the range for interpreting the level of risk for each of the four indicators for mapping biophysical vulnerability. However, the actual map (Figure 3.4) is built from the probability layers and the scale represents the level of risk of the biophysical indicators. In this way, both Figure 3.4 and Table 3.6 are interpreted together. The same applies to Figure 3.6 and Table 3.7 in the subsequent section.

Table 3.6 Level of exposure to biophysical risk

	Biophysical Indicators			
Level of exposure	*Renewable water ($mm^3\ yr^{-1}$)*	*Market access (hours)*	*Tropical Livestock Unit (TLU) (number km^{-2})*	*Population density (number km^{-2})*
High	10,000	<1	>40	<20
Medium	1041–8668	1 to 4	20–40	20–100
Low	0–1041	>4	0–9	100–1000

For all agricultural production systems, the level of risk of vulnerability to biophysical shocks ranges between 3.2 and 3.9. In the pastoral and agro-pastoral systems the levels of risk are 3.5 and 3.7, respectively, and the variation in the level of risk is less than 10 per cent in both systems. The total area covered under this level of risk is 3.9 million ha. In the mixed rain-fed

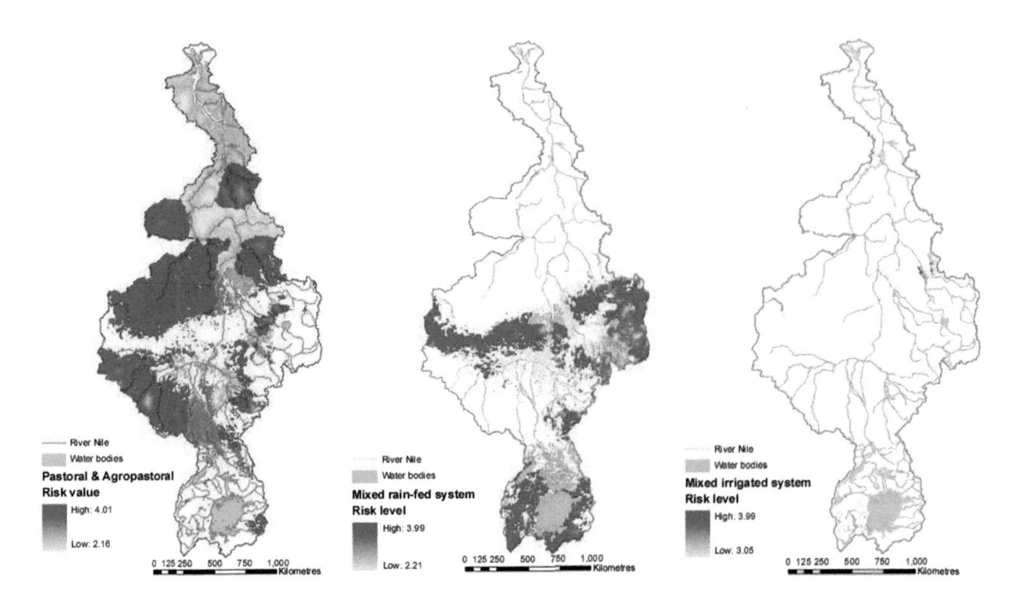

Figure 3.4 Biophysical vulnerability

Source: Kinyangi *et al.*, 2009

systems the level of risk of the four biophysical indicators is 3.7 across 1.1 million ha. In the mixed irrigated system, the total area covered is significantly lower: 0.4 million ha and a level of risk of 3.6. Overall, the entire basin is highly exposed to biophysical vulnerability suggesting negative attributes for human (100 to 1000) and livestock population (>40) density, but positive market access (<1) and internal renewable water resources (10,000). This indicates that the most intensely cultivated agricultural systems are most vulnerable to biophysical shocks (Figure 3.4). Given that these are the systems projected to further undergo agricultural intensification, there is need to focus on providing incentives to build the adaptive capacity of the agricultural communities in these systems. For example, an increase in the density of livestock is likely to exacerbate land degradation from poor management of pasture, pressure on water resources from higher stocking density and insecurity from human conflicts in pastoral systems.

For all agricultural systems, the level of risk of vulnerability to social shocks ranges between 1.5 and 3.2. In the pastoral and agro-pastoral systems the levels of risk are 2.5 and 3.2, respectively, and the variation in the level of risk is between 15 and 30 per cent in both systems. The total area covered under this level of risk is approximately 58.3 million ha. The mean levels of risk for the four biophysical indicators in the mixed rain-fed systems and mixed rain-fed cereals system across 26.4 million ha are 2.9 and 3.3, respectively. In the mixed irrigated system, the total area covered is significantly lower: 0.4 million ha with a risk level of 1.5. The rangelands and mixed rain-fed systems show high exposure to social shocks suggesting that there are significant negative impacts occurring from exposure to human diseases and child malnutrition and development (Figure 3.5).

Water-related risks in agricultural systems

For all agricultural systems, the level of risk of vulnerability to water hazards ranges between 0.79 and 1.89. In the pastoral and agro-pastoral systems the levels of risk are 1.7 and 1.9,

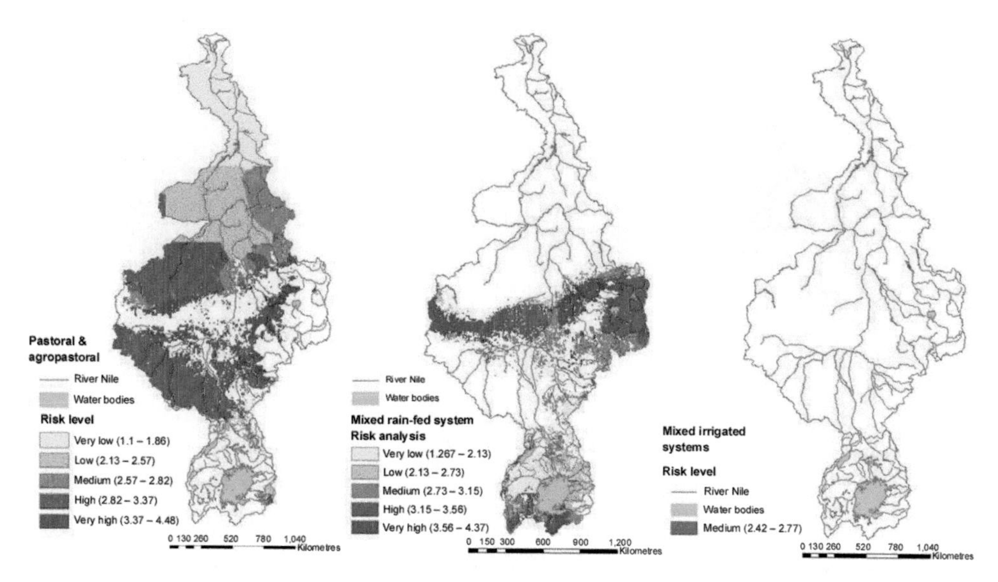

Figure 3.5 Social vulnerability
Source: Kinyangi *et al.*, 2009

respectively, and the variation in the level of risk is 20–25 per cent in both systems. The total area covered for this level of risk is 37 million ha. In the mixed rain-fed systems the level of risk of water-related indicators ranges between 1.5 and 1.6 across 12.2 million ha. In the mixed irrigated system, the total area covered is significantly lower (0.4 million ha), and a level of risk of 1.4 shows that irrigated agricultural production systems provide a lower level of exposure to water-related risks. Overall, mixed irrigated systems show low exposure to vulnerability to water-related hazards, suggesting that negative impacts occur for all four indicators of exposure to these hazards. This indicates that all of the area in pastoral, agro-pastoral and mixed rain-fed agriculture is highly exposed to vulnerability from water-related hazards, while mixed irrigated agricultural systems are less vulnerable to these hazards (Figure 3.6). Because agricultural water is managed in irrigated systems, the severity of impacts of variation in rainfall and the magnitude of loss in the length of the growing period, together with the exposure to drought conditions are well mitigated.

Table 3.7 Level of exposure to water-related risks

	Water-related risk indicators				
Level of exposure	*Coefficient of variation (CV) rain*	*LGP loss*	*LGP gain*	*Drought*	*Floods*
Low	0–20	0–5 (–ve)	0–5 (+ve)	0–1	0–1
Medium	20–40	5–20 (–ve)	5–20 (+ve)	1–2	1–2.5
High	40–233	>20 (–ve)	>20 (+ve)	>2 >2.5	

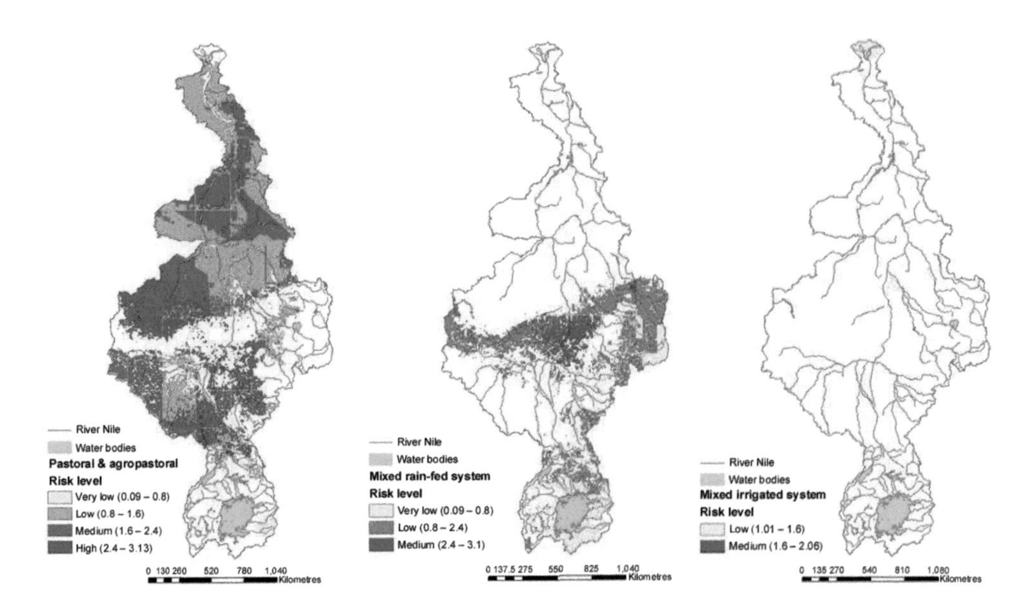

Figure 3.6 Water-related risks

Source: Kinyangi *et al.*, 2009

Conflict and cooperation

Due to the transboundary nature of the Nile, there are formidable obstacles to access to water and productivity. Equitable and effective water management and allocation and environmental protection depend on institutionalized cooperative agreements among riparian countries. Given the low precipitation in countries with high population densities in the sub-basins, sharing agreements are necessary to guarantee present and future access to water resources. For cooperative action, more needs to be done to rehabilitate degraded water catchments upstream, harvest and store water in rangeland and mixed rain-fed agricultural systems and manage flooding risks in irrigated systems downstream.

Conclusions

Intensive agricultural systems are most vulnerable to biophysical shocks. The key drivers of vulnerability to biophysical shocks are the expansion in human population and the intensification of crop-livestock system in hot spots of population growth.

Rangelands and mixed rain-fed systems show a high exposure to social shocks, suggesting negative attributes for human diseases and child malnutrition and development. With high prevalence of poverty incidence, these systems have a weak institutional capacity to cope with the negative impacts of food insecurity and diseases, especially among children and women.

Pastoral, agro-pastoral and mixed rain-fed agriculture is highly exposed to vulnerability from water-related hazards, while mixed irrigated agricultural systems are less vulnerable to them. Low exposure in mixed irrigated systems seems to be a function of better access to agricultural water. The rangelands and mixed agricultural systems rely on rain-fed agriculture and, therefore, these systems are prone to cycles of drought and flooding.

Communities with good access to water can use it for productive purposes, for food production, cottage industries, and so on. When communities or households have poor access to water, their labour supply is reduced due to the time needed to collect water for basic needs. Labour is the biggest asset most people have to earn an income, and its use in water collection reduces income generation potential.

There is a low risk of rainfall variation and changes in length of the growing season in the highlands, as well as in the Lake Victoria sub-basin, but widespread poverty is still unexplained by good market access.

References

Blackden, M. and Wodon, Q. (2006) *Gender, Time Use and Poverty in Sub-Saharan Africa*, World Bank, Washington, DC.

Conway, D. (2000) Some aspects of climate variability in the north east Ethiopian highlands – Wollo and Tigray, *SINET: Ethiopian Journal of Science*, 23, 2, 139–161.

Cook, S. and Gichuki, F. (2006) *Mapping Water Poverty*, Water, Agriculture and Poverty Linkages, BFP Working Paper 3, Challenge Program on Water and Food, Colombo, Sri Lanka.

Cook, S., Fisher, M. Tiemann, T. and Vidal, A. (2011) Water, food and poverty: global- and basin-scale analysis, *Water International*, 36, 1, 1–16.

Corbett, J. (1988) Famine and household coping strategies, *World Development*, 16, 9, 1099–1112.

Dixon, J., Gulliver, A. and Gibbon, D. (2001) *Farming Systems and Poverty: Improving Farmers' Livelihoods in a Changing World*, FAO/World Bank, Rome/Washington, DC.

FAO (Food and Agriculture Organization of the United Nations) (1997) Irrigation potential in Africa: a basin approach, *FAO Land and Water Bulletin*, 4, www.fao.org/docrep/W4347E/W4347E00.htm, accessed 28 December 2011.

Fernández-Rivera, S., Okike, I., Manyong, V., Williams, T., Kruska, R. and Tarawali, S. (2004) *Classification and Description of the Major Farming Systems Incorporating Ruminant Livestock in West Africa*, International Livestock Research Institute, Nairobi, Kenya.

Galvin, K. A., Boone, R. B., Smith, N. M. and Lynn, S. J. (2001) Impacts of climate variability on East African pastoralists: linking social science and remote sensing, *Climate Research*, 19, 161–172.

Herrero, M., Thornton, P. K., Kruska, R. L. and Reid, R. S. (2008) Systems dynamics and the spatial distribution of methane emissions from African domestic ruminants to 2030, *Agriculture, Ecosystems and Environment*, 126, 122–137.

Human Development Report Office (2007) *Climate Change and Human Development in Africa: Assessing the Risks and Vulnerability of Climate Change in Kenya, Malawi and Ethiopia*, Human Development Report Occasional Paper, 2007/2008, United Nations Development Programme, http://hdr.undp.org/en/reports/global/hdr2007-8/papers/IGAD.pdf, accessed 28 December 2011.

ILRI (International Livestock Research Institute) (2002) *Livestock – A Pathway out of Poverty: ILRI's Strategy to 2010*, ILRI, Nairobi, Kenya.

IWMI (International Water Management Institute) (1999) *Water Scarcity and Poverty*, Water Brief 3, IWMI, Colombo, Sri Lanka.

Kasperson, J. X. and Kasperson, R. E. (2001) *A Workshop Summary*, International Workshop on Vulnerability and Global Environmental Change, 17–19 May, Stockholm Environmental Institute, Stockholm, Sweden.

Kristjanson, P., Radeny, M., Baltenweck, I., Ogutu, J. and Notenbaert, A. (2005) Livelihood mapping and poverty correlates at a meso-scale in Kenya, *Food Policy*, 30, 568–583.

Kruska, R.L., Reid, R.S., Thornton, P.K., Henninger, N. and Kristjanson, P.M. (2003) Mapping livestock-orientated agricultural production systems for the developing world, *Agricultural Systems*, 77, 39–63.

Mishra, A. and Hata, T. (2006) A grid-based runoff generation and flow routing model for the Upper Blue Nile basin, *Hydrological Sciences Journal (Journal des Sciences Hydrologiques)*, 51, 191–206.

Mohamed, Y. A., van den Hurk, B. J. J. M., Savenije, H. H. G. and Bastiaanssen, W. G. M. (2005) The Nile Hydro-climatology: results from a regional climate model, *Hydrology and Earth System Sciences*, 9, 263–278.

Molden, D., Murray-Rust, H., Sakthivadivel, R. and Makin, I. (2003) A water-productivity framework for understanding and action, in *Water Productivity in Agriculture: Limits and Opportunities for Improvements*, W. Kijne, R. Barker and D. Molden (eds), CAB International, Wallingford, UK.

Mwebaze, S. (2002) *Pasture Improvement Technologies; Based on an On-farm Study in Uganda*, Working Paper No.18, Regional Land Management Unit, Nairobi, Kenya.
Peden, D., Alemayehu, M., Amede, T., Awulachew, S. B., Faki, F., Haileslassie, A., Herrero, M., Mapezda, E., Mpairwe, D., Musa, M. T., Taddesse, G. and van Breugel, P. (2009) *Nile Basin Livestock Water Productivity*, CPWF Project Report Series, PN37, Challenge Program on Water and Food (CPWF), Colombo, Sri Lanka.
Perry, B. D., McDermott, J. J., Randolph, T. F., Sones, K. R. and Thornton, P. K. (2002) *Investing in Animal Health Research to Alleviate Poverty*, International Livestock Research Institute, Nairobi, Kenya.
Seré, C. and Steinfeld, H. (1996) *World Livestock Production Systems: Current Status, Issues and Trends*, FAO Animal Production and Health Paper 127, Food and Agriculture Organization of the United Nations, Rome, Italy.
Thornton, P. K., Kruska, R. L., Henninger, N., Kristjanson, P. M., Reid, R. S., Atieno, F., Odero, A. and Ndegwa, T. (2002) *Mapping Poverty and Livestock in the Developing World*, International Livestock Research Institute, Nairobi, Kenya.
Thornton, P.K., Galvin, K.A. and Boone, R.B. (2003) An agro-pastoral household model for the rangelands of East Africa, *Agricultural Systems*, 76, 601–622.
Thornton, P. K., Jones, P. G., Owiyo, T. M., Kruska, R. L., Herrero, M., Kristjanson, P., Notenbaert, A., Bekele, N. and Omolo, A., with contributions from Orindi, V., Otiende, B., Ochieng, A., Bhadwal, S., Anantram, K., Nair, S., Kumar, V. and Kulkar, U. (2006) *Mapping Climate Vulnerability and Poverty in Africa*, Report to the Department for International Development, ILRI, Nairobi, Kenya.
Turner, B. L., Kasperson, R. E., Matson, P. A., McCarthy, J. J., Corell, R. W., Christensen, L., Eckley, N., Kasperson, J. X., Luers, A., Martello, M. L., Polsky, C., Pulsipher, A. and Schiller, A. (2003) A framework for vulnerability analysis in sustainability science, *Proceeding of the National Academy of Sciences*, 100, 14, 8074–8079.
Van Koppen, B. (2001) Gender in integrated water management: an analysis of variation, *Natural Resources Forum*, 25, 299–312.
Van Koppen, B., Giordano, M. and Butterworth, J. (2007) *Community-Based Water Law and Resource Management Reform in Developing Countries*, Comprehensive Assessment of Water Management in Agriculture 5, Column Designs Ltd., Reading, UK.
Wilson, R. T. (2007) Perceptions, Practices, principles and policies in the provision of livestock water in Africa, *Agricultural Water Management*, 90, 1–12.
World Bank (2002) *World Development Report 2002: Building Institutions for Markets*, Oxford University Press, New York, NY.
World Bank (2003) *World Development Report 2003: Sustainable Development in a Dynamic World: Transforming Institutions, Growth and Quality of Life*, World Bank, Washington, DC.
World Bank (2005) *World Development Report 2005: A Better Investment Climate for Everyone*, World Bank, Washington, DC.
World Bank (2006) *World Development Report 2006: Equity and Development*, World Bank, Washington, DC.
World Bank (2007) *World Development Report 2008: Development and the Next Generation*, World Bank, Washington, DC.
WRI (World Resources Institute) (2007) *Ideas into Action, Annual Report 2006–2007*, WRI, Washington, DC.
You, L. and Wood, S. (2004) *Assessing the Spatial Distribution of Crop Production Using a Cross-Entropy Method*, IFPRI, EPTD discussion paper no. 126, International Food Policy Research Institute, Washington, DC.

4

Spatial characterization of the Nile Basin for improved water management

Solomon S. Demissie, Seleshi B. Awulachew, David Molden and Aster D. Yilma

Key messages

- Hydronomic (water management) zones are instrumental in identifying and prioritizing water management issues and opportunities in different parts of a river basin. Such zoning facilitates the development of management strategies and informed decision-making during planning and operation.
- Hydronomic zones are identified using various maps of the basin, describing topography, climate, water sources and sinks, soil properties, vegetation types, and environmentally sensitive areas.
- Nineteen hydronomic zones are identified in the Nile Basin. Eighteen of these are identified based on six classes of humidity index and three soil classes. In addition, one environmentally sensitive zone is formed by merging wetlands and protected areas. The identified zones have unique climate and soil properties, and point to the need for distinct water management interventions in each zone.
- Nearly 15 per cent of the Nile Basin falls into water sources zone – where run-off is generated. About 10 per cent of the Nile Basin falls into the environmentally sensitive zone, where conservation and protection of the natural ecosystem should be promoted.

Introduction

The rapid population growth and associated environmental degradation have substantially increased the demand for terrestrial freshwater resources. Different economic sectors and riparian communities sharing river basins are competing for water consumption. The river system also requires an adequate amount of water for preserving its quality and for protecting its ecosystem. Moreover, climate variability and change would affect the availability of water required for human development and ecological functions. The current and anticipated challenges of the overwhelming disparity between water demand and supply could be addressed through managing the scarce freshwater resources in an effective and integrated manner within hydrological domains. However, if the water management practice fails to move away from isolated engagements within administrative boundaries, livelihoods, food security and environmental health would be compromised. The water management system should also focus on interventions that use water efficiently and improves productivity. The Nile River Basin covers expansive areas with greater topographic, climatic and hydro-ecological variability. The water management interventions should be very specific and most adaptable to the different parts of the basin. Therefore, it is essential to characterize the spatial variability of water management drivers in the Nile Basin and to classify the basin into similar water management zones.

Water management zones are instrumental in identifying and prioritizing the water management issues and opportunities in different parts of a river basin. Hence, the information and intervention requirements for addressing the water management issues and harnessing the opportunities in each zone could be exhaustively developed in the water development and monitoring strategies. Generally, classifying the river basin into water management zones facilitates development of management strategies and informed decision-making during planning and operation of water management interventions.

The concept of hydronomic (water management) zones was first developed by Molden *et al.* (2001). They proposed hydronomic zones as indispensable tools for defining, characterizing and developing management strategies for river basin areas with similar characteristics. They illustrated the potential of hydronomic zones in improved understanding of complex water interactions within river basins and assisting the development of water management strategies better tailored to different conditions within basins. They classified hydronomic zones based on the fate of water applied to the irrigation field. Later, Onyango *et al.* (2005) applied the hydronomic concept with that of terranomics (land management) to explore the linkages between water and land management in rain-fed agriculture and irrigation areas in the Nyando Basin, Kenya.

The main purpose of this chapter is to improve understanding of the Nile Basin characteristics using a spatial multivariate analysis of biophysical factors that significantly influence the development, management and protection of water resources of the basin. The relevant biophysical factors are used to classify the basin into similar water management zones that require identical interventions for efficient and sustainable development and management of the scarce water resources.

Hydronomic zones and classification methods

Adaptive and integrated water management of river basins is accepted as the best practice of developing, operating and protecting scarce water resources even under competing demands and climate change conditions. Classification of river basins into similar hydronomic zones facilitates efficient and sustainable application of adaptive and integrated water resources management. Molden *et al.* (2001) have developed and defined a set of six hydronomic zones based on similar hydrological, geological and topographical conditions, and the fate of water flowing from the zone. They demonstrated the concept of hydronomic zoning in four agricultural areas with similar characteristics: the Kirindi Oya Basin in Sri Lanka, the Nile Delta in Egypt, the Bhakra command area in Haryana, India, and the Gediz Basin in Turkey. The six hydronomic zones identified are: water source zone, natural recapture zone, regulated recapture zone, stagnation zone, final use zone and environmentally sensitive zone. In addition, two conditions that influence water management are defined in terms of presence or absence of appreciable salinity or pollution loading and availability or inaccessibility of groundwater for utilization. Generic strategies for irrigation in the four water management areas (the natural recapture, regulated recapture, final use, and stagnation zones) are presented in their analysis.

The water source zone and environmentally sensitive zone are also discussed in terms of their overall significance in basin water use and management.

Different classifications of physical systems have been developed to improve utilization of natural resources and protection of the environment. Koppen climate classification is one of the earliest attempts to classify the physical systems into zones of similar climatic patterns. The Koppen climate classification underwent successive improvements using improved precipitation and temperature records (Peel *et al.*, 2007). This climate classification method adopts different threshold values of parameters derived from monthly precipitation and temperature data sets for different climate zones. The other notable classification of the physical system relevant to water management is agro-ecological zones. The agro-ecological classification follows a GIS-based modelling framework that combines land evaluation methods with socio-economic and multi-criteria analyses to evaluate spatial and dynamic aspects of agriculture (Fischer *et al.*, 2002). The agro-ecological methodology provides a standardized objective framework for characterization of climate, soil and terrain conditions relevant to agricultural production.

The availability of spatial GIS and remote sensing information has contributed towards the advancement of classification methods from experience-based subjective decisions to data-intensive objective frameworks. Fraisse *et al.* (2001) applied principal components and unsupervised classification of topographic and soil attributes to develop site-specific management zones for variable application of agricultural inputs according to unique combinations of potential yield-limiting factors. Muthuwatta and Chemin (2003) developed vegetation growth zones for Sri Lanka through analysis and visual interpretation of remote sensing images of biomass production. They claimed that the vegetation growth zones would have better contribution to water resources planning than the agro-ecological zones since the vegetation growth zones are based on the prevailing environment and have strong linkages to hydrological processes.

Biophysical factors relevant to water management

The water management issues in a river basin are largely driven by the biophysical, socio-economic, institutional and ecological factors. Among these drivers of water management, the biophysical factors (such as climate, topography, soil, vegetation and hydro-ecological structures) are the most dominant. Therefore, these biophysical factors could provide the analytical platform required to objectively define the hydronomic zones. Moreover, the water management classification based on these static drivers of the river basin could provide an insight into the relationship among themselves and with water management indicators (Wagener *et al.*, 2007).

Loucks and Beek (2005) assert that a more complete large-scale perspective of the river system management could be achieved when watershed hydrology is combined with landscape ecology and actions in 'problem sheds'. Therefore, different factors that are related, either adversely or beneficially, to the water management issues of the basin should be exhaustively considered during classification of water management zones. The spatial distribution and disparity between water supply and demand within the basin require appropriate management strategies that consider constraints and opportunities of the basin water resources. Classification of the Nile Basin into hydronomic zones that have similar biophysical attributes would enable to devise adaptive and integrated water management strategies. The biophysical factors relevant to water management could be broadly categorized into climatic, hydrological, topographic, soil, vegetation and environmental factors. The following sub-sections provide brief descriptions and spatial patterns of these major categories of the biophysical attributes of the Nile Basin.

Topographic features

The topography of the river basin dictates the movement of water within the basin. The river basin classification into sub-basins and watersheds is primarily based on the altitude of the topography. Crop production and land suitability for agriculture are largely affected by topographic attributes. A high-gradient slope exposes the landscape for soil erosion and land degradation. The undulating topography also influences rainfall generating mechanisms in the mountainous areas. The aspect of sloping land surface could distinguish the rain-shadow part of mountain areas.

The upper parts of the Nile Basin have a ridged topography with steep slopes as depicted in Figure 4.1a, b. The central and downstream parts of the basin are predominantly flat areas. The impact of topography on movement of water within the basin and on the wetness of the underlying land surface could be characterized by a compound topographic index. The compound topographic index at the grid point in the basin is evaluated from its slope and the area that contribute flow to the grid point (USGS, 2000). The compound topographic index map of the Nile Basin in Figure 4.1c shows that flat areas of the basin that receive water from large upstream catchments have greater values of the topographic index. Such areas of the basin would have greater chances of becoming wet if the upstream catchments receive a substantial amount of precipitation.

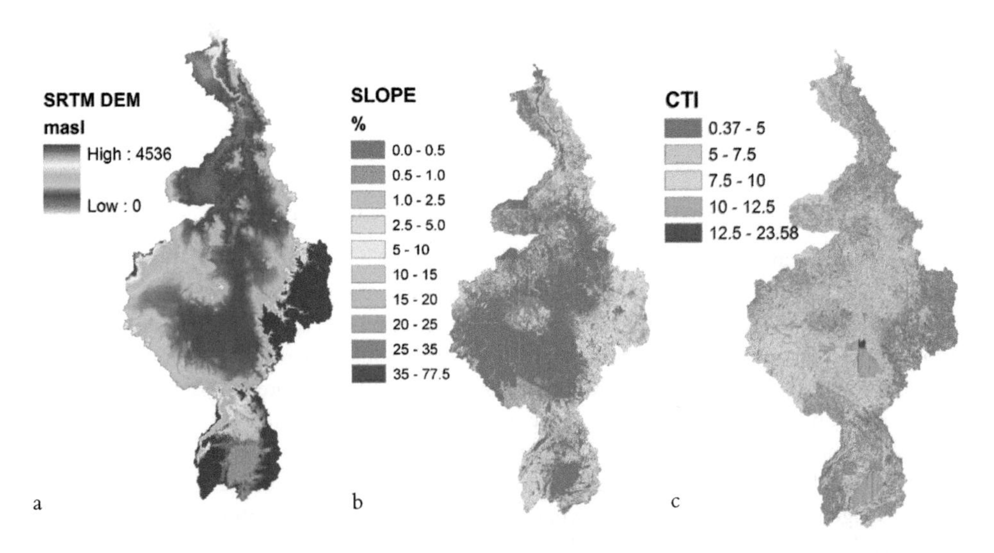

Figure 4.1 Topographic patterns of the Nile Basin: (a) Shuttle Radar Topography Mission Digital Elevation Model (metres above sea level), (b) slope (%) and (c) compound topographic index

Climatic and hydrological factors

The climate system is the major sources and sinks of water for river basins. While the climate system provides precipitation for the river basin, it takes away water in the form of evapotranspiration. The climate of the Nile Basin is largely driven by latitudinal contrasts of about 36°

from the southern (upstream) to the northern (downstream) ends. The Nile Basin climate can be broadly classified as arid, temperate and tropical. The Koppen–Geiger climate classification (Figure 4.2a) shows that the greater part of the basin is either arid desert hot or tropical savannah. The humidity index, the ratio of mean annual precipitation to potential evapotranspiration, characterizes the aridity or humidity of the basin. According to the humidity index derived from IWMI's Climate Atlas (Figure 4.2b), about half of the Nile Basin falls under the arid category. The Ethiopian Highland plateaus and equatorial lakes region below the Sudd wetlands are classified as humid zones.

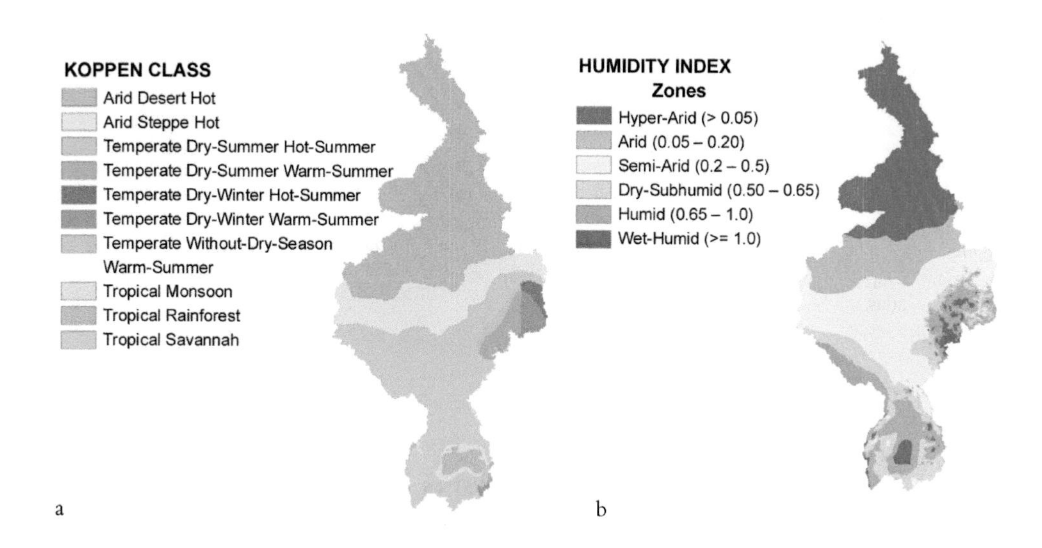

Figure 4.2 Climatic patterns of the Nile Basin from (a) the Koppen–Geiger climate classification and (b) humidity zones derived from the IWMI climate atlas

The hydrological cycle interrelates the physical processes and feedback mechanisms between the hydrological, atmospheric and lithospheric systems. The main sources and sinks of water in the river basin are precipitation and evapotranspiration, respectively. These climate variables exhibit temporal and spatial variability in the Nile Basin as depicted in Figure 4.3a, b, and this has resulted in very low average annual run-off, about 30 mm over the entire basin, as compared with the size of the basin, which is about 3 million km^2 (Sutcliffe and Parks, 1999). Despite their greater spatial variability, precipitation and evapotranspiration are some of the major factors that determine water availability within the river basin. Therefore, water source and deficit zones in the river basin can be identified by analysing differences between these climatic variables. The difference between mean annual precipitation and potential evapotranspiration in the Nile Basin (Figure 4.3c) reveals that most parts of the basin, particularly the central and downstream parts, are predominantly water-deficit zones. The water source zones are located in the Ethiopian Highland plateaus and the equatorial lakes region.

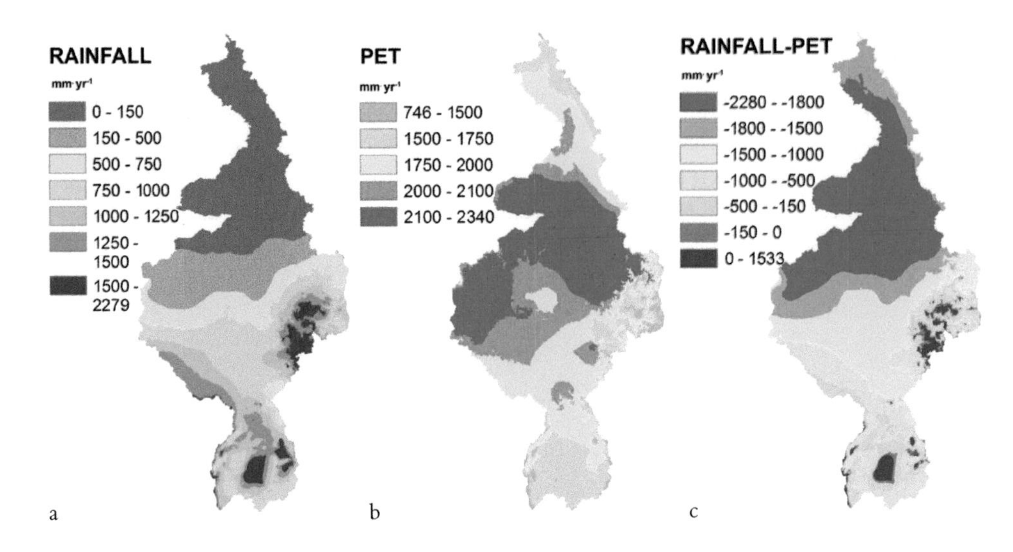

Figure 4.3 Water sources and sinks in the Nile Basin: (a) rainfall distribution, (b) potential evapotranspiration and (c) run-off production potentials derived from the IWMI climate atlas

Soil characteristics

The suitability of landscape for crop production largely depends on the soil properties of the landscape. Like slope, soil is one of the major factors for classifying lands for rain-fed and irrigation farming systems. Among the soil properties, texture, drainage, bulk density, available water content, electrical conductivity and calcium carbonate content could potentially describe the impact of soil on water resources management. These soil factors are obtained from the ISRIC-WISE derived data set (Batjes, 2006). The spatial patterns of the selected soil factors for the Nile Basin are illustrated in Figure 4.4.

Vegetation indices

The vegetation cover of the river basin has significant influence on the proportion of rainfall converted into direct run-off. Similarly, it also influences the infiltration rate of rainwater. Moreover, the degree of soil erosion and land degradation is largely related to vegetation cover. The degraded highland plateaus are producing substantial amounts of sediment that impair water storage facilitations and irrigation infrastructures in downstream parts of the basin. The Normalized Differenced Vegetation Index (NDVI) evaluated from the red and near-infrared reflectance of remotely sensed images characterizes the vegetation cover of the land surface. The United States Geological Survey (USGS) land use land cover map and the average annual SPOT NDVI plots in Figure 4.5 show that the spatial vegetation patterns in the Nile Basin are very similar to the climate patterns shown in Figure 4.2.

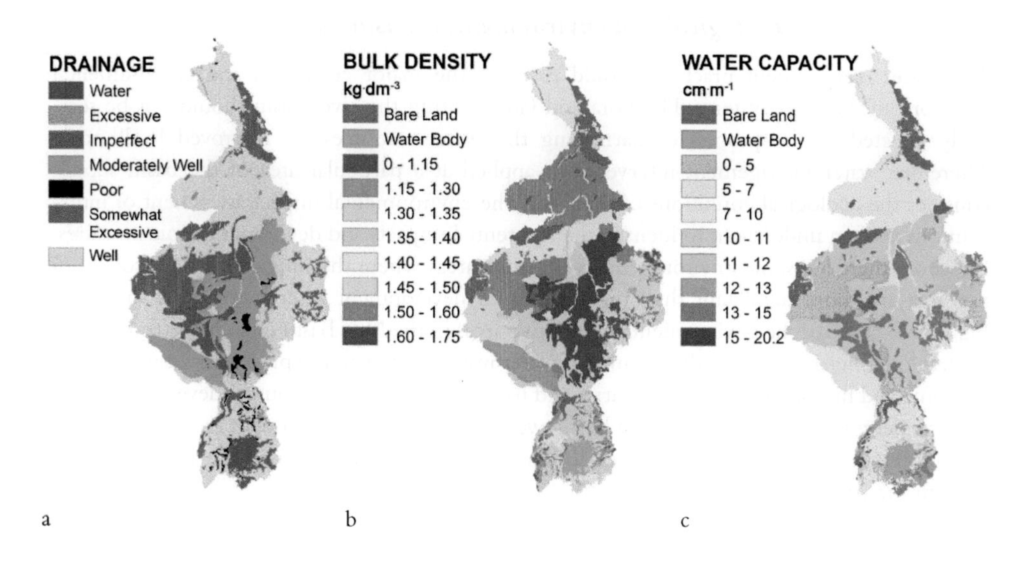

Figure 4.4 Soil properties in the Nile Basin: (a) drainage class, (b) bulk density (kg dm^{-3}) and (c) available water capacity (cm m^{-1}) derived from ISRIC-WISE data

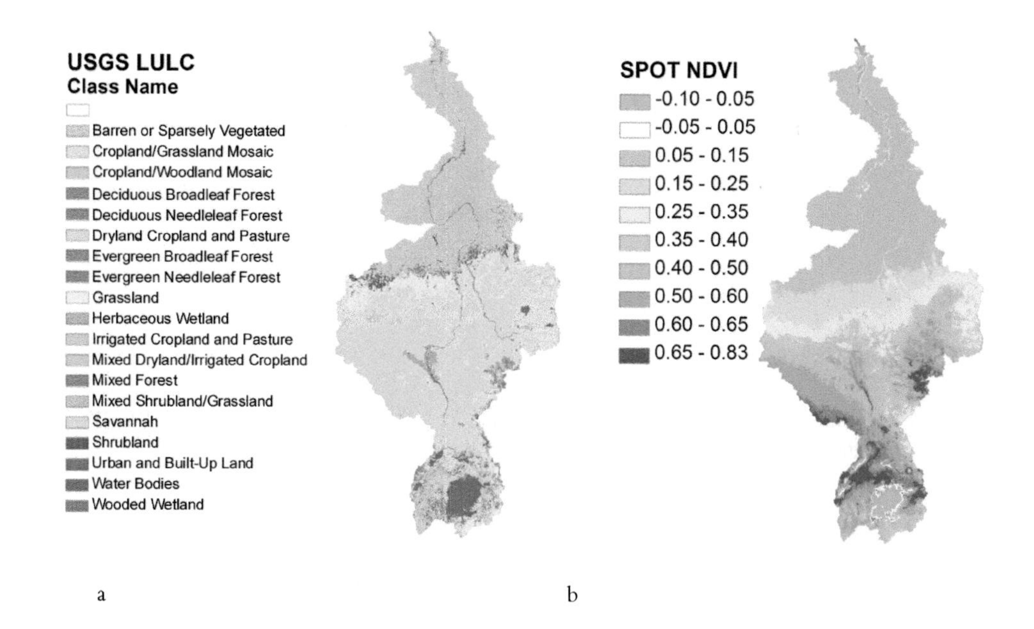

Figure 4.5 Vegetation profiles in the Nile Basin: (a) USGS land use land cover and (b) average SPOT NDVI (mean annual from 1999 to 2006)

Ecological and environmental considerations

The water management practices should preserve the major ecological and environmental functions of the river systems. The flora and fauna within the river basin should not be seriously affected in the process of harnessing the water resources for improved livelihoods. Therefore, water management interventions applied at a particular area of the basin should consider the ecological conditions of that area. The environmental impact assessment of interventions is often undertaken to identify their potential impacts and devise mitigation measures. However, there are some environmentally sensitive areas where the impacts on the ecology of the area are more important than the benefits of development interventions. As shown in Figure 4.6, some of the environmentally sensitive areas in the Nile Basin include wetlands, flood plains along the river course, the vicinity of water impoundments, and protected areas for natural, game and hunting reserves, sanctuaries and national parks. Water resources development and management interventions should not be allowed in such ecological hot-spot areas of the basin. Therefore, the water management zone should clearly delineate the environmentally sensitive areas in the basin.

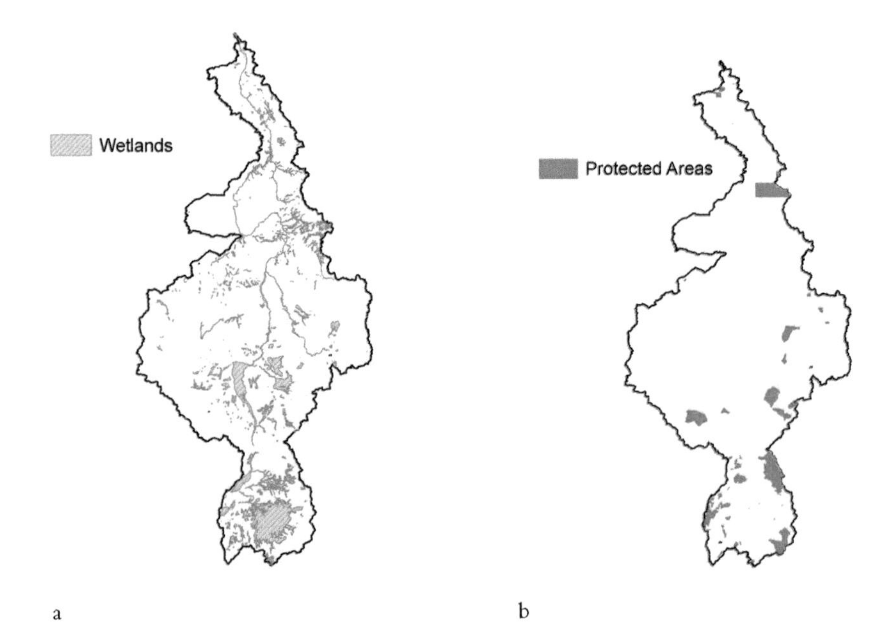

Figure 4.6 Environmentally sensitive areas: (a) wetlands and (b) protected areas compiled from IWMI's Integrated Database Information System Basin Kits

Multivariate analysis of basin characteristics

The biophysical factors of water management discussed in the previous section are obviously related to one another. For example, the climate and vegetation factors have similar spatial patterns in the Nile Basin. In fact, the Koppen climate classification was initially derived from vegetation cover since observed climate variables in the early twentieth century were very limited (Peel *et al.*, 2007). In order to use these biophysical factors for classification of water management zones, the interdependency between the factors should be removed. Moreover, the relative importance of the biophysical factors should be known to minimize the numbers of relevant factors used for classification of water management zones.

Principal components analysis (PCA) is a multivariate statistical technique that transforms interdependent multidimensional variables into significant and independent principal components of the variables with fewer dimensions. The PCA tool is employed for removing interdependency and reducing the dimensions of the biophysical factors of water management. After a preliminary analysis, the following six biophysical factors that represent climatic, topographic, soil and vegetation features of the basin are selected for principal components analysis: humidity index, landscape slope, compound topographic index, soil bulk density, available soil water content and normalized differenced vegetation index.

The selected biophysical factors are standardized by their respective means and standard deviations in order to comply with the Gaussian assumption of PCA and to give equal opportunity to factors with large and small numerical differences. The linear correlation matrix of the selected factors in Table 4.1 shows that the selection process has minimized the interdependency between the factors. The highest correlation was obtained between landscape slope and compound topographic index. The PCA transformation will remove these correlations between the selected factors.

Table 4.1 Linear correlation matrix of the relevant biophysical factors

	HI	*Slope*	*CTI*	*SBD*	*SWC*	*NDVI*
HI	1.00					
Slope	0.17	1.00				
CTI	–0.03	–0.49	1.00			
SBD	–0.19	–0.17	0.14	1.00		
SWC	0.00	0.14	–0.09	–0.22	1.00	
NDVI	0.40	0.09	–0.03	–0.27	–0.11	1.00

Note: HI = humidity index, Slope = landscape slope, CTI = compound topographic index, SBD = soil bulk density, SWC = available soil water content, NDVI = normalized differenced vegetation index

The principal component analysis of the standardized factors is performed using the selected six biophysical factors. The PCA evaluates the eigenvalues and eigenvectors of the covariance matrix of the standardized biophysical factors. The eigenvalue is literally the variance of the normalized factors explained by the corresponding principal component. The transpose of the eigenvectors provides the coefficients (weights) of the normalized factors for each principal component. The amount of the total variances of the normalized factors, which is equal to the number of variables (6), explained by each principal component, and the coefficients (weights) of the factors for each principal component are provided in Table 4.2. While the first principal component has explained half of the total variances of the six biophysical factors, the first three principal components have explained about 99 per cent of the total variance. Therefore, principal components would enable us to reduce the dimensions of the factors from six to two or three without losing significant spatial information.

Table 4.2 The percentage of variance of the biophysical factors explained by each principal component and the weights (coefficients) of the factors for the principal components

Principal components	*% of variance*	*HI*	*Slope*	*CTI*	*SBD*	*SWC*	*NDVI*
PC1	50.21	–0.254	–0.316	0.240	0.321	–0.022	–0.821
PC2	37.52	–0.052	0.545	–0.531	–0.145	0.471	–0.418
PC3	11.72	–0.032	0.311	–0.408	0.348	–0.780	–0.072
PC4	0.36	–0.103	–0.189	–0.094	–0.858	–0.383	–0.248
PC5	0.14	0.305	0.627	0.664	–0.127	–0.149	–0.187
PC6	0.06	0.910	–0.278	–0.211	0.039	0.000	–0.221

Note: HI = humidity index, Slope = landscape slope, CTI = compound topographic index, SBD = soil bulk density, SWC = available soil water content, NDVI = normalized differenced vegetation index

The weights of the biophysical factors, which linearly transform the relevant factors to the principal components, reveal that vegetation (NDVI), topographic (Slope and CTI) and soil (SWC) attributes are the most dominant factors for the first, the second and the third principal components, respectively. The graphical patterns of the principal components (Figure 4.7) are very similar to the corresponding biophysical factors.

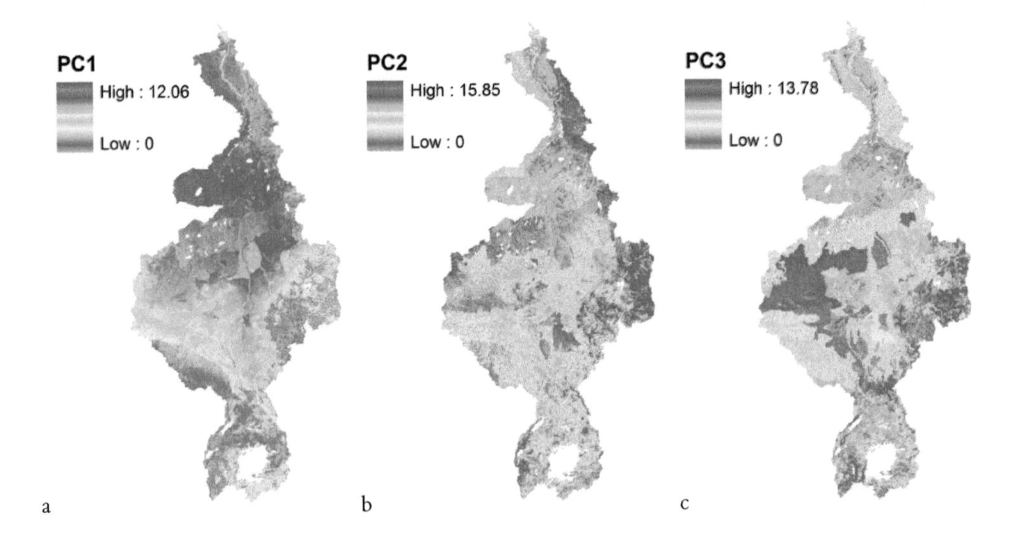

Figure 4.7 The dominant principal components of the biophysical factors: (a) PC1, (b) PC2 and (c) PC3

Classification of hydronomic zones

The similarity patterns of the biophysical factors discussed and the results of the principal components analysis are used to develop a classification framework for hydronomic zoning of the Nile Basin. Both subjective and objective approaches are employed in setting out the classification framework. The assessment of the biophysical factors indicated that climatic and vegetation attributes have similar spatial patterns in the Nile Basin. However, the principal components analysis of the relevant biophysical factors revealed that vegetation (NDVI) is the most dominant factor for water management classification, followed by topographic (Slope and CTI) and soil (SWC) attributes. The unsupervised classification of the first three principal components provided indicative patterns of the water management zones. But these zones are very patchy and often mixed up, since the analysis was performed at 1 km resolution. Therefore, the climatic factor (humidity index) that has distinctive zones is used as the primary (first-level) classification factor instead of NDVI since both factors have similar patterns. The humidity index in Figure 4.8a has six unique zones: hyper-arid (Ha), arid (Aa), semi-arid (Sa), dry subhumid (Ds), humid (Hh) and wet humid (Wh).

The topographic factors have greater spatial variability and could not provide distinct classes for the entire basin. These factors could provide better classification for sub-basins and catchments as suggested by the principal components analysis. Consequently, the soil attribute (SBD) is used for secondary (second-level) classification. The soil bulk density was divided into three classes: light soil (Ls), medium soil (Ms) and dense soil (Ds), as shown in Figure 4.8b. Hence, for each of the primary six classes defined by humidity index, there are three classes of soil attributes, which classify the basin into eighteen water management zones.

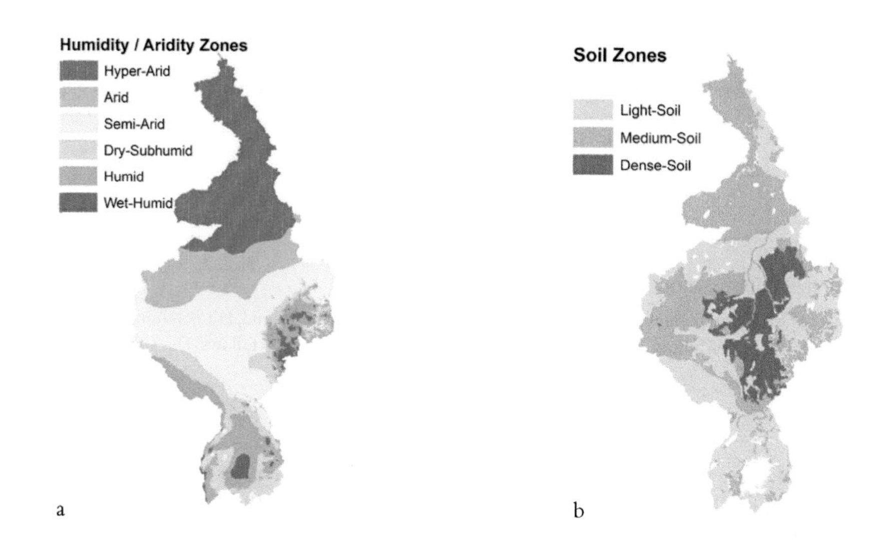

Figure 4.8 Water management classification framework for the Nile Basin: (a) humidity/aridity zones and (b) soil zones

Following the works of Molden *et al.* (2001), the environmentally sensitive (EnSe) zone was formed by merging the wetland and protected areas in Figure 4.6. The final hydronomic zones of the Nile Basin are developed by superimposing the EnSe zone over the eighteen identified zones (Figure 4.9).

The developed hydronomic zones of the Nile Basin have 19 distinct zones in which similar water management interventions could be applied. The hydronomic zoning includes

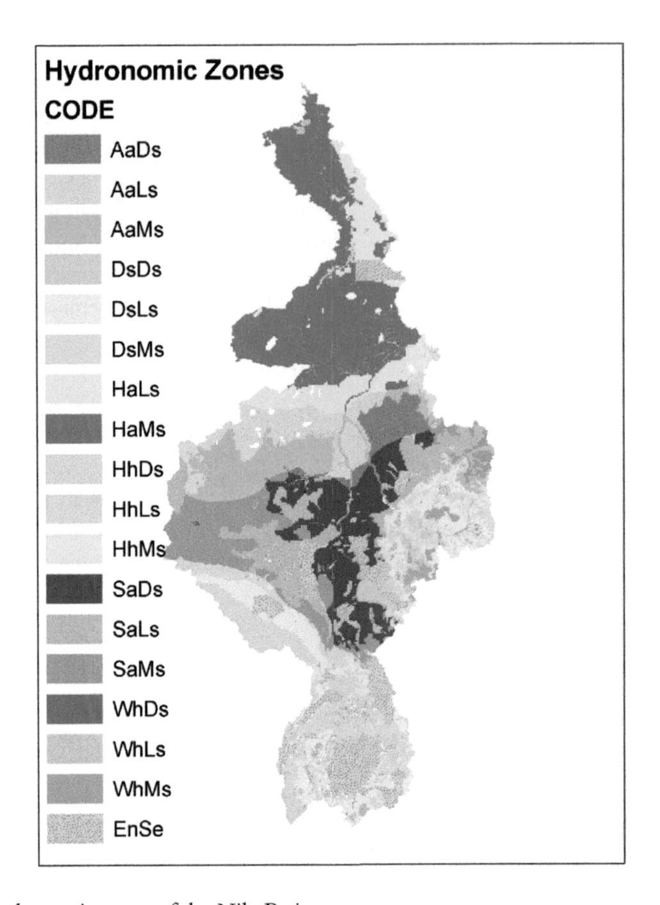

Figure 4.9 The hydronomic zones of the Nile Basin

Note: The first part of each label defines the zone, as follows: Aa = arid, Ds = dry subhumid, Hh = humid, Ha = hyper-arid, Sa = semi-arid, Wh = wet humid. The second part defines the soil bulk density, as follows: Ds = dense soil, Ls = light soil, Ms = medium soil

different aspects of water management. For example, the water source areas of the basin can be easily identified as humid and wet humid zones (HhLs, HhMs, HhDs, WhLs, WhMs and WhDs) where the humidity index is greater than 0.65.

The classes of the developed hydronomic zones could be increased to 37 by including two classes of topographic attribute as a third classification factor for applications at sub-basin or watershed levels.

Discussions and concussions

The spatial patterns of the biophysical factors relevant to the water management of the Nile Basin are examined for the purpose of identifying potential attributes for classification of water management zones. The principal component analysis of the selected biophysical factors indicated that the vegetation (NDVI) attribute has the greatest spatial variability followed by the topographic indices (Slope and CTI) and the soil variable (SWC). These identified biophysical factors have greater spatial variability in the Nile Basin. Hence, the water management zones obtained through unsupervised classification of the dominant principal components have shown greater variation across the basin. Attaching physical names for such a detailed classification requires extensive ground observation; and this may not be applicable to large basins like the Nile. However, the observed patterns of the biophysical factors indicated that the vegetation indices have a similar spatial pattern with the humidity index, and the variability of the soil bulk density is much smoother than, but has similar patterns with, the topographic indices. Therefore, the humidity index and the soil bulk density are used for setting a classification framework for water management zones.

Eighteen water management zones are identified from six classes of humidity index and three classes of the soil factor. In addition, one environmentally sensitive zone is formed by merging wetland and protected areas. The proportional areas of the 19 water management zones are listed in Table 4.3. About 10 per cent of the Nile Basin falls under the environmentally sensitive zone. In this zone, water development interventions should not be permitted. Rather, conservation and protection of the natural ecosystem should be promoted.

The humid and wet humid zones are the water source zones of the Nile Basin. The water source zones account for less than 15 per cent of the basin area. This fact complies with the low specific run-off of the Nile Basin. Since the identified zones have unique climate and soil properties, the water management interventions required to address issues in each zone should also be unique. Therefore, developing a water management strategy for the Nile Basin should commence by mapping potential water management interventions at basin and regional scales within such similar hydronomic zones.

Table 4.3 The proportional areas of the hydronomic zones in the Nile Basin

Name of zone	*Zone code*	*Zone area (million km²)*	*Percentage of basin area*
Hyper arid – light soil	HaLs	537.45	17.22
Hyper arid – medium soil	HaMs	0.00	0.00
Hyper arid – dense soil	HaDs	179.45	5.75
Arid – light soil	AaLs	196.29	6.29
Arid – medium soil	AaMs	188.26	6.03
Arid – dense soil	AaDs	78.24	2.51
Semi-arid – light soil	SaLs	276.41	8.86
Semi-arid – medium soil	SaMs	265.43	8.51
Semi-arid – dense soil	SaDs	280.94	9.00
Dry subhumid – light soil	DsLs	189.30	6.07
Dry subhumid – medium soil	DsMs	85.21	2.73
Dry subhumid – dense soil	DsDs	23.52	0.75
Humid – light soil	HhLs	296.99	9.52
Humid – medium soil	HhMs	80.76	2.59
Humid – dense soil	HhDs	4.11	0.13
Wet humid – light soil	WhLs	23.56	0.75
Wet humid – medium soil	WhMs	27.87	0.89
Wet humid – dense soil	WhDs	0.09	0.003
Environmentally sensitive	EnSe	351.49	11.26
Unclassified		35.24	1.13
Total		3120.59	100.00

References

Batjes, N. M. (2006) *ISRIC-WISE Derived Soil Properties on a 5 by 5 Arc-minutes Global Grid*, Report 2006/02, ISRIC-World Soil Information, Wageningen, The Netherlands.

Fischer, G., van Velthuizen, H., Shah, M. and Nachtergaele, F. (2002) *Global Agro-ecological Zones Assessment for Agriculture for the 21st Century: Methodology and Results*, Research Report 02, International Institute for Applied Systems Analysis, Laxenburg, Austria.

Fraisse, C. W., Sudduth, K. A. and Kitchen, N. R. (2001) Delineation of site-specific management zones by unsupervised classification of topographic attributes and soil electrical conductivity, *Transactions of American Society of Agricultural Engineers*, 44, 1, 155–166.

Loucks, D. P. and van Beek, E. (2005) *Water Resources Systems Planning and Management: An Introduction to Methods, Models and Application*, Studies and Reports in Hydrology, UNESCO Publishing, Turin, Italy.

Molden, D. J., Keller, J. and Sakthivadivel, R. (2001) *Hydronomic Zones for Developing Basin Water Conservation Strategies*, Research Report 56, International Water Management Institute, Colombo, Sri Lanka.

Muthuwatta, L. and Chemin, Y. (2003) Vegetation growth zonation of Sri Lanka for improved water resources planning, *Agricultural Water Management*, 58, 123–143.

Onyango, L., Swallow, B. and Meinzen-Dick, R. (2005) *Hydronomics and Terranomics in the Nyando Basin of Western Kenya*, Proceedings of International Workshop on African Water Laws, Plural Legislative Frameworks for Rural Water Management in Africa, Gauteng, South Africa.

Peel, M. C., Finlayson, B. L. and McMahon, T. A. (2007) Updated world map of the Köppen-Geiger climate classification, *Hydrology and Earth System Sciences*, 11, 1633–1644.

Sutcliffe, J. V. and Parks, Y. P. (1999) *The Hydrology of the Nile*, IAHS Press, Wallingford, UK.

USGS (United States Geological Survey) (2000) HYDRO1k elevation derivative database, http://edc.usgs.gov/products/elevation/gtopo30/hydro, accessed 25 September 2009.

Wagener, T., Sivapalan, M., Troch, P. and Woods, R. (2007) Catchment classification and hydrologic similarity, *Geography Compass*, 1, 4, 901–931.

5

Availability of water for agriculture in the Nile Basin

Robyn Johnston

Key messages

- Rain-fed agriculture dominates water use in the Nile Basin outside Egypt, with more than 70 per cent of the total basin rainfall depleted as evapotranspiration from natural systems partially utilized for pastoral activities, and 10 per cent from rain-fed cropping, compared with less than 1 per cent depleted through irrigation. There is a potential to considerably expand and intensify rain-fed production in upstream areas of the basin without significantly reducing downstream water availability.
- Proposals for up to 4 million ha of additional irrigation upstream of the Aswan Dam are technically feasible if adequate storage is constructed. However, if implemented, they would result in significant reduction of flows to Egypt, offset, to some extent, by reduction in evaporative losses from Aswan. Increasing irrigation area in Sudan will have a much greater impact on flows at Aswan than comparable increases in Ethiopia, due to more favourable storage options in Ethiopia. Expansion of irrigation in the Equatorial Lakes Region by up to 700,000 ha would not significantly reduce flows to Aswan, due to the moderating effects of Lake Victoria and the Sudd wetlands.
- Uncertainties in estimates of both irrigation demand and available flows within the basin are so high that it is not possible to determine from existing information the stage at which demand will outstrip supply in Egypt. Higher estimates suggest that Egypt is already using 120 per cent of its nominal allocation and is dependent on 'excess' flows to Aswan which may not be guaranteed in the longer term; and thus it is vulnerable to any increase in upstream withdrawals.
- Managing non-beneficial evaporative losses through a coordinated approach to construction and operation of reservoirs is an urgent priority. Total evaporative losses from constructed storage in the basin are more than 20 per cent of flows arriving at Aswan. By moving storage higher in the basin, security of supply in the upper basin would be improved, and evaporative losses reduced to provide an overall increase in available water. This can only be achieved through transboundary cooperation to manage water resources at the basin scale.
- Conversely, proposals to reduce evaporation by draining wetlands should be approached with caution, since the gains are relatively small and the Nile's large wetland systems provide important benefits in terms of both pastoral production and biodiversity.

- Projected changes in rainfall due to climate change are mostly within the envelope of existing rainfall variability, which is already very high. However, temperature increases may reduce the viability of rain-fed agriculture in marginal areas, and increase water demand for irrigation.

Introduction

Rapid population growth and high levels of food insecurity in the Nile Basin mean that increasing agricultural production is an urgent imperative for the region. In much of the basin, agriculture is dominated by subsistence rain-fed systems with low productivity and high levels of risk due to variable climate. Egypt's highly productive, large-scale irrigation is seen as a model for agricultural development in other Nile Basin countries, but there are concerns that irrigation development in upstream countries could jeopardize existing production in Egypt. The Nile Basin Initiative (NBI) was launched in 1999 as a mechanism to share the benefits of the Nile waters more equitably. A critical question for the NBI is the extent to which upstream agricultural development will impact on water availability in the lower basin. This chapter synthesizes evidence from several of the studies presented in this book to examine current and future water availability for agriculture in the Nile Basin.

A distinction must be made between water availability (the total amount of water present in the system) and water access (ease of obtaining and using it). Availability is generally fixed by climate and hydrology, while access can be improved through infrastructure and/or enabling institutional mechanisms. In much of Africa, access to water is a more pressing constraint on livelihoods, and a contributor to high levels of poverty. In the Nile Basin, there is the apparently contradictory situation that access to water is often poor in the highland areas where water is abundant; but in arid Egypt, access has been significantly enhanced due to well-developed infrastructure. The chapter will examine only water availability (the nexus between water, poverty and vulnerability is discussed in Chapter 3).

Nile Basin overview

The Nile basin covers 3.25 million km^2 in nine countries, and is home to a population of around 200 million. The Nile comprises five main subsystems. There have been a number of different delineations of the extent of the Nile Basin and its component sub-basins. This study adopts the delineation currently used by NBI, amalgamating some sub-basins to eight larger units. Reference is also made to results of Kirby *et al.* (2010), who used a set of 25 sub-basins, which nest within the NBI units. Figure 5.1 illustrates the major tributaries and sub-basins of the Nile basin, which are:

- The White Nile sub-basin, divided into three sections:
  - headwaters in the highlands of the Equatorial Lakes Region (ELR), including Lake Victoria;
  - middle reaches in western and southern Sudan, where the river flows through the lowland swamps of the Sudd (Bahr el Jebel) and Bahr el Ghazal; and
  - Lower White Nile (LWN) sub-basin in central Sudan south of Khartoum.
- The Sobat-Baro-Akobo sub-basin, including highlands of southern Ethiopia and Machar Marshes and lowlands of southeast Sudan.
- The Blue Nile (Abay) sub-basin, comprising the central Ethiopian plateau and Lake Tana, and the arid lowlands of western Ethiopia and eastern Sudan, including the major irrigation area at Gezira where the Blue Nile joins the White Nile near Khartoum.

- The Atbara–Tekeze sub-basin, comprising highlands of northern Ethiopia and southern Eritrea and arid lowlands of northeast Sudan.
- The Main Nile system, divided into two distinct sections:
  - Main Nile in Sudan above the Aswan Dam; and
  - Egyptian Nile below Aswan, including the Nile Valley and Delta.

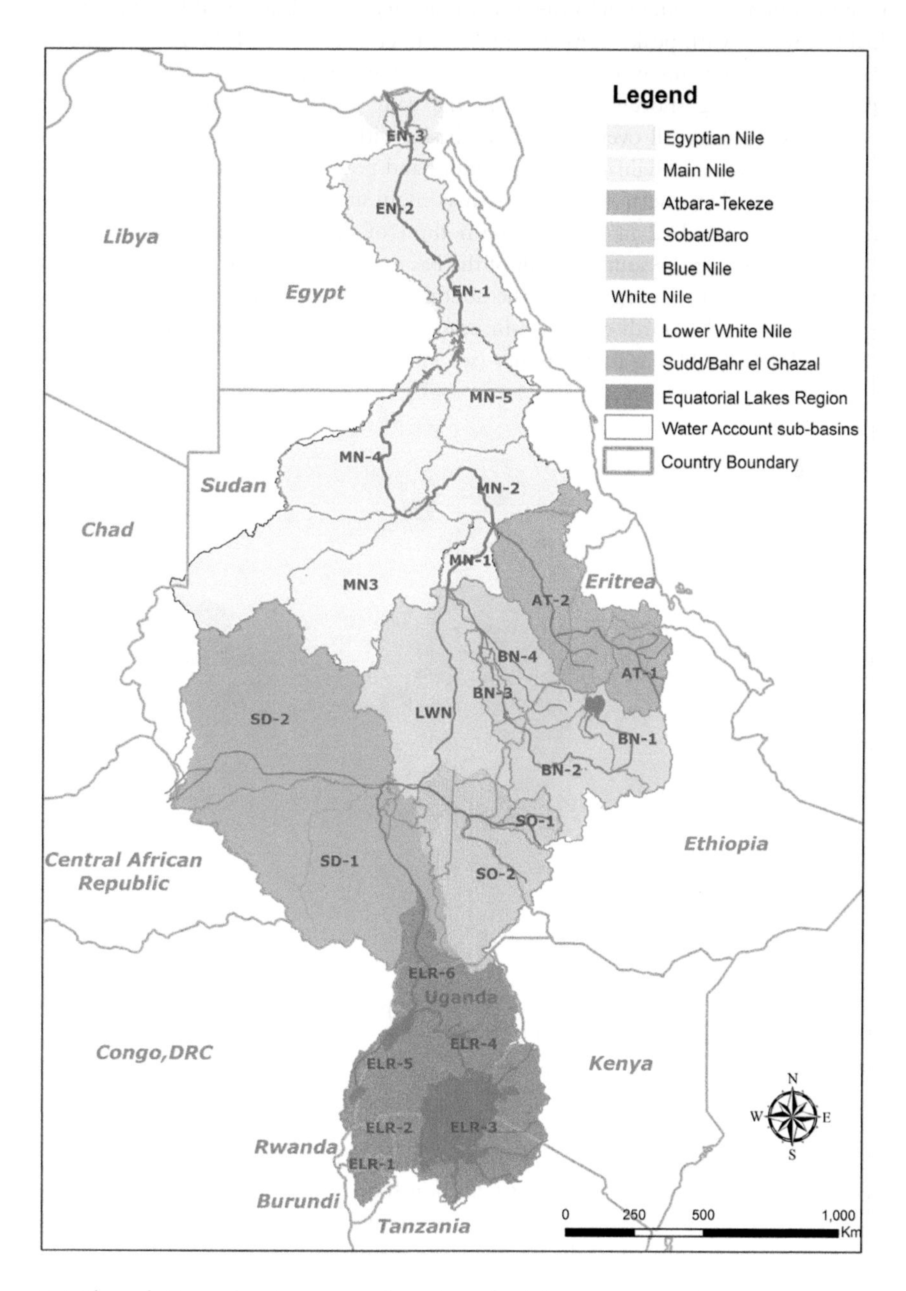

Figure 5.1 The Nile Basin, showing major tributaries and sub-basins. Smaller sub-catchments used in the water accounting framework of Kirby *et al.* (2010) are also shown

Climate

The climate of the Nile Basin has strong latitudinal and topographic gradients. Mean annual precipitation (MAP) decreases from the highlands of the south and east to the lowland deserts in the north, and ranges from more than 2000 mm around Lake Victoria and in the Ethiopian highlands to less than 10 mm in most of Egypt. Rainfall in the basin is strongly seasonal, although the timing and duration of the wet season vary. In the Equatorial Lakes Region there is a dual wet season with peaks in April and November; parts of the Ethiopian Highlands also experience a weak second wet season. In most of the basin, the wet season peaks around July–August, becoming shorter and later in the eastern and northern parts of the basin. Evaporation exceeds rainfall over most of the basin, with the exception of small areas in the equatorial and Ethiopian Highlands. Temperatures and potential evapotranspiration (PET) are highest in central and northern Sudan, where maximum summer temperatures rise above 45°C and annual PET exceeds 2 m. The northern third of the basin is classified as hyper-arid (MAP/PET <0.05); but the southern half of the basin is semi-arid to humid with MAP above 600 mm. In the equatorial regions in the south, temperatures and PET vary only slightly through the year; in the north of the basin, seasonal changes in temperature are reflected in PET, which almost doubles in mid-summer (see Figure 5.2).

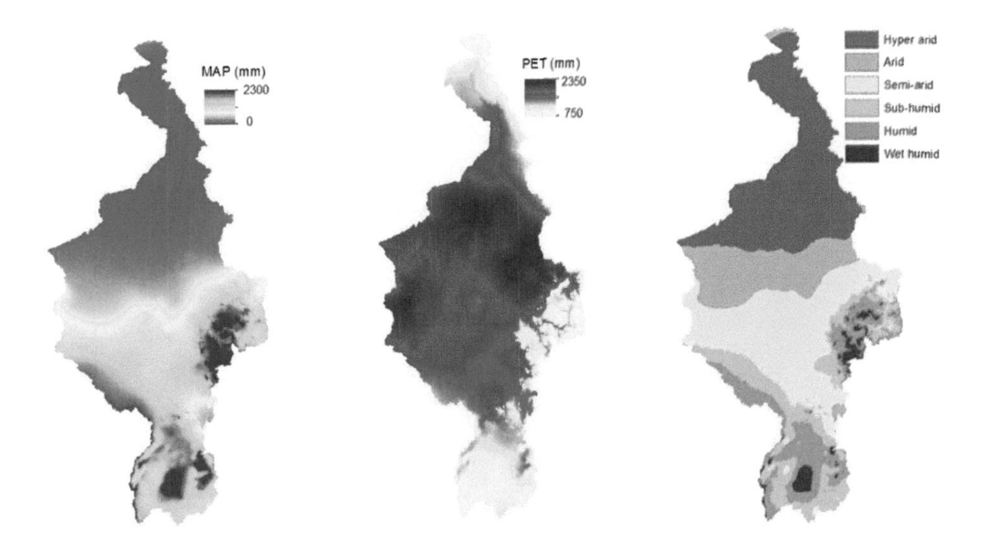

Figure 5.2 Mean annual precipitation (MAP), mean annual potential evapotranspiration (PET) and humidity index for the Nile Basin

Source: World Climate Atlas (http://waterdata.iwmi.org)

Inter-annual rainfall variability is very high, related to the movements of the Inter-tropical Convergence Zone and to the Southern Oscillation Index, with low rainfall in El Niño years. Camberlin (2009) describes long-term rainfall variability on the scale of decades, with different patterns in different regions. Over the period 1951–2000, the northern belt (15–16° N) experienced high rainfall in the 1950s and 1960s, with dryer conditions from 1970 to 2000; the African Sahel (7–14° N) saw a severe downward trend in rainfall through the 1970s and 1980s, with partial recovery in the 1990s; and equatorial regions experienced a wet period in the 1960s, but rainfall was otherwise stable. Seasonal and inter-annual climate variability has a more significant impact on river flows than long-term climatic trends (Awulachew *et al.*, 2008; Di Baldassarre *et al.*, 2011). For temperature over the same period, a majority of warming trends are observed, with a rise of 0.4° C reported for 1960–2000 across East Africa (Hulme *et al.*, 2001).

Hydrology

The hydrology of the Nile Basin is reviewed in detail by Sutcliffe and Parks (1999) and Sutcliffe (2009). A schematic of flow distribution in the Nile system based on average flows is shown in Figure 5.3. The total annual flow arriving at the Aswan High Dam varies between 40 and 150 km^3, averaging around 85 km^3; of this total, the White Nile, Sobat and Atbara systems each contributes about one-seventh, with the Blue Nile contributing four-sevenths (Blackmore and Whittington, 2008). The Main Nile is a losing system, with high transmission losses due to evaporation and channel infiltration (NWRC, 2007). The volume and distribution of flows in the various Nile sub-basins vary markedly from year to year and over decades, reflecting variability in rainfall. Table 5.1 compares discharge from the main sub-basins for periods in the first and second half of the twentieth century. In the south of the basin, there was a marked increase in flows but in the north, flows decreased by around 20 per cent. These differences reflect a complex interplay of climate variability and human modification of the river system, and do not necessarily represent continuing trends.

Table 5.1 Variability of Nile flows; comparison of long-term average flows over different time periods

Sub-basin	*Station*	*Annual discharge (km^3)*				*Change*	*Data source*
		Pre-1960		*Post-1960*			
White Nile/ Equatorial Lakes Region	Lake Victoria	1901–1960	20.6	1961–1990	37.5	1.8	Sutcliffe and Parks, 1999
White Nile above Sudd	Mongalla	1905–1960	26.8	1961–1983	49.2	1.8	
White Nile below Sudd	Sudd outflow	1905–1960	14.2	1961–1983	20.8	1.5	
Sobat	Doleib Hill	1905–1960	13.5	1961–1983	13.7	1.0	
LWN above Jebel Aulia	Malakal	1905–1960	27.6	1961–1995	32.8	1.2	
LWN below Jebel Aulia	Mogren	1936–1960	23.1	1961–1995	28.1	1.2	
Blue Nile	Khartoum	1900–1960	52.8	1961–1995	48.3	0.9	
Atbara	Atbara at mouth	1911–1960	12.3	1961–1994	8.6	0.7	
Main Nile above Aswan	Dongola	1911–1960	86.1	1961–1995	73.1	0.8	
Egyptian Nile below Aswan	Aswan	1952–1960	89.7	1970–1984	56.9	0.6	Dai and Trenberth, 2003
Egyptian Nile Delta	–	before 1970	32.4	after 1970	4.5	0.1	El-Shabraway, 2009

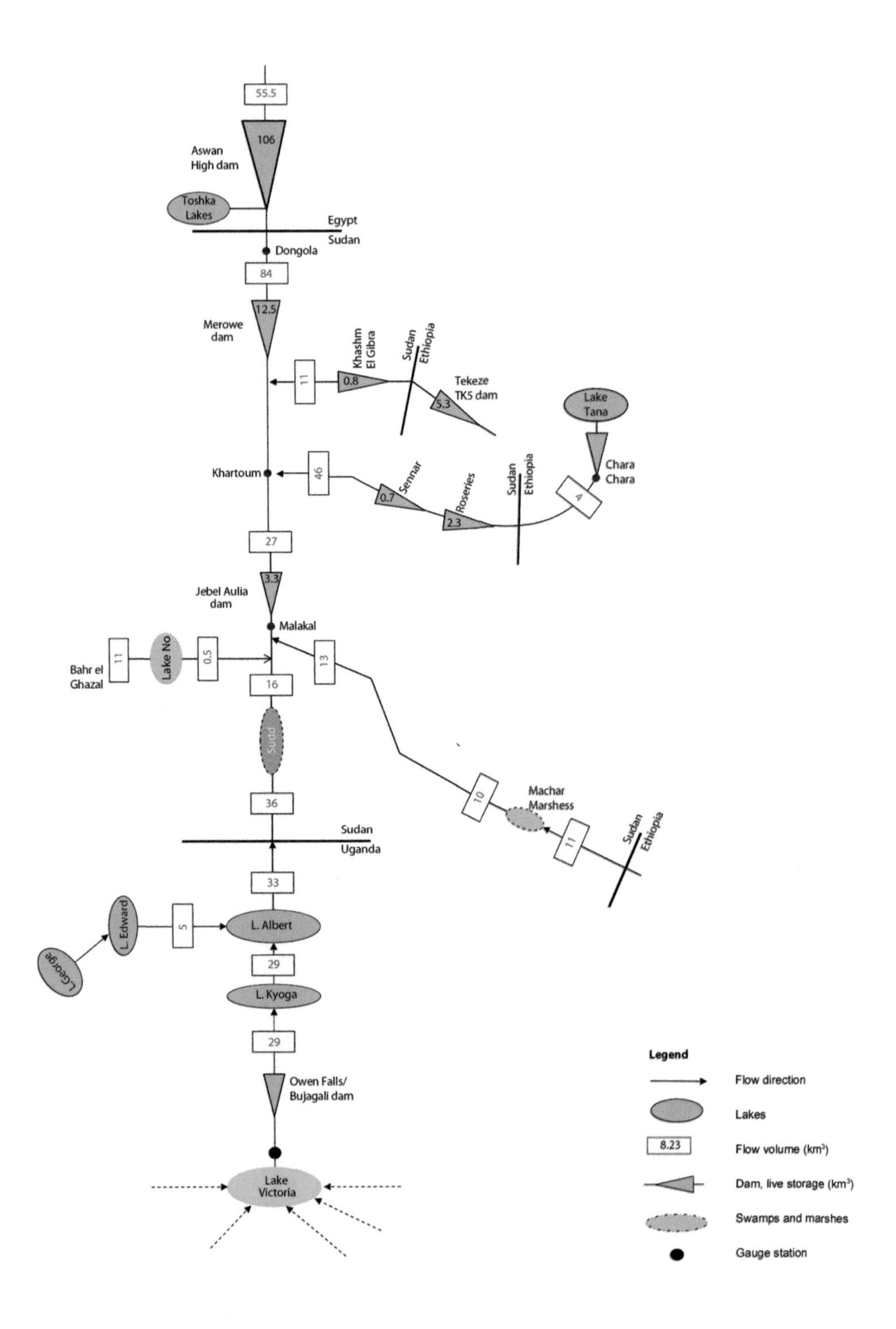

Figure 5.3 Schematic of Nile flows

Source: modified from Awulachew *et al.*, 2010

Flows in the Nile are highly seasonal, with different patterns in different parts of the basin, in response to rainfall distribution. The White Nile, fed mainly from Lake Victoria and moderated by the vast wetlands of the Sudd, has a relatively constant flow throughout the year. In contrast, the tributaries rising in the Ethiopian Highlands have pronounced peak flows during the wet season with little base flow. As a result, the Main Nile below Khartoum has peak flows (from the Blue Nile, Sobat and Atbara) superimposed on modest base flows (from the White Nile). Peak flows are substantially reduced below the Aswan High Dam (AHD) as releases are timed to meet agricultural demand. Figure 5.4 illustrates the spatial distribution of seasonal variability in flows, based on calculated discharge from 25 sub-basins for the period 1952–2000 (Kirby *et al.*, 2010).

	White Nile									Sobat		Blue Nile				Atbara-Tezeke		Main Nile above Aswan					Egyptian Nile		
	Equatorial Lakes Region						Sudd		LWN																
	ELR-1	ELR-2	ELR-3	ELR-4	ELR-5	ELR-6	SD-1	SD-2	LWN	SO-1	SO-2	BN-1	BN-2	BN-3	BN-4	AT-1	AT-2	MN-1	MN-2	MN3	MN-4	MN-5	EN-1	EN-2	EN-3
Jan	8%	7%	8%	8%	7%	7%	7%	7%	8%	1%	2%	1%	2%	1%	5%	6%	0%	6%	5%	5%	5%	5%	5%	6%	8%
Feb	9%	9%	8%	8%	7%	7%	7%	6%	8%	0%	1%	1%	2%	1%	5%	5%	0%	5%	4%	4%	4%	5%	6%	7%	8%
Mar	9%	9%	8%	8%	8%	8%	7%	5%	7%	1%	1%	1%	1%	0%	4%	4%	0%	4%	3%	3%	3%	5%	7%	8%	8%
Apr	10%	10%	8%	9%	9%	9%	7%	5%	6%	2%	2%	1%	1%	0%	4%	3%	0%	3%	3%	3%	3%	6%	9%	9%	8%
May	10%	10%	9%	9%	9%	9%	8%	6%	6%	5%	4%	2%	2%	1%	3%	3%	0%	3%	3%	3%	3%	8%	13%	12%	8%
Jun	11%	11%	9%	9%	9%	8%	8%	7%	7%	9%	8%	4%	5%	5%	6%	6%	0%	6%	5%	5%	5%	8%	11%	10%	8%
Jul	10%	10%	9%	9%	8%	8%	9%	9%	9%	15%	13%	23%	17%	19%	12%	11%	33%	11%	14%	14%	14%	12%	11%	10%	8%
Aug	8%	9%	9%	9%	9%	9%	10%	11%	12%	19%	18%	32%	26%	29%	19%	19%	54%	19%	23%	23%	23%	17%	10%	10%	8%
Sep	6%	6%	8%	8%	9%	9%	10%	12%	13%	20%	19%	21%	22%	24%	18%	18%	10%	18%	17%	17%	17%	13%	9%	9%	8%
Oct	6%	6%	8%	8%	9%	9%	9%	12%	9%	17%	17%	10%	12%	12%	11%	11%	1%	11%	10%	10%	10%	9%	8%	8%	8%
Nov	6%	7%	8%	8%	9%	9%	9%	11%	8%	8%	11%	3%	6%	5%	7%	8%	1%	8%	7%	7%	7%	6%	6%	7%	8%
Dec	6%	6%	8%	8%	7%	7%	8%	9%	8%	3%	6%	2%	3%	2%	6%	7%	0%	7%	6%	6%	6%	5%	5%	6%	8%
CV	32%	30%	26%	27%	27%	23%	22%	18%	25%	21%	25%	36%	25%	28%	24%	25%	88%	25%	28%	29%	29%	35%	26%	30%	43%

Proportion of total annual discharge in month: >40% | 20-40% | 10-20% | 5-10% | 1-5% | 1%

Figure 5.4 Spatial patterns of seasonal flow in the Nile sub-basins, displayed as proportion of annual flow in each calendar month

Note: Averages are for 1952–2000, except for the Egyptian Nile, which are 1970–2000 (i.e. after construction of Aswan High Dam). Coefficient of variation (CV) is for annual flows, 1952–2000

Source: Based on CRU data compiled by Kirby *et al.*, 2010

In response to this variability, dams were constructed prior to 1970 on the White Nile (Jebel Aulia), Blue Nile (Roseries and Sennar) and Atbara (Khasm el Girba) with total storage of 7.7 km^3, to retain peak flows for irrigation in Egypt and Sudan. In 1970, AHD was constructed to provide over-year storage to safeguard flows to Egypt. The reservoir was designed for 'century storage', to guarantee a supply equal to the mean inflow over a period of 100 years. Total storage is 162 km^3, almost twice the mean annual flow. Prior to its construction, Egypt and Sudan signed an agreement which divided the expected yield of the project between the two countries. Based on the long-term annual flow of 84 km^3 and estimated evaporative losses of 10 km^3, the remaining amount of 74 km^3 was apportioned as 55.5 km^3 to Egypt and 18.5 km^3 to Sudan. The upstream Nile Basin countries were not party to the agreement, and do not recognize it.

Agriculture in the Nile Basin

Farming systems in the Nile Basin are diverse, including a range of pastoral, agro-pastoral and cropping system (for more detail, see Chapter 8). Livestock are an important component of agricultural systems throughout the basin. Irrigation from the Nile and its tributaries has allowed development of agriculture in otherwise arid regions of Egypt and Sudan. Otherwise, land use reflects climatic gradients, with a transition from rain-fed mixed agriculture in the humid south and east, through agro-pastoral systems in the semi-arid central regions, to low-intensity rangelands and desert in the arid north. The length and timing of the growing season (defined as months where precipitation/PET >0.5, depicted in Figure 5.5) exercise a primary control on land use.

	White Nile									Sobat		Blue Nile				Atbara-Tezeke		Main Nile above Aswan					Egyptian Nile		
	Equatorial Lakes Region						Sudd		LWN																
	ELR-1	ELR-2	ELR-3	ELR-4	ELR-5	ELR-6	SD-1	SD-2	LWN	SO-1	SO-2	BN-1	BN-2	BN-3	BN-4	AT-1	AT-2	MN-1	MN-2	MN3	MN-4	MN-5	EN-1	EN-2	EN-3
Jan	0.9	0.5	0.6	0.2	0.3	0.1	0.0	0.0	0.0	0.1	0.0	0.1	0.1	0.0	0.0	0.0	0.0	0.0	0.0	0.0	0.0	0.0	0.0	0.0	0.2
Feb	0.9	0.6	0.7	0.3	0.4	0.1	0.0	0.0	0.0	0.2	0.1	0.2	0.1	0.0	0.0	0.1	0.0	0.0	0.0	0.0	0.0	0.0	0.0	0.0	0.1
Mar	1.1	0.9	0.8	0.6	0.7	0.4	0.2	0.0	0.0	0.4	0.2	0.3	0.2	0.0	0.0	0.2	0.0	0.0	0.0	0.0	0.0	0.0	0.0	0.0	0.1
Apr	1.4	1.2	1.4	1.0	1.1	0.8	0.4	0.1	0.1	0.6	0.4	0.5	0.4	0.0	0.0	0.2	0.0	0.0	0.0	0.0	0.0	0.0	0.0	0.0	0.0
May	0.8	0.9	1.1	1.1	0.8	0.9	0.8	0.2	0.2	1.2	0.7	0.5	0.8	0.2	0.1	0.3	0.1	0.0	0.0	0.0	0.0	0.0	0.0	0.0	0.0
Jun	0.2	0.2	0.5	[illegible]	0.6	0.9	1.1	0.4	0.4	1.7	0.9	0.9	1.5	0.5	0.4	0.5	0.2	0.0	0.0	0.0	0.0	0.0	0.0	0.0	0.0
Jul	0.1	0.1	0.4	0.8	0.5	1.0	1.2	[illegible]	0.8	2.1	1.1	2.6	2.5	1.0	0.8	1.9	0.7	0.2	0.0	0.2	0.0	0.0	0.0	0.0	0.0
Aug	0.2	0.3	0.4	[illegible]	0.8	1.1	1.3	1.0	0.9	2.2	1.1	2.5	2.4	1.1	1.1	1.9	0.7	0.3	0.1	0.3	0.0	0.0	0.0	0.0	0.0
Sep	0.5	0.5	0.5	[illegible]	0.9	0.9	1.0	0.6	0.6	1.7	0.9	1.3	1.6	0.7	0.5	0.7	0.3	0.1	0.0	0.1	0.0	0.0	0.0	0.0	0.0
Oct	0.7	0.6	0.6	0.8	1.0	0.9	0.7	0.2	0.3	0.9	0.5	0.5	0.7	0.2	0.2	0.3	0.1	0.0	0.0	0.0	0.0	0.0	0.0	0.0	0.0
Nov	1.2	1.0	1.0	0.7	1.0	0.5	0.1	0.0	0.0	0.5	0.3	0.2	0.2	0.0	0.0	0.1	0.0	0.0	0.0	0.0	0.0	0.0	0.0	0.0	0.1
Dec	1.0	0.7	0.7	0.3	0.5	0.1	0.0	0.0	0.0	0.2	0.1	0.1	0.1	0.0	0.0	0.1	0.0	0.0	0.0	0.0	0.0	0.0	0.0	0.0	0.2
CV	10%	11%	14%	14%	10%	13%	9%	16%	15%	11%	20%	13%	9%	17%	16%	25%	25%	55%	68%	44%	76%	66%	54%	38%	33%

Rainfall/PET: >1 | 0.7-1.0 | 0.5-0.7 | 0.2-0.5 | <0.2

Figure 5.5 Monthly variation in humidity index (rainfall/PET) for Nile sub-basins 1951–2000, illustrating spatial variability of timing and duration of growing season

Note: CV is for annual rainfall 1951–2000

Source: Based on CRU data compiled by Kirby *et al.*, 2010

Dominant land use in the basin in terms of area is low-intensity agro-pastoralism, with grasslands and shrublands interspersed with small-scale cropping, covering more than a third of the basin (see Figure 5.6). Extensive seasonally flooded wetlands in South Sudan (Machar Marshes, Sudd and Bahr el Ghazal) support large livestock herds: it is estimated that there are over one million head of cattle in the Sudd (Peden *et al.*, 2009). In the semi-arid to arid zones of central Sudan, sparse grasslands are utilized for low-intensity agro-pastoral production and extensive grazing. In the northern parts of the basin, low and variable rainfall jeopardizes availability of feed in the rangelands, and demand for drinking water for stock exceeds supply in most areas (Awulachew *et al.*, 2010). Water productivity in these areas is generally very low; in most areas, rainfall cannot meet crop water demands resulting in low yields. At Gezira and New Halfa, development of irrigation has significantly improved productivity (Karimi *et al.*, 2012) and further large expansion of irrigation is proposed in these areas.

The total area of rain-fed cropping in the basin is estimated at over 33 million ha (FAO, 2010; Table 5.2). Almost half of this area is in the ELR, where major crops are bananas, maize and wheat, with some commercial cultivation of coffee, sugar cane and cotton. Natural swamps and marshes are used extensively for agriculture, with over 230,000 ha of cultivation in wetlands and valley bottoms in Burundi, Rwanda and Uganda (FAO, 2005). Despite mostly adequate rainfall, yields and productivity in this region are low to moderate (Karimi *et al.*, 2012).

Sudan has almost 15 million ha of rain-fed cropping, mainly subsistence cultivation of cereals, groundnut and soybean. During the 1960–1980s, the Sudan government promoted

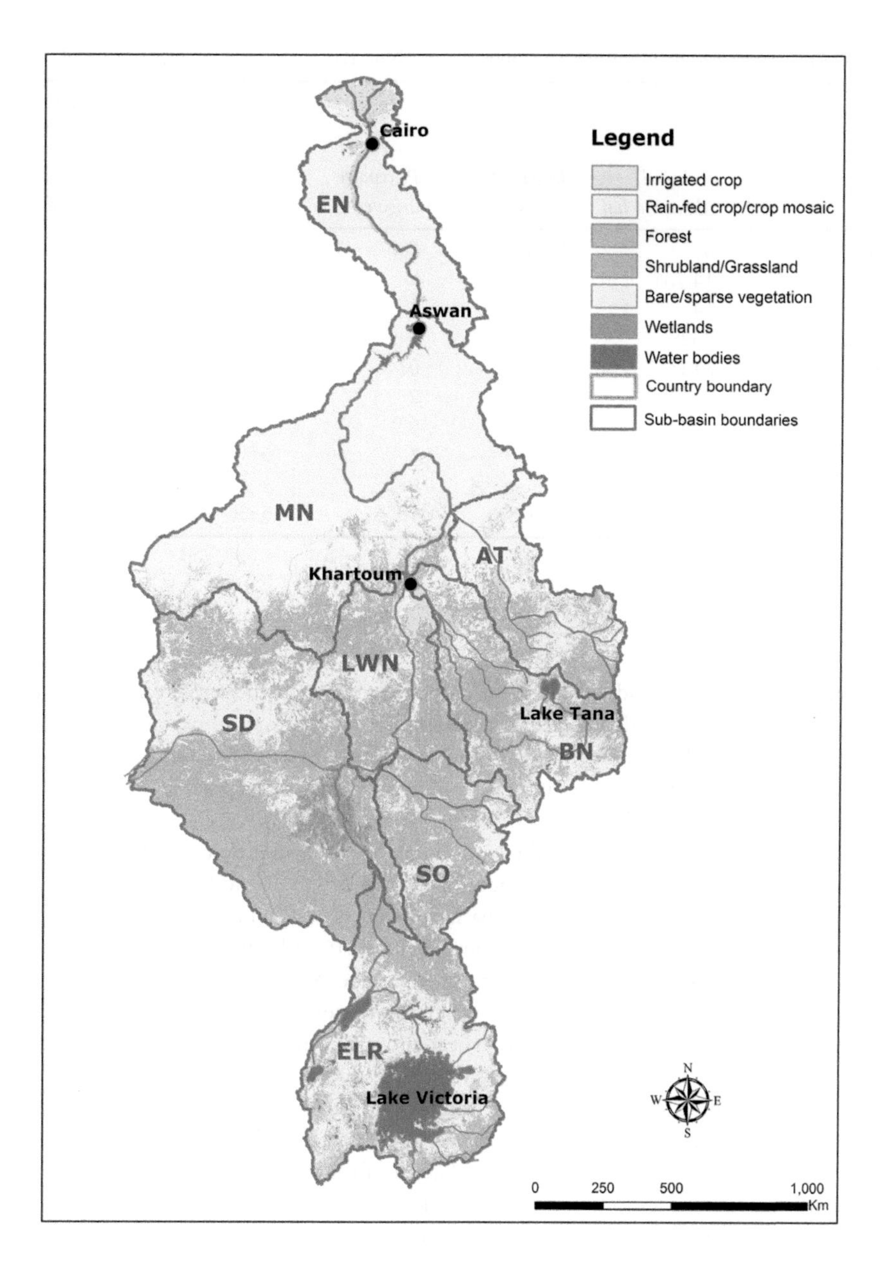

Figure 5.6 Land cover in the Nile Basin

Source: Globcover 2009, © ESA 2010 & UCLouvain, http://ionia1.esrin.esa.int

mechanized rain-fed agriculture, designed to utilize the fertile cracking clay soils that were not suited to traditional cultivation practices. Over 0.75 million ha were cultivated under official schemes or informally, with sorghum, groundnut and sugar cane. Initial yields were high, but unsustainable farming practices, drought and civil war meant that, by the mid-1990s, much of the land had been abandoned (Mongabay, 1991; UNEP/GRID, 2002).

Table 5.2 Areas of irrigated and rain-fed cropping in the Nile Basin reported by different studies

Country	*FAO (2010)*			*Chapter 15, this volume*	*Bastiaanssen and Perry, 2009*
	Rain-fed (thousand ha)	*Irrigated* (thousand ha)*	*Percentage irrigated*	*Irrigated (thousand ha)*	*Irrigated (thousand ha)*
Egypt	0	5117	100	3324	2963
Sudan	14,785	1207	8	2176	1749
Eritrea	64	5	7	–	–
Ethiopia	3328	15	0	16	91
Uganda	8123	33	0	9	25
Kenya	2153	42	2	6	34
Tanzania	2593	0	0	0	7
Rwanda	1375	21	1	5	18
Burundi	808	0	0	–	14
Total	33,229	6440	16	5536	4901

Note: *Includes multiple cropping

In the Ethiopian Highlands, a variety of crops are grown, including cereals (wheat, barley, maize), enset root crops, coffee, teff and sorghum; livestock are an important component of farming systems. Double-cropping is possible in some areas. Erosion from steep cultivated lands is a major problem, reducing agricultural productivity and causing rapid sedimentation in downstream reservoirs (Awulachew *et al.*, 2008).

Estimates of total irrigated area in the basin range from 4.9 million to 6.4 million ha (Table 5.2). Large areas of formal irrigation are developed only in Egypt and Sudan. In Sudan, irrigation schemes totaling around 1.5 million ha have been developed at Assalaya and Kenana on the Lower White Nile (0.08 million ha), New Halfa (0.16 million ha) on the Atbara downstream of Khasm el Gibra Dam, and Gezira on the Blue Nile (1.25 million ha). The Gezira scheme, one of the largest in Africa, draws water from reservoirs at Roseires and Sennar. The major irrigated crops are sorghum, cotton, wheat and sugar. In addition, small-scale pump irrigation occurs along the main Nile channel. Most irrigation in Sudan overlaps at least a part of the wet season, with little irrigation in the winter dry period. Generally, low productivity in Sudan's irrigation areas is attributed to a range of factors including poor farming practices, problems with water delivery resulting from siltation of reservoirs and lack of flexibility due to the requirements of releases for hydropower, poor condition of canals, drainage problem, salinization and an unfavourably hot climate (Bastiaanssen and Perry, 2009).

In Egypt, total agricultural area in the Nile Valley and Delta exceeds 3 million ha; double-cropping means that over 5 million ha are planted annually (Bastiaanssen and Perry, 2009; FAO, 2010; Chapter 15, this volume). Water is provided by the AHD and seven barrages diverting water into an extensive network of canals (32,000 km of canals) with complementary drainage systems. There are three agricultural seasons: winter (October to February), when main crops are wheat, fodder and berseem; summer (March–June), when the main crops are maize, rice and cotton; and *nili* (July–September), when the main crops are rice, maize, pulses, groundnut and vegetables. Sugar cane, citrus, fruits and oil crops are grown all year. Because rainfall is so low, virtually all agriculture is irrigated, although there may be opportunistic rain-fed cropping in some years. In the last 10 years, new irrigation areas have been developed at the 'New Valley' irrigation project near the Toshka lakes, drawing water from Lake Nasser via a pumping station to irrigate around 0.25 million ha, with total water requirements of 5.5 km^3 when fully operational (NWRC, 2007; Blackmore and Whittington, 2008). Withdrawals from the Nile are also used outside the basin in the Sinai irrigation development, where 0.168 million ha are to be irrigated using 3 km^3 of water derived from 50 per cent Nile water mixed with 50 per cent recycled water (NWRC, 2007).

Karimi *et al.* (2012) assessed water productivity in the Nile Basin, and found that overall productivity in irrigated systems in Egypt is high, with intensive irrigation, high yields and high-value crops. Bastiaanssen and Perry (2009) point out that there is great variability in productivity between different irrigation districts, with some functioning very poorly, but some of the systems in the Delta ranking among the best in the world.

Water balance/water account

The Nile River flows constitute only a very small proportion of total water resources in the basin. On average, a total of around 2000 km^3 of rain falls over the basin annually, but annual flow in the Nile at Aswan is <5 per cent (around 85 km^3). In order to assess availability of water for agriculture, water accounts have been constructed for the basin at different spatial and temporal scales to illustrate the way rainfall is distributed, stored and depleted in the basin.

Kirby *et al.* (2010) developed a dynamic water use account for the Nile Basin, based on 25 sub-catchments at monthly time steps, with hydrological and evapotranspiration (ET) components. The sub-catchments used in the analysis are shown in Figure 5.1. A hydrological account of inflows, storages, depletion by ET and outflows is based on lumped partitioning of rainfall into run-off and infiltration; downstream flows are calculated using a simple water balance. ET is estimated from PET and the surface water store, and partitioned between land uses based on the ratio of their areas, using crop coefficients to scale ET relative to other land uses. The account uses a 50-year run of climate data (1951–2000; Climate Research Unit, University of East Anglia; CRU_TS_2.10), and provides information on both seasonal and inter-annual variability in rainfall, flows and ET. It does not account for changes in land use over the period, but assumes static land cover, derived from 1992–1993 Advanced Very High Resolution Radiometer (AVHRR) data. The results were validated against available flow data from 20 gauging stations in the basin from the ds552.1 data set (Dai and Trenberth, 2003). For this study, some corrections and adjustments were made to the Kirby *et al.* model and the results were recalculated.

The results (summarized in Figure 5.7) indicate that, at the basin scale, only a small fraction of total rainfall is depleted as ET from managed agricultural systems (10% from rain-fed cropping and 3% from irrigation). A large proportion of rainfall (70%) is used by grasslands, shrubland and forest that are not actively managed, although a large proportion of this area is used for pastoral activities at different intensities. Significant run-off volumes are generated only from a few catchments, mainly in the highlands of the ELR and Ethiopia. Calculated values for local run-off, totaling 7 per cent of basin rainfall (163 km^3), are higher than net discharge through the river system, since flows are depleted by channel losses and evaporation.

Over 8 per cent of total basin water resources are depleted through evaporation from water bodies (lakes, dams and open water swamps). Of this, around half is from Lake Victoria (ELR-3 in Figure 5.7), and another third from the major wetland systems of the Sudd and Bahr el Ghazal (SD-1 and SD-2). Excluding these, the water account indicates that total annual evaporative losses from man-made reservoirs exceed 15 km^3.

Karimi *et al.* (2012) constructed a water account for the Nile at basin scale using remotely

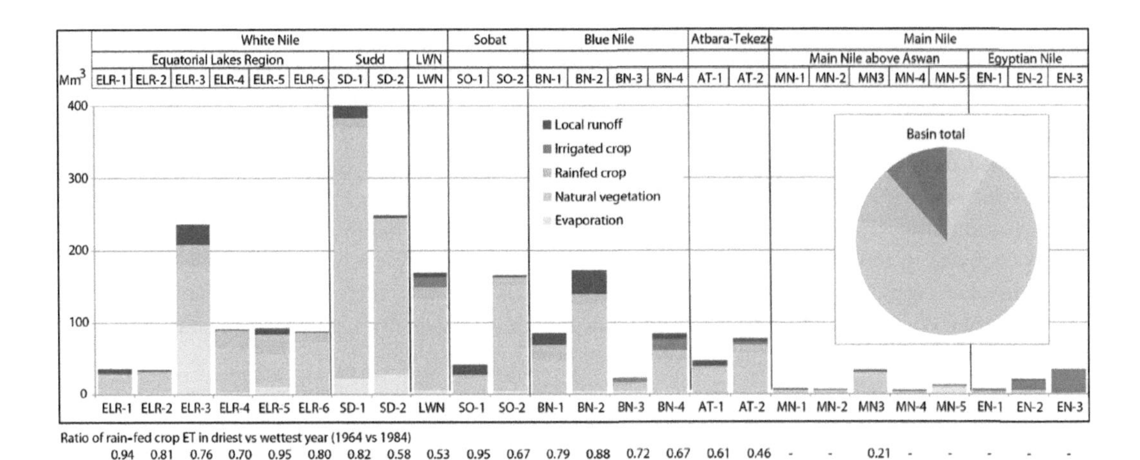

Figure 5.7 Water account for the Nile, showing partitioning of rainfall into ET (by land use category) and locally generated run-off for each sub-catchment and the basin as a whole

sensed data on rainfall, land use and evapotranspiration (ET) for a single year (2007). They estimate that of the 2045 km³ of rainfall delivered to the basin, only 13 per cent was depleted from crops (10% rain-fed cropping and 3% irrigation). Surface run-off represented only 4.5 per cent of rainfall, with outflows to the seas less than 2 per cent of total inputs. These results correspond well with those from the modelled water account above.

Availability of water for rain-fed agriculture

Four important points about availability of water for rain-fed agriculture in the Nile Basin emerge from the water accounts. First, the proportion of rainfall in the basin that is currently depleted from cropping is very small – around 13 per cent in total, 10 per cent in rain-fed cropping and only around 3 per cent in irrigated production. Most of the rainfall is depleted from natural grasslands and woodlands used for extensive pastoralism, often with very low productivity (Karimi *et al.*, 2012). This is despite the fact that in the subhumid regions in the southern basin (particularly in South Sudan and central Sudan), average rainfall is more than sufficient to support rain-fed cropping. There is significant opportunity to extend more intensive and productive cropping and agro-pastoral land uses into areas currently dominated by low productivity grazing through improved water management, including rainwater harvesting and storage for small-scale irrigation.

While the proportion of ET from rain-fed crops remains relatively stable between years, the absolute amount varies very significantly, from 180 to 256 km³, representing a large difference in potential crop production between years and illustrating the risks associated with rain-fed agriculture in the region. The variability is higher in low rainfall areas: the ratio of rain-fed crop ET between the driest and the wettest years is around 0.7–0.9 in the humid uplands, but falls to around 0.5 in the semi-arid catchments of central Sudan and the Atbara basin. In terms of food security, this annual variability is exacerbated by the occurrence of multi-year droughts. Under these conditions, opportunistic cropping in wet years (routinely practised in dryland production in semi-arid regions in Australia) may be a viable strategy commercially, although it is difficult to reconcile it with the need for smallholders to produce a crop every year to ensure food security. Small-scale agricultural water management techniques, such as rainwater harvesting, small-scale storages and groundwater have a potentially important role in securing rain-fed crops in these regions. Araya and Stroosnijder (2011) found that in northern Ethiopia, where crops failed in more than a third of years in the period 1978–2008, one month of supplementary irrigation at the end of the wet season could avoid 80 per cent of crop yield losses and 50 per cent of crop failures. Other strategies used in the area to manage erratic rainfall include supplementary irrigation to establish crops (to avoid false starts to the wet season), postponement of sowing until adequate soil moisture is available, and growing quickly maturing cash crops such as chickpea at the end of the growing period, to utilize unused soil water reserves.

Estimates of net discharge indicate that the majority of flow is generated from only a small number of sub-catchments in the ELR and Ethiopian Highlands, which constitute water source zones (see Chapter 4). Land use in areas outside of these zones will have little impact on downstream flows. Thus, there are significant areas where rainwater harvesting, intensification of cropping and conversion of natural vegetation to agriculture are unlikely to significantly reduce downstream water availability.

In the ELR below Lake Victoria, the Sudd, Bahr el Ghazal and lower Sobat sub-basins, locally generated run-off is high, but net discharge from the sub-basin constitutes only a small fraction of total run-off (<20%). These areas retain water in natural sinks (wetlands, shallow groundwater) that are used to secure rain-fed production. There is cultivation of over 230,000ha in wetlands and valley bottoms in the Nile Basin in Burundi, Rwanda and Uganda (FAO, 2005). The wetlands of South Sudan are very important for livestock grazing: over one million head of cattle are estimated to be in the Sudd (Awulachew *et al.*, 2010). Better understanding of the hydrology of these regions, and particularly the connections between surface water and groundwater, could open up new opportunities for agriculture, capitalizing on annually recharged wetlands and shallow aquifers.

There is at this stage no compelling evidence that climate change will significantly alter total rainfall across the Nile Basin in the next 50 years. Different studies project both increased and decreased rainfall (Di Baldassarre *et al.*, 2011; Kim *et al.*, 2008). Most projected changes are within the envelope of existing rainfall variability. However, projected temperature increases of 2–4° C by 2090–2099 (IPCC, 2007) will increase ET and water stress, and may reduce the viability of rain-fed agriculture in marginal areas. Shifts in seasonality or variability of rainfall may also increase risks to rain-fed production, even if total rainfall remains constant. Climate adaptation efforts that focus on local agricultural water management interventions to reduce risks to agriculture from rainfall variability constitute 'no-regrets' options that can help in addressing current as well as future vulnerability.

Availability of water for irrigated agriculture

Current demands for water for irrigation in the upstream Nile countries (excluding Egypt and Sudan) are very low compared with available resources. In the White Nile sub-system, total estimated annual demand in the ELR is less than 1 km^3 in total (see Table 15.4, page 306), representing only a small fraction of the average flow of around 32 km^3 leaving Uganda. Available resources are more than adequate to supply current demand, and constraints on irrigation development are about infrastructure and access, rather than about water availability. Similarly, in Ethiopia current demand is minimal compared with available resources.

In Sudan, however, current irrigation demand is much higher, though difficult to estimate precisely. Reported irrigated area ranges from 1.2 million to 2.2 million ha (see Table 5.2);

depending on assumptions made about per hectare application rates, channel losses and return flows, total demand exceeds 12.5 km^3 and could be as high as 22.4 km^3 annually. The higher figure is considerably in excess of the 18.5 km^3 allocated to Sudan under the 1959 agreement, but represents only a quarter of total annual flows in northern Sudan. Water shortages do occur in both Gezira and New Halfa, but these are due to inadequate seasonal storage and siltation of reservoirs (Bastiaanssen and Perry, 2009). At this stage, there is no indication that Sudan's withdrawals have resulted in flows to Egypt becoming less than the 55.5 km^3 annual allocation; in fact, during the early 2000s, arrivals at Aswan were beyond the capacity of the reservoir, and releases were made through the Toshka Escape (see below). It seems likely that actual withdrawals in Sudan have been at the lower end of this range, although that may now be changing. Since 2009, there have been additional evaporative losses from Merowe Dam, estimated at 2.1 km^3 annually (Blackmore and Whittington, 2008); and the current expansion of Roseires Dam will also result in increased evaporative losses (DIU, 2011).

Availability of water for agriculture in Egypt is determined predominantly by the volumes arriving at, and released from, the AHD, but it is surprisingly difficult to get a clear idea of the exact volumes. Since these numbers are critical for determining how much is available for downstream use and how much additional withdrawal can be made upstream without jeopardizing downstream commitments, it is worthwhile examining the problem in some detail.

The first problem is which measurement best represents water available to Egypt. Long-term records are available from stations at Dongola in Sudan and at Aswan. The Dongola station is a composite record, which has been moved upstream twice since the 1920s to accommodate changes in lake level with construction of dams at Aswan and is now at Dongola, 430 km south of the border. At Aswan itself, flows are computed from reservoir levels and sluice discharges, not measured directly. Flows are reported as 'water arriving at Aswan' (derived by adding change in reservoir contents to downstream discharge), 'natural river at Aswan' (corrected for water abstracted upstream and for some evaporative losses) and 'flows below Aswan' (outflows from Aswan reservoir). Sutcliffe and Parks (1999) conclude that 'water arriving at Aswan' is the more reliable of the upstream records, since the basis for calculating 'natural river at Aswan' has changed over time. However, comparison of records for 'water arriving at Aswan' with flows at Dongola shows marked inconsistencies, with apparent high losses in most years. Long-term average for 'water arriving at Aswan' is 85.4 km^3 (1869–1992), while flows in the main Nile at Dongola averaged 88.1 km^3 over the same period. Suttcliffe and Parks (1999) attribute these differences to channel losses, measurement errors, the shift in the Dongola station and, since 1970, to evaporative losses from Aswan. To compound the uncertainty, records of Nile flows below Khartoum have not been released publicly beyond 1990, although flow measurements have been collected within government agencies; and (as discussed above) the volume of extractions within Sudan is not well constrained.

Under the Nile Agreement, the average annual flow in the Nile was agreed to be 84.5 km^3, based on the long-term average flows at Aswan and it is on this basis that 55.5 km^3 yr^{-1} were allocated to Egypt. However, reported annual flows are very variable, ranging from 150 to 40 km^3 (Blackmore and Whittington, 2008). A decline in flows is observed after the mid-1960s (despite higher flows in the White Nile in this period; see Table 5.1), attributed partly to increased abstraction and evaporative losses from reservoirs, and partly to the decline in rainfall in the Blue Nile and Atbara catchments. In the period 1970–1984, reported flows at Dongola averaged only 69 km^3. Though considerably below the notional long-term average on which the Agreement was predicated, this volume exceeds the average of 65.5 km^3 expected at Aswan if Sudan used its entire entitlement.

During the 1990s, flows to Aswan were much higher. Figures for 'water arriving at Aswan' have not been published, but around 1998 Egypt began diverting water through the Toshka Flood Escape, a canal leading to a depression in the western desert about 250 km above Aswan. By 2002 an area of more than 1500 km^2 had been flooded with a calculated stored volume in 2002 of 23 km^3 to form the Toshka Lakes (El-Shabrawy and Dumont, 2009). Since annual PET in the region is about 2 m, over 3 km^3 must have been pumped from Lake Nasser each year in order to maintain the lakes. Imagery from May 2011 shows the lakes still very much in evidence (Earth Snapshot, 2011). In addition, releases below Aswan were considerably above the nominated 55.5 km^3 – between 1998 and 2001, they averaged around 65 km^3 (Blackmore and Whittington, 2008). AHD operated at close to maximum level through 1995–2005. What is not clear in this equation is the extent to which the surplus can be attributed to low withdrawals by Sudan (below their allocation of 18.5 km^3).

Even during the extreme drought of 1984–85, releases from Aswan were maintained at above 53 $km^3\ yr^{-1}$, confirming the success of the 'century storage' concept. Thus, since the construction of the AHD, availability of water for Egypt has not only not fallen below the nominated 55.5 $km^3\ yr^{-1}$, but has, in many years, considerably exceeded it. The extent of the 'surplus' is not clear, but is probably about 5–10 km^3.

Demand within Egypt is similarly poorly quantified. The volume of irrigation withdrawals and return flows in the Nile Valley and Delta are complex and difficult to account, particularly given widespread use of recycled water and shallow pumped groundwater mixed with river water. FAO (2005) estimated total withdrawals in Egypt as 68.3 $km^3\ yr^{-1}$, with 59 km^3 diverted to agriculture. A later study by FAO (2010) estimated annual water use for irrigation in the Nile Basin in Egypt at 65.6 km^3. Blackmore and Whittington (2008) report Egypt's current total water use as 55.5 km^3 based on government estimates. A much lower agricultural demand of around 43.2 km^3 is estimated in Chapter 15 of this volume.

In addition, at least 6–8 km^3 of flow from the delta to the Mediterranean are needed to mitigate saline intrusion and preserve the salt balance of the Nile Delta (El-Arabawy, 2002). The actual extent of discharge from the Nile to the Mediterranean is highly debated. Because of the complexity of the delta, direct measurement of outflows is not possible, and calculation on the basis of upstream flows is hampered by incomplete data. Various modelling studies (Bonsor *et al.*, 2010; Karimi *et al.*, 2012; Chapter 15, this volume) estimate outflows of about 28 km^3. However, Hamza (2009) estimated that total annual releases are as low as 2–4 km^3; similarly, El-Shabraway (2009) reports that annual outflows decreased from about 32 km^3 before construction of AHD to 4.5 km^3 after.

The uncertainty surrounding estimates of total water use within Egypt thus remains very high. In planning terms, the difference between estimates is highly significant. Given that Egypt's nominal total allocation (for all uses including irrigation) under the Nile Agreement is 55.5 km^3, the lower estimate indicates that Egypt has room for substantial proposed increases from current levels of irrigation, within the supply limit guaranteed by the agreement. In contrast, the higher estimates suggest that Egypt is already overusing its allocation by 10 km^3 (20%) and is dependent on 'excess' flows to Aswan which may not be guaranteed in the longer term; and is thus potentially vulnerable to any increase in upstream withdrawals.

Future development

All Nile countries have ambitious proposals to expand irrigated agriculture to meet growing food demands and boost economic development. Based on national planning documents (Chapter 15), the overall increase will be more than 10 million ha in the Nile Basin, doubling from current levels of around 5 million ha. The question is whether such plans are feasible, where the limits to development lie in terms of the balance between demand and availability of water for irrigation in the basin, and at what stage withdrawals by upstream countries will impact upon Egypt.

Modelling studies (Blackmore and Whittington, 2008; McCartney *et al.*, 2010; Chapters 14–15) have examined the potential impact on flows of large-scale developments in the basin, for a range of scenarios with different levels of irrigation development. In the upper basin, much of the proposed irrigation development is part of multi-purpose schemes with an emphasis on hydropower generation, particularly in Ethiopia where dams with a total storage of over 167 km^3 have been proposed (McCartney *et al.*, 2010). These large-scale developments, which include water transfer schemes as well as storage, would profoundly change patterns of water availability.

Results from all studies indicate that absolute shortage of water is not limiting for proposed development in the upper basin countries. Proposed expansion of irrigation in the ELR will have no significant impact on downstream flows – increasing irrigated area in the ELR by 0.7 million ha would decrease average outflows from the Sudd by less than 1.5 km^3. In contrast, proposed development in Ethiopia and Sudan will cause large net reductions in flow in the Main Nile above Aswan, estimated at between 15 and 27 km^3 for expansion of total irrigation above Aswan by 2.2 million and 4.1 million ha, respectively (Chapter 15). If adequate storage is constructed, extractions of this magnitude are physically feasible (setting aside downstream impacts for the moment). Blackmore and Whittington (2008) also concluded that deficits in Ethiopia and Sudan would be negligible or small under even the most extreme of the scenarios they modelled. In reality, of course, the need to ensure downstream flows cannot be neglected, but these results reinforce that, to a large extent, limits to water availability for irrigation in the upstream Nile Basin countries are likely to be political rather than physical.

Uncertainties in estimates of total demand and supply within the basin are so high that it is difficult to draw firm conclusions about the stage at which irrigation demand in Egypt will outstrip supply – or whether this has already happened. The model results presented in Chapter 15 illustrate the trends resulting from increased withdrawals, but until current flows to Aswan and usage within Egypt are better constrained, it is not clear where along these trends Egypt sits. However, the results confirm three important points. First, withdrawals in the White Nile system upstream of the Sudd only have limited impact on water availability for Egypt. Second, irrigation development in Sudan will have a larger impact on water availability for Egypt than comparable increases in irrigation areas in Ethiopia. For example, under the 'long-term' scenario, expanding irrigation by 0.47 million ha in the Blue Nile Basin in Ethiopia resulted in an annual decrease compared with current conditions of only 2 km^3 at the Sudan Border; while an increase of 0.89 million ha in Sudan resulted in a decrease in projected flows in the Blue Nile at Khartoum of almost 10 km^3. This is due in large part to favourable options for storage in Ethiopia that can reduce evaporative losses, which are very high in Sudan. Third, model results illustrate the potential gains from managing evaporative losses from Aswan. Under the current and medium-term scenarios, where Aswan operates at relatively high levels, projected losses at Aswan are 11 km^3 while under the long-term scenario, Aswan operates at close to minimum levels, with evaporative losses reduced to 2 km^3, providing a significant offset to the increased withdrawals upstream of the dam.

Transboundary and environmental flow requirements

In estimating availability of water for irrigation, consideration must be given to requirements for environmental flows to maintain the ecosystems of the river, and to the obligations to maintain flow levels under international treaties and agreements. Since the late 1800s, international treaties and agreements have been in place to prevent upstream developments or withdrawals that would reduce Nile flows. These agreements were negotiated by Britain as a colonial power, and their validity is now contested. In the early 1950s, on construction of the Ripon Falls Dam on the outflow of the White Nile from Lake Victoria, an agreement was signed between Egypt and Uganda to ensure that releases from the lake retained their natural pattern (the 'Agreed Curve'). In 1959, prior to the construction of Aswan Dam, the Nile Agreement was signed which allocated the water of the Nile between Egypt (55.5 km^3) and Sudan (18.5 km^3). Ethiopia and the upstream countries were not signatories to the treaty, and do not recognize its validity. In 1999, the NBI was inaugurated to develop the river in a cooperative manner, without binding rules on flow management. In 2010–2011, six of the nine Nile Basin countries (Burundi, Ethiopia, Kenya, Rwanda, Tanzania, Uganda) signed the Entebbe Agreement on Nile River, which calls for Nile Basin countries to modify the existing agreement and reallocate water shares. The Entebbe Agreement has been strongly opposed by Egypt and Sudan (Swain, 2011; Ibrahim, 2011).

No comprehensive assessment of environmental flow requirements for the Nile has been conducted, but the various transboundary agreements have, to some extent, acted to secure environmental flows. The Agreed Curve governing releases from Lake Victoria based on lake levels formed the basis for releases from the 1950s to 2001 (except for a period in the 1960s when lake levels rose above the limit of gauging). When the Kiira Power Station (an extension of the Owen Falls hydro scheme) began operation in 2002, water levels fell. To meet demand for hydropower, Uganda has released 55 per cent more than the Agreed Curve (Kull, 2006); combined with several years of low rainfall this reduced lake levels to an 80-year low (Awange *et al.*, 2008). Lake levels have recovered since 2007, but manipulation of the lake level for hydropower generation remains controversial. Adverse impacts are felt primarily around the lake itself, since the additional flows were released from Kiiraare, moderated by the Sudd.

Studies in the Sudd on the feasibility of the proposed Jonglei Canal examined potential impacts of diversions on the extent of flooding in the wetlands. Sutcliffe and Parks (1999) estimated that diversion of 20–25m^3 day^{-1} (about 20–25% of flows on an annualized basis) would reduce the area of permanent swamp by more than a third, and of seasonal swamp by around 25 per cent; the decrease in seasonally flooded area could be mitigated to some extent by varying withdrawals according to season. Related proposals to regulate the inflows to the Sudd through storage in Lake Albert or Lake Victoria would also reduce the seasonal flooding. Seasonally flooded grasslands are a vital component of the grazing cycle for the herds of the Nuer and Dinka, while the permanent swamps are an important dry-season refuge for wildlife, including large populations of elephants. South Sudan has recently declared the Sudd a national reserve, with plans to develop eco-tourism in the area.

McCartney *et al.* (2009) conducted a study to determine environmental flow requirements (both high and low flows) for the Blue Nile downstream of CharaChara weir on Lake Tana. They estimate that an average annual allocation of 22 per cent of the mean annual flow (862Mm^3) is needed to maintain the basic ecological functioning in this reach, with an absolute minimum mean monthly allocation not less than approximately 10 million m^3.

Construction of the AHD and large withdrawals for irrigation have modified the ecosystem of the Delta through reductions in both flow and sediment, and exacerbated the decline in water quality from agricultural, urban and industrial uses (Hamza, 2009). A minimum level of flow (around 6–8 km^3) is critical in a number of different contexts: to prevent intrusion of salt water into the agricultural systems of the Delta; to flush other salts and pollutants from the system; and to maintain the coastal ecosystems of the Delta and the fringing lakes (El-Arabawy,

2002). Continued degradation of the water quality and ecosystems of the Delta suggests that these requirements are not being met.

Finding more water

It has long been recognized that there is potential to increase total water availability in the Nile Basin by reducing 'losses' of water through evaporation and infiltration to groundwater from both natural and man-made water bodies and irrigation schemes. There are three mechanisms proposed to achieve this: diversion of flows from natural wetlands; management of man-made storage to minimize evaporation; and improving the efficiency of irrigation. In addition, there is potential to expand the role of groundwater resources in supplementing surface supplies.

Since the 1930s, there have been proposals to divert flows from the floodplains of southern Sudan through canals to reduce evaporative losses. The best known and most ambitious is the Jonglei Canal project, diverting water for 360 km around the Sudd to gain around 4 km^3 yr^{-1} in flows (Sutcliffe and Parks, 1999). Construction began in 1978, but was interrupted by civil unrest and suspended from 1982, and has never been completed. A similar proposal for the Bahr el Ghazal would divert flow to the White Nile using collector canals; Hurst *et al.* (1978) suggested that as much as 8 km^3 yr^{-1} could be diverted. Such schemes should be approached with caution, however. The dynamics of the wetland systems are not well understood, and potential impacts on both ecology and livelihoods are very high. Sutcliffe and Parks (1999) concluded that the Jonglei diversion would reduce the area of both permanent swamps and seasonally flooded regions in the Sudd, with potential impacts on livelihoods of the local populations.

There are similar proposals to reduce evaporative losses from the Sobat-Baro either by regulating peak flows with upstream storage to reduce floodplain spillage, or by diverting flows from the Machar Marshes (Hurst, 1950). However, an analysis by WaterWatch (reported in Blackmore and Whittington, 2008) indicates that more than 90 per cent of water in the marshes is derived from local rainfall, with only 1 km^3 from overbank spills. They conclude that gains from diversions would be small, but that potential impacts on local livelihoods and ecology could be severe.

Total evaporative losses from constructed storage in the basin are now estimated at around 15 – 20 km^3 yr^{-1}, more than 20 per cent of flows arriving at Aswan (Blackmore and Whittington, 2008; Kirby *et al.*, 2010). Annual evaporative losses from Lake Nasser/Lake Nubia (impounded by AHD) vary with water level, from around 5 km^3 at 160 m to more than 10 km^3 at 180 m (El-Shabrawy, 2009). Blackmore and Whittington (2008) estimate somewhat higher losses of 14.3 km^3 yr^{-1} at maximum levels, compared with 10 km^3 under 'normal' operations. Other areas where significant evaporative losses occur include the recently completed Merowe hydropower dam on the Main Nile in Sudan (more than 1.3 km^3 yr^{-1}); Toskha Lakes (2–3 km^3 yr^{-1}); and Jebel Aulia Dam, upstream of Khartoum, with losses estimated at 2.1 km^3 yr^{-1} (WaterWatch, 2011) to 3.45 km^3 yr^{-1} (Blackmore and Whittington, 2008). Jebel Aulia was originally constructed in 1937 to prolong the natural recession for irrigation downstream. Since construction of the AHD, its primary function has become redundant, and it has been operated mainly to optimize costs for pumped irrigation from the river downstream of Khartoum. Removal of the dam would provide significant evaporative gains for very little cost compared, for example, to the cost of the Jonglei scheme.

As demonstrated in Chapter 14, by shifting storage from broad shallow reservoirs in arid downstream areas (Aswan, Merowe, Jebel Aulia) to deep reservoirs in upstream areas with lower rates of evaporation, overall losses from reservoir evaporation can be reduced very significantly.

Evaporative losses from reservoirs outweigh putative gains from proposed canal systems at Jonglei and Bahr el Ghazal designed to reduce ET from natural systems. Evaporation from reservoirs is entirely non-beneficial, while ET from natural wetlands provides important benefits in terms of both pastoral production and biodiversity. Managing non-beneficial evaporative losses through a coordinated approach to construction and operation of reservoirs is a much more urgent priority as a water-saving measure than draining wetlands.

Irrigation demand in the basin could be substantially reduced by improving the efficiency of irrigation systems (see Chapter 15). The estimates suggest that if efficiency of water use could be increased from 50 to 80 per cent, total demand in the long-term (high development) scenario would be reduced by 40 km^3. Bastiaanssen and Perry (2009) provide a comprehensive review of the productivity of large-scale irrigation schemes within the Nile Basin, including criteria related to water use efficiency (crop water consumption and beneficial fraction). In Sudan, beneficial fraction emerged as one of the lowest scoring criteria in almost half of the irrigation schemes studied, and crop water consumption scored very poorly in Gezira, the largest irrigation area. In Egypt, beneficial fraction scored well, but crop water consumption was very high in many of the schemes studied. These results confirm that there are significant gains to be made in terms of efficiency of water use in existing schemes; and that measures to ensure efficiency of water use in new schemes are an important priority.

Role of groundwater

Shallow groundwater systems, seasonally recharged from local run-off, exist over large areas of the southern half of the basin where rain-fed cropping dominates. These aquifers generally have low to moderate flow rates, and are not suitable for large-scale irrigation, but could provide an accessible source of supplementary irrigation to reduce risk in rain-fed cropping. Calow and MacDonald (2009) concluded that groundwater has the potential to provide limited irrigation across wide areas of sub-Saharan Africa, and could increase food production, raise farm incomes and reduce vulnerability. Surface seepage from shallow groundwater systems is already utilized in wetland and valley bottom cultivation in ELR and the Ethiopian Highlands. In many cases, shallow groundwater and surface waters are connected, and function as a single system. This is true both in the Ethiopian Highlands, where groundwater supply baseflow for rivers in the dry season and in the alluvial aquifers of the Nile Valley, which are directly linked to the river. In these cases, groundwater must be accounted as part of the surface water system, though there may still be advantages in drawing water from the subsurface in terms of evaporative losses and local access. In southern and eastern Sudan, local to regional aquifer systems recharged from adjacent highlands and swamps are not directly linked to the rivers, and potentially constitute a very significant resource that Sudan could exploit with no impact on downstream flows.

In rangeland pastoral systems in Ethiopia, Sudan and Uganda, access to drinking water for stock can be limiting even when fodder is available. If surface water sources dry up in the dry season, alternatives are to travel long distances to permanent sources, to harvest and store rainwater, or to provide watering points from groundwater. Shallow wells are the mainstay for provision of human and animal drinking water in the arid zone. Provision of local watering points can significantly improve livestock productivity, both by reducing energy requirements for additional travel to water and by reducing grazing pressure around surface water sources (Peden *et al.*, 2009).

Currently, groundwater is used for large-scale irrigation only in Egypt, where it accounts for 11 per cent of irrigated agricultural production (FAO, 2005), though a substantial proportion of this is from shallow aquifer systems linked directly to the Nile. The Nubian Sandstone Aquifer System (NSAS) supplies water for large-scale irrigation in Egypt and Libya, and there is potential for similar developments in northern Sudan and areas of Egypt within the Nile Basin. The aquifer is an enormous resource of 375,000 km^3, but recharge is very limited, and extraction is essentially mining fossil waters. Given the size of the resource, properly managed extractions could support irrigation for hundreds, or even thousands of years; but experience in Chad, Egypt and Libya has demonstrated that over-extraction can lead to rapid drawdown locally. In addition, irrigation-induced salinization must be carefully managed; and both establishment costs and energy costs for pumping are high (IAEA, 2011).

Evaporative losses from man-made reservoirs account for a large and increasing fraction of Nile flows. Internationally, sub-surface water storage is being explored as one option for managing evaporative losses using managed aquifer recharge (MAR), storage and recovery techniques – for example, in Menindee Lake in Australia (Geoscience Australia, 2008). The Egyptian government has investigated options for injecting excess water from Lake Nasser into the Nubian Sandstone Aquifer for subsequent recovery by pumping to provide water for new irrigation for land development. Preliminary studies indicate feasibility of the approach, but caution that irrigation combined with local groundwater mounds from injection could lead to waterlogging and salinization (Kim and Sultan, 2002). MAR requires a detailed knowledge of aquifer structures and properties, and may not be feasible in the short term; but in the longer term, the potential evaporative savings mean that it deserves serious technical appraisal.

Groundwater use in the Nile Basin is, to a large extent, buffered from the impacts of climate change. NSAS draws on fossil water, and so will not be affected by changes in recharge. In low rainfall areas (<200 mm) recharge is minimal, and groundwater sustainability is determined by the balance between withdrawals and local drawdown, and is unlikely to be affected by climate change. Current levels of use for domestic and livestock supplies require very little recharge to be sustainable: Calow and MacDonald (2009) estimate 3 mm yr^{-1} across much of Africa. In the complex basement hydrogeology of the southern basin, low-yielding aquifers are to some extent self-regulating – excessive withdrawals result in local drawdown and decreased yield. Large regional aquifers are at more risk of depletion. Dependence on groundwater is likely to increase as populations grow, and increases in demand will be more significant than any changes in overall supply due to climate change. Groundwater resources in the Nile Basin are discussed in more detail in Chapter 10.

Conclusions

Although the debates about availability of agricultural water in the Nile Basin are usually framed primarily around irrigation, rain-fed agriculture dominates production in the basin outside of Egypt. The dominant land use in the basin is low productivity agro-pastoralism. More than 70 per cent of rain falling in the basin is depleted as ET from natural systems partially utilized for pastoral activities. Rain-fed cropping, which constitutes around 85 per cent of total cropped area, uses only about 10 per cent of total basin rainfall. In the subhumid to humid regions of the southern Nile Basin, there is significant potential to intensify rain-fed production and to increase the area of rain-fed cropping at the expense of grasslands and savannahs, transferring at least some of this component to more productive, higher-value uses. In the semi-arid zones of the central basin, there is capacity to reduce risks and improve productivity of rain-fed agriculture using supplementary irrigation from small-scale storages and groundwater, and improved soil water management and agronomic practices. Since most of the flow is generated from only a small number of highland sub-catchments in the ELR and Ethiopian Highlands, rainwater harvesting, intensification of cropping and conversion of natural vegetation to agriculture are unlikely to significantly reduce downstream water availability.

Current demand for irrigation water in the upstream Nile countries (excluding Egypt and Sudan) constitutes only a small fraction of available resources, and constraints on irrigation development are about infrastructure and access, rather than about water availability. In Sudan, estimates of current extractions vary, but may be as high as 20 km^3, exceeding Sudan's nominal allocation under the 1959 Agreement. However, at this stage there is no indication that Sudan's withdrawals have resulted in reduced flows to Egypt. In fact, evidence from several sources suggests that for most of the last 15 years, flows to Egypt have considerably exceeded the nominal allocation of 55.5 $km^3 yr^{-1}$. The extent to which the surplus can be attributed to withdrawals in Sudan below their nominal allocation is not clear.

Demand within Egypt is similarly poorly quantified, with estimates varying from 43 to 65 $km^3 yr^{-1}$. High releases from Aswan in at least some years, and low outflows to the Mediterranean (~4 km^3) suggest that usage within Egypt is at least in the middle range of estimates. In planning terms, the difference between estimates is highly significant. The lower estimate indicates that Egypt has scope to expand water use for irrigation, within its nominated allocation under the 1959 Agreement. The higher estimates suggest that Egypt is already over-using its nominal allocation by up to 10 km^3 (20%) and is dependent on 'excess' flows to Aswan, which may not be guaranteed in the longer term, and is thus potentially vulnerable to any increase in upstream withdrawals.

All the upstream Nile countries have ambitious plans to expand irrigation to meet growing food demands and boost economic development. Modelling studies indicate that absolute shortage of water is not limiting for proposed development in the upper basin countries. If adequate storage is constructed, proposed expansion of an additional 4 million ha of irrigation upstream of Aswan is technically feasible but would result in significant reduction of flows to Egypt, though this would be offset to some extent by reduction in evaporative losses from Aswan. Due to the moderating effect of the Sudd on peak flows, development of up to 0.7 million ha of irrigation in ELR would not significantly reduce outflows from the White Nile system. Increasing irrigation area in Sudan will have a much greater impact on flows at Aswan than comparable increases in Ethiopia, due to more favourable storage options in Ethiopia.

While it is clear that upstream development will cause a reduction in flows to Egypt, uncertainties in estimates of total irrigation demand and available flows within the basin are so high that it is not possible to determine from existing information the stage at which demand will outstrip supply in Egypt, or even whether this has already happened. Shortages already occur in some years, but high flows in others present opportunities to expand irrigation through increased upstream storage, better management of flow variability and improved efficiency of use. A more flexible approach to management of high flows, including over-year storage in upstream areas, could provide an overall increase in available water, but requires management of water resources at the basin scale. Total evaporative losses from constructed storage in the basin are now estimated at around 15–20 $km^3 yr^{-1}$, more than 20 per cent of flows arriving at Aswan. By supplementing Aswan with storage higher in the basin, options for managing and storing high flows could be extended, security of supply in the upper basin improved, and evaporative losses reduced to provide an overall increase in available water, but this can only be achieved through transboundary cooperation to manage water resources at the basin scale.

Evaporative losses from reservoirs outweigh putative gains from proposed canal systems at Jonglei and Bahr el Ghazal, designed to reduce ET from wetland systems. Evaporation from reservoirs is entirely non-beneficial, while ET from natural wetlands provides important benefits in terms of both pastoral production and biodiversity. Managing non-beneficial evaporative losses through a coordinated approach to construction and operation of reservoirs is a much more urgent priority as a water-saving measure than draining wetlands.

Groundwater has a potential role in increasing water availability in the Nile Basin in four different contexts: small-scale supplementary irrigation in rain-fed zones; improving productivity of rangeland pastoral systems; large-scale irrigation in arid areas; and a potential role in reducing evaporative losses in storage through the use of managed aquifer recharge.

Climate change projections for the basin are equivocal, with no compelling evidence for large changes in rainfall. Projected changes mostly fall within the envelope of existing rainfall variability, but increases in temperature increases may reduce the viability of rain-fed agriculture in marginal areas and increase water demands for irrigation.

References

Araya, A. and Stroosnijder, L. (2011) Assessing drought risk and irrigation need in northern Ethiopia, *Agricultural and Forest Meteorology,* 151, 425–436.

Awange, J., Sharifi, M., Ogonda, G., Wickert, J., Grafarend, E. and Omulo, M. (2008) The falling Lake Victoria water level: GRACE, TRIMM and CHAMP satellite analysis of the lake basin, *Water Resources Management*, 22, 775–796.

Awulachew, S. B., McCartney, M., Steenhuis, T. S. and Ahmed, A. A. (2008) *A Review of Hydrology, Sediment and Water Resource Use in the Blue Nile Basin,* IWMI Working Paper 131, International Water Management Institute, Colombo, Sri Lanka.

Awulachew, Seleshi, Rebelo, Lisa-Maria and Molden, David (2010) The Nile Basin: tapping the unmet agricultural potential of Nile waters, *Water International,* 35, 5, 623–654.

Bastiaanssen, W. and Perry, C. (2009) *Agricultural Water Use and Water Productivity in the Large Scale Irrigation Schemes of the Nile Basin,* Report for the Efficient Water Use in Agriculture Project, Nile Basin Initiative, Entebbe, Uganda.

Blackmore, D. and Whittington, D. (2008) *Opportunities for Cooperative Water Resources Development on the Eastern Nile: Risks and Rewards,* Report to the Eastern Nile Council of Ministers, Nile Basin Initiative, Entebbe, Uganda.

Bonsor, H. C., Mansour, M. M., MacDonald, A. M., Hughes, A. G., Hipkin, R. G. and Bedada, T. (2010) Interpretation of GRACE data of the Nile Basin using a groundwater recharge model, *Hydrology and Earth System Sciences Discussions,* 7, 4501–4533.

Calow, R. and MacDonald, A. (2009) *What Will Climate Change Mean for Groundwater Supply in Africa?* ODI Background Note, Overseas Development Institute, London.

Camberlin, P. (2009) Nile Basin climates, in *The Nile,* H. J. Dumont (ed.), Monographiae Biologicae 89, 307–334, Springer, Dordrecht, The Netherlands.

Dai, A. and Trenberth, K. E. (2003) *New Estimates of Continental Discharge and Oceanic Freshwater Transport,* Proceedings of the Symposium on Observing and Understanding the Variability of Water in Weather and Climate, 83rd Annual AMS Meeting, American Meteorological Society, Long Beach, CA.

Di Baldassarre, G., Elshamy, M., Griensven, A. van,Soliman, E., Kigobe, M., Ndomba, P., Mutemi, J., Mutua, F., Moges, S., Xuan, Y., Solomatine, D. and Uhlenbrook, S. (2011) Future hydrology and climate in the River Nile Basin: a review, *Hydrological Sciences Journal*, 56, 2, 199–211.

DIU (Dam Implementation Unit) (2011) *Alrrusairis Dam,* Ministry of Electricity and Dams, Khartoum, Sudan, http://diu.gov.sd/roseires/en/about_rosirs.htm, accessed May 2011.

Earth Snapshot (2011) *Crops of the New Valley Project West of Toshka Lakes and Lake Nasser, Egypt,* www.eosnap.com/?s=toshka+lakes, accessed December 2011.

El-Arabawy, M. (2002) *Water Supply-Demand Management for Egypt: A Forthcoming Challenge,* Proceedings Nile Conference 2002, Nairobi, Kenya, 28–31 January.

El-Shabrawy, G. I. (2009) Lake Nasser-Nubia, in *The Nile,* H. J. Dumont (ed.), Monographiae Biologicae 89, 125–155, Springer, Dordrecht, The Netherlands.

El-Shabrawy, G. I. and Dumont, H. (2009) The Toshka Lakes, in *The Nile,* H. J. Dumont (ed.), Monographiae Biologicae 89, 157–162, Springer, Dordrecht, The Netherlands.

FAO (Food and Agriculture Organization of the United Nations) (2005) *FAO Water Report No. 29,* FAO, Rome, Italy.

FAO (2010) Areas of rainfed and irrigated cropping in the Nile Basin, unpublished data.

Geoscience Australia (2008) *Assessment of Groundwater Resources in the Broken Hill Region,* Professional Opinion 2008/05, Department of Environment, Water, Heritage and the Arts, Canberra, Australia.

Hamza, W. (2009) The Nile delta, in *The Nile*, H. J. Dumont (ed.), Monographiae Biologicae 89, 75–94, Springer, Dordrecht, The Netherlands.

Hulme, M., Doherty, R., Ngara, T., New, M. and Lister, D. (2001) African climate change: 1900–2100, *Climate Research*, 17, 145–168.

Hurst, H. E. (1950) *The Hydrology of the Sobat and White Nile and the Topography of the Blue Nile and Atbara*, The Nile Basin, VIII, Government Press, Cairo, Egypt.

Hurst, H. E., Black, R. P. and Simaika, Y. M. (1978) *The Hydrology of the Sadd el Aali*, The Nile Basin, XI, Government Press, Cairo, Egypt.

Ibrahim, A. M. (2011) The Nile Basin Cooperative Framework Agreement: The beginning of the end of Egyptian hydro-political hegemony, *Missouri Environmental Law and Policy Review*, 18, 282.

IPCC (Intergovernmental Panel on Climate Change) (2007) *Climate Change 2007: The IPCC Fourth Assessment Report*, IPCC, Geneva, Switzerland, www.ipcc.ch/publications_and_data/publications_and_data.htm, accessed April 2011.

Karimi, P., Molden, D., Bastiaanssen, W. G. M. and Cai, X. (2012) Water accounting to assess use and productivity of water – Evolution of the concept and new frontiers, in *International Approaches to Policy and Decision-Making*, Godfrey, J. M. and Chalmers, K. (eds), Edward Elgar Publishing, Cheltenham, UK.

Kim, J. and Sultan, M. (2002) Assessment of the long-term impacts of Lake Nasser and related irrigation projects in Southwestern Egypt, *Journal of Hydrology*, 262, 68–83.

Kim, U., Kaluarachchi, J. J. and Smakhtin, V. U. (2008) *Climate Change Impacts on Water Resources of the Upper Blue Nile River Basin, Ethiopia*, IWMI Research Report 126, IWMI, Colombo, Sri Lanka.

Kirby, M., Mainuddin, M. and Eastham, J. (2010) *Water-Use Accounts in CPWF Basins: Model Concepts and Description*, CPWF Working Paper BFP01, CGIAR Challenge Program on Water and Food, Colombo, Sri Lanka.

Kull, D. (2006) *Connection between Recent Water Level Drop in Lake Victoria, Dam Operation and Drought*, www.irn.org/programs/nile/pdf/060208vic.pdf, accessed May 2011.

McCartney, M., Alemayehu, T., Shiferaw, A. and Awulachew, S. B. (2010) *Evaluation of Current and Future Water Resources Development in the Lake Tana Basin, Ethiopia*, IWMI Research Report 134, IWMI, Colombo, Sri Lanka.

Mongabay. (1991) *Sudan – Rainfed Agriculture*, www.mongabay.com/history/sudan/sudan-rainfed_agriculture.html, accessed May 2011.

NWRC (National Water Research Centre). (2007) *Nile River Basin Baseline*, revised report, National Water Research Centre, Cairo, Egypt.

Peden, D., Taddesse, G. and Haileslassie, A. (2009) Livestock water productivity: implications for sub-Saharan Africa, *The Rangeland Journal*, 31, 187–193.

Sutcliffe, J. (2009) The hydrology of the Nile Basin, in *The Nile*, H. J. Dumont (ed.), Monographiae Biologicae 89, 335–364, Springer, Dordrecht, The Netherlands.

Sutcliffe, J. V. and Parks, Y. P. (1999) *The Hydrology of the Nile*, IAHS Special Publication 5, International Association of Hydrological Sciences, Wallingford, UK.

Swain, A. (2011) Challenges for water sharing in the Nile basin: changing geo-politics and changing climate, *Hydrological Sciences Journal*, 56, 4, 687–702.

UNEP/GRID (United Nations Environment Programme/GRID-Arendal) (2002) *Global Environment Outlook 3*, Earthscan, London.

WaterWatch (2011) *Determination of Water Surface Area and Evaporation Losses from Jebel Aulia Reservoir*, www.waterwatch.nl/fileadmin/bestanden/Project/Africa/0136_SD_2006_JebelAulia.pdf, accessed May 2012.

6
Hydrological processes in the Blue Nile

Zachary M. Easton, Seleshi B. Awulachew, Tammo S. Steenhuis, Saliha Alemayehu Habte, Birhanu Zemadim, Yilma Seleshi and Kamaleddin E. Bashar

Key messages

- While we generally have a fundamental understanding of the dominant hydrological processes in the Blue Nile Basin, efforts to model it are largely based on temperate climate hydrology. Hydrology in the Blue Nile Basin is driven by monsoonal climate, characterized by prolonged wet and dry phases, where run-off increases as the rainy season (which is also the growing season) progresses. In temperate climates, run-off typically decreases during the growing season as plants remove soil moisture. In the Blue Nile Basin there is a threshold precipitation level needed to satisfy soil-moisture capacity (approximately 500 mm) before the basin begins to generate run-off and flow.
- Not all areas of the basin contribute equally to Blue Nile flow. Once the threshold moisture content is reached, run-off generation first occurs from localized areas of the landscape that become saturated or are heavily degraded. These saturated areas are often found at the bottom of large slopes, or in areas with a large upslope-contributing area, or soils with a low available soil moisture storage capacity. As the monsoonal season progresses, other areas of the basin with greater soil moisture storage capacity begin to contribute to run-off. By the end of the monsoon, more than 50 per cent of the precipitation can end up as run-off. This phenomenon is termed 'saturation excess run-off' and has important implications for identifying and locating management practices to reduce run-off losses.
- The Soil and Water Assessment Tool (SWAT) model is modified to incorporate these run-off dynamics, by adding a landscape-level water balance. The water balance version of SWAT (SWAT-WB) calculates the water deficit (e.g. available soil moisture storage capacity) for the soil profile for each day, and run-off is generated once this water deficit is satisfied. We show that this conceptualization better describes hydrological processes in the Blue Nile Basin.
- Models that include saturation excess (such as our adaptation to the SWAT model) are not only able to simulate the flow well but also good in predicting the distribution of run-off in the landscape. The latter is extremely important when implementing soil and water conservation practices to control run-off and erosion in the Blue Nile Basin. The SWAT-WB model shows that these practices will be most effective if located in areas with convergent topography.

Overview

This chapter provides an analysis of the complexity of hydrological processes using detailed studies at scales from the micro-watershed to the Blue Nile Basin (BNB). Data collected from various sources include long-term Soil Conservation Reserve Program (SCRP) data (Hurni, 1984), and consist of both hydrological data in the form of streamflow and stream sediment concentrations and loads. Data collected by students in the SCRP watersheds include piezometric water table data and plot studies of run-off dynamics. Governmental and non-governmental sources provided meteorological data, as well as data on land use, elevation and soil inputs for modelling analysis. In the small SCRP watersheds, the analysis focused on piezometric water table data and run-off losses as they relate to the topographic position in the watershed. In parallel, basin-scale models were used to enhance understanding of rainfall run-off and erosion processes and the impact of management interventions on these processes in the BNB.

Introduction

A better understanding of the hydrological processes in the headwaters of the BNB is of considerable importance because of the trans-boundary nature of the BNB water resources. Ethiopia has abundant, yet underutilized, water resource potential, and 3.7 million ha of potentially irrigable land that can be used to improve agricultural production (MoWR, 2002; Awulachew *et al.*, 2007). Yet only 5 per cent of Ethiopia's surface water (0.6% of the Nile Basin's water resource) is being currently utilized by Ethiopia (Arseno and Tamrat, 2005). Sudan receives most of the flow leaving the Ethiopian Highlands, and has considerable infrastructure in the form of reservoirs and irrigation schemes that utilize these flows. Ethiopian Highlands are the source of more than 60 per cent of the Nile flow (Ibrahim, 1984; Conway and Hulme, 1993). This proportion increases to almost 95 per cent during the rainy season (Ibrahim, 1984). However, agricultural productivity in Ethiopia lags behind other similar regions, which is attributed to unsustainable environmental degradation mainly from erosion and loss of soil fertility (Grunwald and Norton, 2000). In addition, there is a growing concern about how climate and human-induced degradation will impact the BNB water resources (Sutcliffe and Parks, 1999), particularly in light of limited hydrological and climatic studies in the basin (Arseno and Tamrat, 2005).

One characteristic of Ethiopian BNB hill slopes is that most have infiltration rates in excess of the rainfall intensity. Consequently, most run-off is produced when the soil saturates (Ashagre, 2009) or from shallow, degraded soils. Engda (2009) showed that the probability of rainfall intensity exceeding the measured soil infiltration rate is 8 per cent. This is not to imply that infiltration excess, or Hortonian flow (Horton, 1940), is not present in the basin, but that it is not the dominant hydrological process. Indeed, Steenhuis *et al.* (2009) and Collick *et al.* (2009) not only note the occurrence of infiltration excess run-off but also state that it is predominantly found in areas with exposed bedrock or in extremely shallow and degraded soils. Most of the models utilized to assess the hydrological response of basins such as the BNB are arguably incorrect (or at the very least incomplete) in their ability to adequately simulate the complex interrelations of climate, hydrology and human impacts. This is not because the models are poorly constructed but often because they were developed and tested in very different climates, locations and hydrological regimes. Indeed, most hydrological models have been developed in, and tested for, conditions typical of the United States or Europe (e.g. temperate climate, with an even distribution of rainfall), and thus lack the fundamental understanding of how regions dominated by monsoonal conditions (such as the BNB) function hydrologically. Thus, a paradigm shift is needed on how the hydrological community conceptualizes hydrological processes in these data-scarce regions.

In monsoonal climates a given rainfall volume at the onset of the monsoon produces a different run-off volume than the same rainfall at the end of the monsoon (Lui *et al.*, 2008). Lui *et al.* (2008) and Steenhuis *et al.* (2009) showed that the ratio of discharge to precipitation minus evapotranspiration (Q/(P – ET)) increases with cumulative precipitation from the onset of the monsoon and, consequently, the watersheds behave differently depending on the amount of stored moisture, suggesting that saturation excess processes play an important role in the watershed run-off response. Other studies in the BNB or nearby catchments have suggested that saturation excess processes control overland flow generation (Collick *et al.*, 2009; Ashagre, 2009; Engda, 2009; Tebebu, 2009; Easton *et al.*, 2010; Tebebu *et al.*, 2010; White *et al.*, 2011) and that infiltration-excess run-off is rare (Liu *et al.*, 2008; Engda, 2009).

Many of the commonly used watershed models employ some form of the Soil Conservation Service curve number to predict run-off, which links run-off response to soils, land use and five-day antecedent rainfall (AMC), and not the cumulative seasonal rainfall volume. The Soil and Water Assessment Tool (SWAT) model is a basin-scale model where run-off is based on land use and soil type (Arnold *et al.*, 1998), and not on topography. As a result, run-off and sediment transport on the landscape are only correctly predicted for infiltration excess of overland flow, and not when saturation excess of overland flow from variable source areas (VSA) dominates. Thus, critical landscape sediment source areas might not be explicitly recognized.

The analysis in this chapter utilizes existing data sets to describe the hydrological characteristics of the BNB with regard to climatic conditions, rainfall characteristics and run-off process across various spatial scales in the BNB. An attempt is made to explain the processes governing the generation of run-off at various scales in the basin, from the small watershed to the basin level and to quantify the water resources at these scales. Models used to predict the source, timing and magnitude of run-off in the basin are reviewed, the suitability and limitations of existing models are described, and approaches and results of newly derived models are presented.

Much of the theory of run-off production that follows is based on, and corroborated by, studies carried out in the SCRP watersheds. These micro-watersheds are located in headwater catchments in the basin and typify the landscape features of much of the highlands, and are thus somewhat hydrologically representative of the basin. A discussion of the findings from these SCRP micro-watersheds is followed by work done in successively larger basins (e.g. watershed, sub-basin and basin) and, finally, by an attempt to integrate these works using models across the various scales.

Rainfall run-off processes

Micro-watershed hydrological processes

SCRP watersheds have the longest and most accurate record of both rainfall and run-off data available in Ethiopia. Three of the sites are located in the Amhara region either in or close to the Nile Basin: Andit Tid, Anjeni and Maybar (SCRP, 2000). All three sites are dominated by agriculture, with control structures built for soil erosion to assist the rain-fed subsistence farming (Table 6.1).

Table 6.1 Location, description and data span from the three SCRP research sites

Site	*Watershed centroid (region)*	*Area (ha)*	*Elevation range (masl)*	*Precipitation (mm yr⁻¹)*	*Length of record*
Andit Tid	39°43' E, 9°48' N (Shewa)	477.3	3040–3548	1467	1987–2004 (1993, 1995–1996 incomplete)
Anjeni	37°31' E, 10°40' N (Gojam)	113.4	2407–2507	1675	1988–1997
Maybar	37°31' E, 10°40' N (South Wollo)	112.8	2530–2858	1417	1988–2001 (1990–1993 incomplete)

The Andit Tid Research Unit covers a total area of 481 ha with an elevation of 3040–3548 m, with steep and degraded hill slopes (Bosshart, 1997), resulting in 54 per cent of the long-term precipitation becoming run-off (Engda, 2009). Conservation practices including terraces, contour drainage ditches and stone bunding have been installed to promote infiltration and reduce soil loss. In addition to long-term rainfall run-off and meteorological measurements, plot-scale measures of soil infiltration rate at 10 different locations throughout the watershed were taken and geo-referenced with a geographic positioning system (GPS). Soil infiltration was measured using a single-ring infiltrometer of 30 cm diameter.

The Anjeni watershed is located in the Amhara Region of the BNB. The Anjeni Research Unit covers a total area of 113 ha and is the most densely populated of the three SCRP watersheds, with elevations from 2400 to 2500 m. The watershed has extensive soil and water conservation measures, mainly terraces and small contour drainage ditches. From 1987 to 2004 rainfall was measured at five different locations, and discharge was recorded at the outlet and from four run-off plots. Of the rainfall, 45 per cent becomes run-off. During the 2008 rainy season the soil infiltration rate was measured at ten different locations throughout the watershed using a single-ring infiltrometer of 30 cm diameter. In addition, piezometers were installed in transects to measure the water table depths.

The 112.8 ha Maybar catchment was the first of the SCRP research sites, characterized by rugged topography with slopes ranging between 2530 and 2860 m. Rainfall and flow data were available from 1988 to 2004. Discharge was measured with a flume installed in the Kori River using two methods: float-actuated recorder and manual recording. The groundwater table levels were measured with 29 piezometers located throughout the watershed. The saturated area in the watershed was delineated and mapped using a combination of information collected using a GPS, coupled with field observation and groundwater-level data.

Analysis of rainfall discharge data in SCRP watersheds

To investigate run-off response patterns, discharges in the Anjeni, Andit Tid and Maybar catchments were plotted as a function of effective rainfall (i.e. precipitation minus evapotranspiration, P – E) during the rainy and dry seasons. In Figure 6.1a an example is given for the Anjeni catchment. As is clear from this figure, the watershed behaviour changes as the wet season progresses, with precipitation later in the season generally producing a greater percentage of run-off. As rainfall continues to accumulate during the rainy season, the watershed eventually reaches a threshold point where run-off response can be predicted by a linear relationship with effective precipitation, indicating that the proportion of the rainfall that became run-off was constant during the remainder of the rainy season. For the purpose of this study, an approximate threshold of 500 mm of effective cumulative rainfall (P – E) was determined after iteratively examining rainfall/run-off plots for each watershed. The proportion Q/(P – E) varies within a relatively small range for the three SCRP watersheds, despite their different characteristics. In Anjeni, approximately 48 per cent of late season effective rainfall became run-off, while ratios for Andit Tid and Maybar were 56 and 50 per cent, respectively (Liu *et al.*, 2008). There was no correlation between biweekly rainfall and discharge during the dry seasons at any of the sites.

Despite the great distances between the watersheds and the different characteristics, the response was surprisingly similar. The Anjeni and Maybar watersheds had almost the same run-off characteristics, while Andit Tid had more variation in the run-off amounts but, on average, the same linear response with a higher intercept (Figure 6.1b). Linear regressions were generated for all three watersheds (Figures 6.1). The regression slope does not change significantly, but this is due to the more similar values in Anjeni and Maybar dominating the fit (note that these regressions are only valid for the end of rainy seasons when the watersheds are wet).

Why these watersheds behave so similarly after the threshold rainfall has fallen is an interesting characteristic to explore. It is imperative to look at various time scales, since focusing on just one type of visual analysis can lead to erroneous conclusions. For example, looking only at storm hydrographs of the rapid run-off responses prevalent in Ethiopian storms, one could conclude that infiltration excess is the primary mechanism generating run-off. However, looking at longer time scales it can be seen that the ratio of Q/(P – E) is increasing with cumulative precipitation and, consequently, the watersheds behave differently depending on how much moisture is stored in the watershed, suggesting that saturation excess processes play an important role in the watershed run-off response. If infiltration excess was controlling run-off responses, discharge would only depend on the rate of rainfall, and there would be no clear relationship with antecedent cumulative precipitation, as is clearly the case as shown in Figure 6.1.

Infiltration and precipitation intensity measurements

To further investigate the hydrological response in the SCRP watersheds, the infiltration rates are compared with rainfall intensities in the Maybar (Figure 6.2a) and Andit Tid (Figure 6.2b) watersheds, where infiltration rates were measured in 2008 by Bayabil (2009) and Engda (2009), and rainfall intensity records were available from the SCRP project for the period 1986–2004. In Andit Tid, the exceedance probability of the average intensities of 23,764 storm events is plotted in Figure 6.2b (blue line). These intensities were calculated by dividing the rainfall amount on each day by the duration of the storm. In addition, the exceedance probability for instantaneous intensities for short periods was plotted as shown in Figure 6.2b (red line). Since there are often short durations of high-intensity rainfall within each storm, the rainfall intensities for short periods exceeded those of the storm-averaged intensities as shown in Figure 6.2b.

The infiltration rates for 10 locations in Andit Tid measured with the diameter single-ring infiltrometer varied between a maximum of 87 cm hr^{-1} on a terraced eutric cambisol in the bottom of the watershed to a low of 2.5 cm hr^{-1} on a shallow sandy soil near the top of the hill slope. This low infiltration rate was mainly caused by the compaction of freely roaming animals for grazing. Bushlands, which are dominant on the upper watershed, have significantly

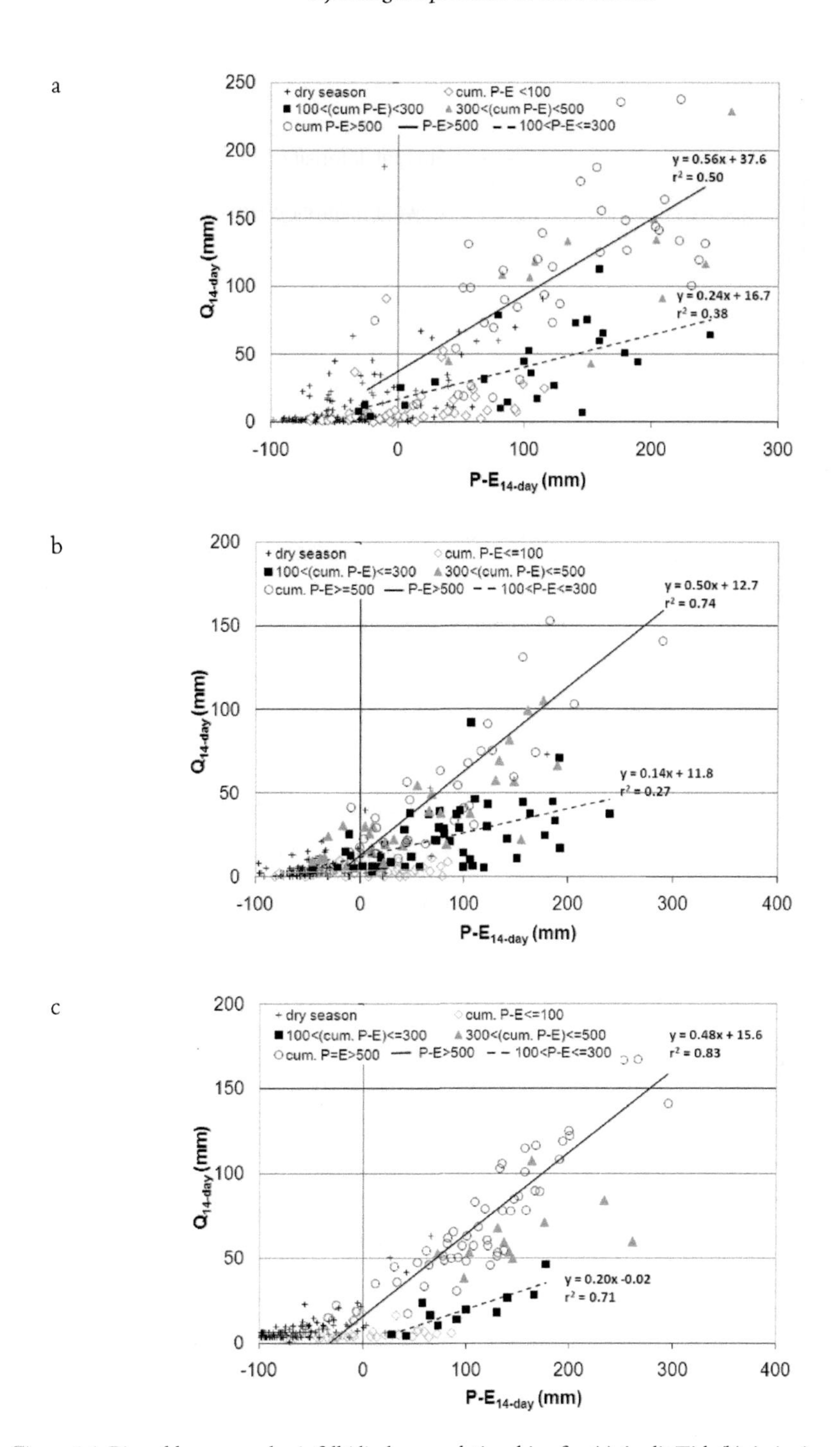

Figure 6.1 Biweekly summed rainfall/discharge relationships for (a) Andit Tid, (b) Anjeni and (c) Maybar. Rainy season values are grouped according to the cumulative rainfall that had fallen during a particular season, and a linear regression line is shown for the wettest group in each watershed

Source: Lui *et al.*, 2008

a

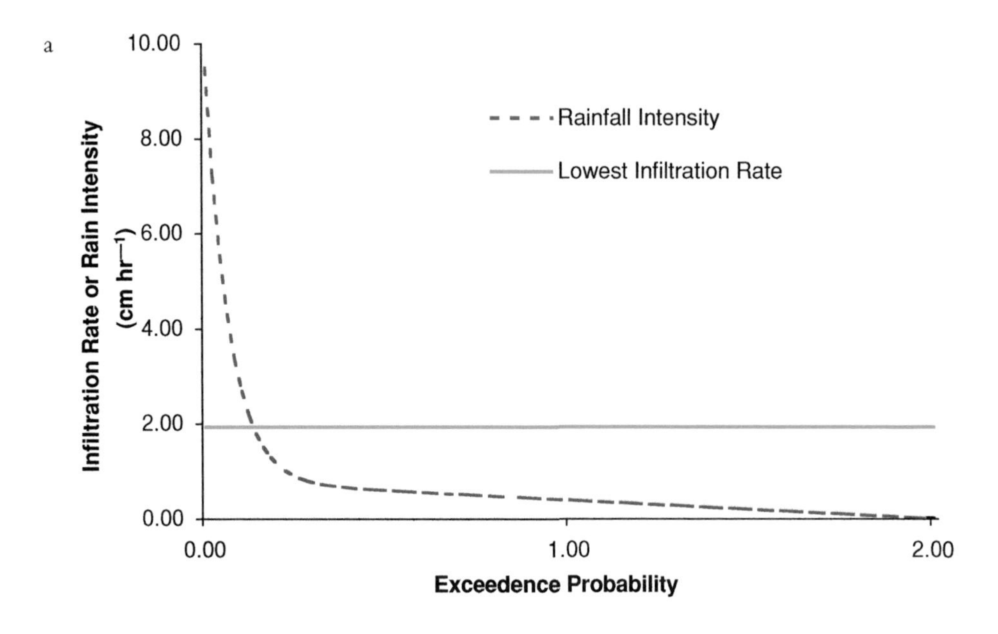

b

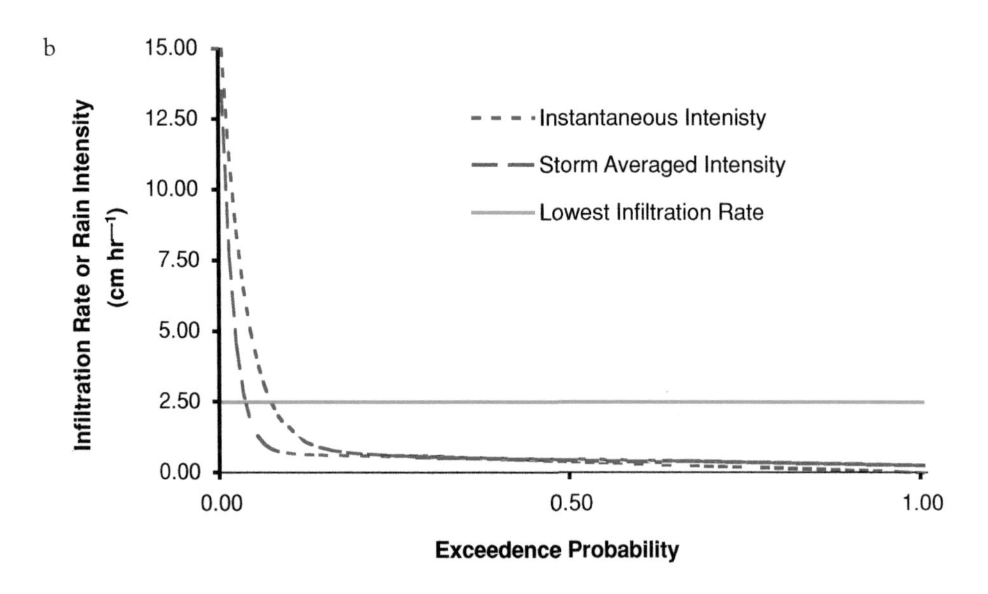

Figure 6.2 Probability of soil infiltration rate being exceeded by a five-minute rainfall intensity for the (a) Andit Tid and (b) Anjeni watersheds. Horizontal lines indicate the lowest measured infiltration rate

higher infiltration rates. In general, terraced and cultivated lands also have higher infiltration rates. The average infiltration rate of all ten measurements of the storm intensities was 12 cm hr^{-1}, and the median 4.3 cm hr^{-1}. The median infiltration rate of 4.3 cm hr^{-1} is the most meaningful value to compare with the rainfall intensity since it represents a spatial average. This median intensity has an exceedance probability of 0.03 for the actual storm intensities and 0.006 for the storm-averaged intensities. Thus the median intensity was exceeded only 3 per cent of the time and for less than 1 per cent of the storms. Storms with greater intensities were all of short duration with amounts of less than 1cm of total precipitation except once when almost 4 cm of rain fell over a 40-minute period. The run-off generated during short-duration intense rainfall can infiltrate into the soil in the subsequent period in the soil down slope when the rainfall intensity is less or the rain has stopped.

A similar analysis was performed in the Maybar watershed (Derib, 2005), where 16 infiltration rates were measured and even greater infiltration rates than in Andit Tid were found. The final steady state infiltration rates ranged from 1.9 to 60 cm hr^{-1}, with a median of 17.5 cm hr^{-1} (Figure 6.2a). The steady state infiltration rates in the Maybar watershed (Derib, 2005) ranged from 19 to 600 mm hr^{-1}. The average steady state infiltration rate of all 16 measurements was 24 cm h^{-1} and the median was 18 cm hr^{-1}. The median steady state infiltration rate (or geometric mean) was 18 cm hr^{-1}. The average daily rainfall intensity for seven years (from 1996 to 2004) was 8.5 mm hr^{-1}. A comparison of the geometric mean infiltration rate with the probability that a rainfall intensity of the same or greater magnitude occurs showed that the median steady state infiltration rate is not exceeded, while the minimum infiltration rate is exceeded only 9 per cent of the time. Thus, despite the rapid observed increase in flow at the outlet of the watershed during a rainstorm, it is unlikely that high rainfall intensities caused infiltration excess run-off, and more likely that saturated areas contributed the majority of the flow. Locally, there can be exceptions. For example, when the infiltration rate is reduced or in areas with severe degradation, livestock traffic can cause infiltration excess run-off (Nyssen *et al.*, 2010). Thus, the probability of exceedence is approximately the same as in Andit Tid, despite the higher rainfall intensities.

These infiltration measurements confirm that infiltration excess run-off is not a common feature in these watersheds. Consequently, most run-off that occurs in these watersheds is from degraded soils where the topsoil is removed or by saturation excess in areas where the upslope interflow accumulates. The finding that saturation excess is occurring in watersheds with a monsoonal climate is not unique. For example, Bekele and Horlacher (2000), Lange *et al.* (2003), Hu *et al.* (2005) and Merz *et al.* (2006) found that saturation excess could describe the flow in a monsoonal climate in southern Ethiopia, Spain, China and Nepal, respectively.

Piezometers and groundwater table measurements

In all three SCRP watersheds, transects of piezometers were installed to observe groundwater table in 2008 during the rainy reason and the beginning of the dry season.

Both Andit Tid and Maybar had hill slopes with shallow- to medium-depth soils (0.5–2.0 m depth) above a slow sloping permeable layer (either a hardpan or bedrock). Consequently, the water table height above the slowly permeable horizon (as indicated by the piezometers) behaved similarly for both watersheds. An example is given for the Maybar watershed, where groundwater table levels were measured with 29 piezometers across eight transects (Figure 6.3). The whole watershed was divided into three slope ranges: upper steep slope (25.1–53.0°), mid-slope (14.0–25.0°) and relatively low-lying areas (0–14.0°). For each slope class the daily perched groundwater depths were averaged (i.e. the height of the saturated layer above the restricting layer). The depth of the perched groundwater above the restricting layer in the steep and upper parts of the watershed is very small and disappears if there is no rain for a few days. The depth of the perched water table on the mid-slopes is greater than that of the upslope areas. The perched groundwater depths are, as expected, the greatest in relatively low-lying areas. Springs occur at the locations where the depth from the surface to the impermeable layer is the same as the depth of the perched water table and are the areas where the surface run-off is generated.

The behaviour of water table is consistent with what one would expect if interflow is the dominant conveyance mechanism. *Ceteris paribus*, the greater the driving force (i.e. the slope of the impermeable layer), the smaller the perched groundwater depth needed to transport the same water because the lateral hydraulic conductivity is larger. Moreover, the drainage area and the discharge increase with the downslope position. Consequently, one expects the perched groundwater table depth to increase with the downslope position as both slopes decrease and drainage area increases.

These findings are different from those generally believed to be the case – that the vegetation determines the amount of run-off in the watershed. Plotting the average daily depth of the perched water table under the different crop types (Figure 6.3a) revealed a strong correlation between perched water depth and crop type. The grassland had the greatest perched water table depth, followed by croplands, while bushlands had the lowest groundwater level. However, some local knowledge was needed to interpret these data. For instance, the grasslands are mainly located in the often-saturated lower-lying areas (too wet to grow a crop), while the croplands are often located in the mid-slope (with a consistent water supply but not saturated) and the bushlands are found on the upper steep slope areas (too droughty for good yield). Since land use is related to slope class (Figure 6.3b), the same relationship between crop type and soil water table height is not expected to be seen as between slope class and water table height. Thus there is an indirect relationship between land use and hydrology. The landscape determines the water availability, and thus the land use.

In the Anjeni watershed, which had relatively deep soils and no flat-bottom land, the only water table found was near the stream. The water table level was above the stream level, indicating that the rainfall infiltrates first in the landscape and then flows laterally to the stream. Although more measurements are needed it seems reasonable to speculate that there was a portion of the watershed that had a hardpan at a shallow depth with a greater percolation rate than in either Andit Tid or Maybar allowing recharge but, at the same time, causing interflow and saturation excess overland flow.

Perhaps the most interesting interplay between soil water dynamics and run-off source areas can be observed in the Maybar catchment. For instance, transect 1 illustrates a typical water table response across slope classes in the catchment. This and other transects have a slow permeable layer (either a hardpan or bedrock) and the water tends to pond above this layer.

At the beginning of August (the middle of the rainy season) the water table at the most down-slope location, P1, increased and reached the surface of the soil on 17 August (Figure 6.4). On a few dates, it was located even above the surface indicating surface run-off at the time. The water table started declining at the end of September, when precipitation ceased. The water level in P2, located upslope of P1, reached its maximum on 29 August, and the level remained below the surface. It decreased around the beginning of September, when rainfall storms were less frequent (Figure 6.4). The water table in P3 responded quickly and decreased rapidly. Thus, unlike P1 and P2, the water did not accumulate there and flowed rapidly as interflow downslope. Finally, the response time in the most upstream piezometer, P4, was probably around the duration of the rainstorm and was not recorded by our manual measurements.

Thus, the piezometric data in this and other transects indicate that the rainfall infiltrates on the hillsides and flows laterally as interflow down slope. At the bottom of the hillside where the slope decreases, the water accumulates and the water table increases. When the water intersects

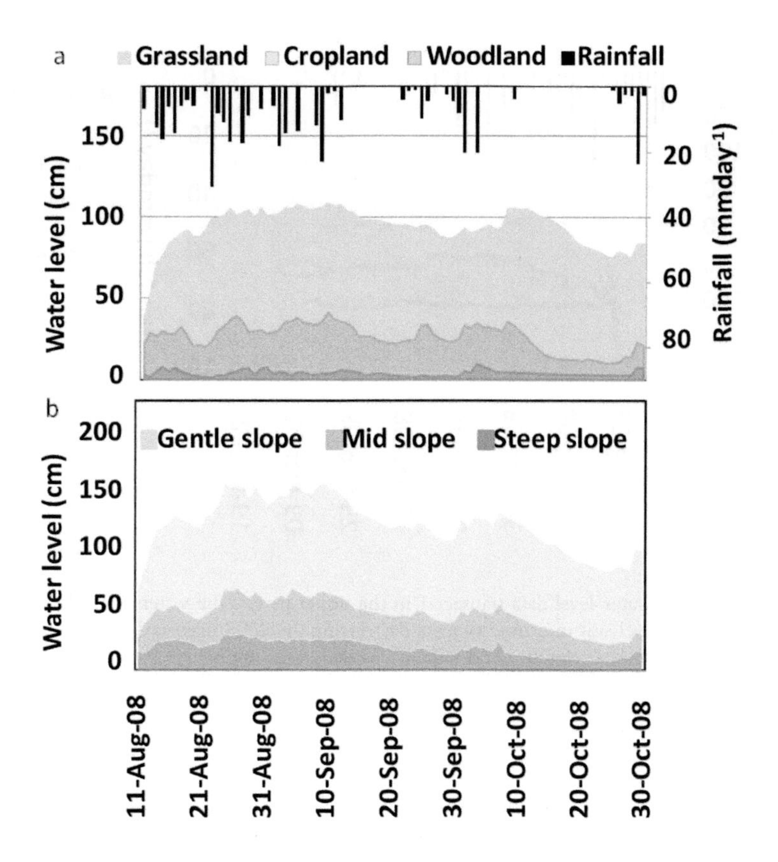

Figure 6.3 Average daily water level for three land uses (i.e. grasslands, croplands and woodlands) calculated above the impermeable layer superimposed (a) with daily rainfall and (b) for three slope classes in the Maybar catchment

with the soil surface, a saturated area is created. Rainfall on this saturated area becomes overland flow. In addition, rainfall at locations where the water table remains steady, such as P2, also becomes run-off; otherwise it would rise to the surface. Natural soil pipes that rapidly convey water from the profile and that have been seen in many places in this watershed might be responsible for this process (Bayabil, 2009).

These findings indicate that topographic controls are important to consider when assessing watershed response. However, when the average daily depth of the perched water table is plotted against the different crop types (e.g. Figure 6.3a), there was also a strong correlation of perched water depth with crop type. The grasslands at the bottom of the slope had the greatest perched water table depth, followed by croplands and woodlands with the lowest groundwater level. Thus, it seems that similar to the plot data, both ecological factors and topographical factors play a role in determining the perched water table height. Since land use is related to slope class, the same relationship between crop type and soil water table height as slope class and water table height is expected. Thus, there is an indirect relationship between land use and hydrology. The landscape determines the water availability and thus the land use.

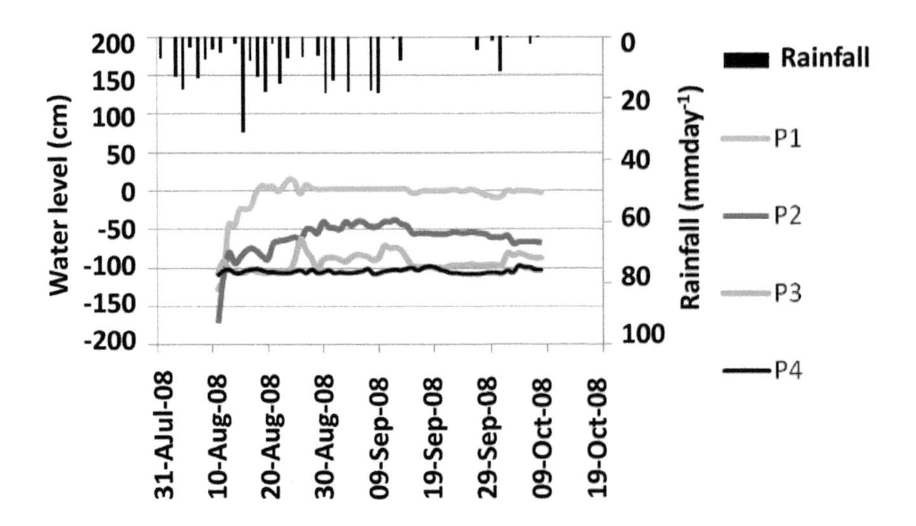

Figure 6.4 Piezometric water-level data transect 1 in the upper part of the watershed, where the slope is even. Water level was measured twice a day during the 2008 main rainy season using the ground surface as a reference and rainfall as a daily measurement

The results of this study are similar to those of the May Zeg watershed, which is in a much dryer area in Tigrai, with an average annual rainfall of around 600 mm yr^{-1} (Nyssen *et al.*, 2010). In this study, the water table in the valley bottom was measured with a single piezometer. Nyssen *et al.* (2010) observed that increasing infiltration on the hillside resulted in a faster increase in water tables in the valley bottom, which is similar to what was observed in the Maybar catchment where water moved via the subsurface, which increased the water levels where the slope decreased.

Thus, although there is a relationship between run-off potential and crop type, the relationship is indirect. The saturated areas are too wet for a crop to survive and these areas are often left as grass. The middle slopes have sufficient moisture (and do not saturate) to survive the dry spells in the rainy season. The steep slopes, without any water table, are likely to be droughty for a crop to survive during dry years and are therefore mainly forest or shrub.

These findings are consistent with the measurements taken by McHugh (2006) in the Lenche Dima watershed near Woldea, where the surface run-off of the valley bottom lands were much greater than the run-off (and erosion) from the hillsides.

Run-off from test plots

The rainfall run-off data collected from run-off plots in the Maybar watershed for the years 1988, 1989, 1992 and 1994 allow further identification of the dominant run-off processes in the watershed. The average annual run-off measured on four plots showed that plots with shallower slopes had higher run-off losses than those with steeper slopes (Figure 6.5a). The run-off coefficients ranged from 0.06 to 0.15 across the slope classes (Figure 6.5a). Nyssen *et al.* (2010) compiled the data of many small run-off plots in Ethiopia and showed an even larger range of run-off coefficients across slope classes. Run-off from plots in the Andit Tid catchment showed a very similar slope response (Figure 6.5b) as the Maybar plots, with shallower slopes producing more run-off. These results indicate that the landscape position plays an important role in the magnitude of the run-off coefficients as well. Indeed, it is commonly accepted that, ceteris paribus, a greater slope causes an increase in the lateral hydraulic conductivity of the soils, and thus these soils maintain a greater transmissivity than shallower slopes, and are able to conduct water out of the profile faster, reducing run-off losses.

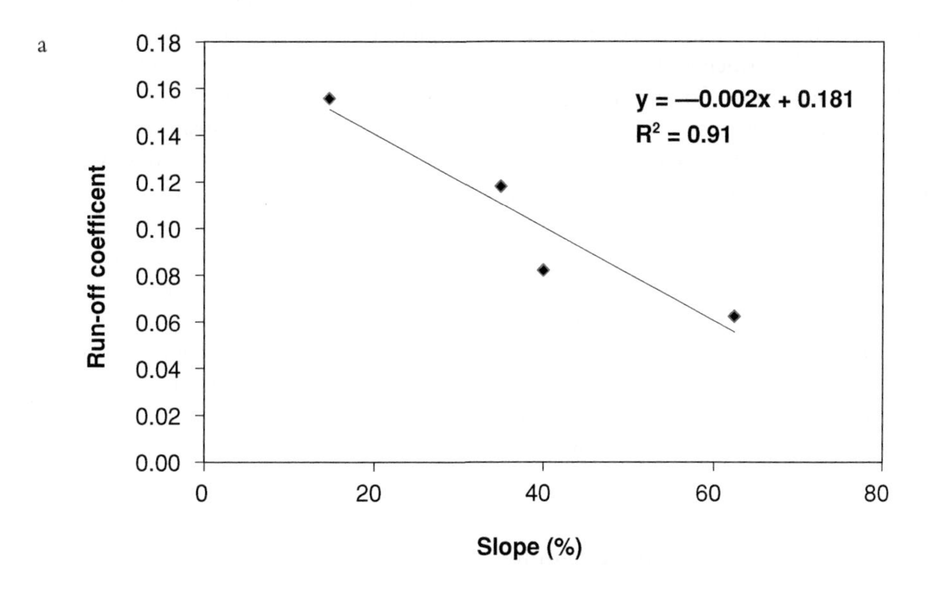

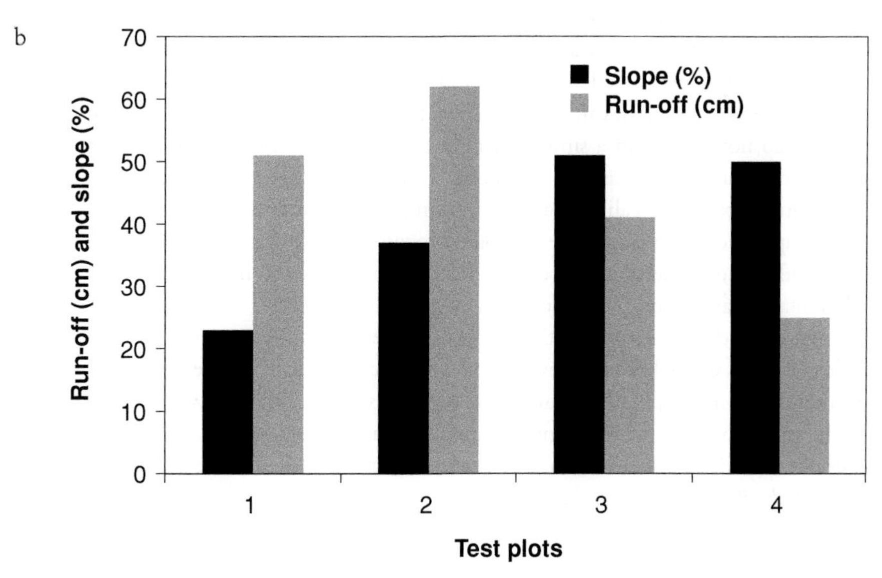

Figure 6.5 Plot run-off coefficient computed from daily 1988, 1989, 1992 and 1994 rainfall and run-off data for different slopes in (a) the Maybar catchment and (b) run-off depths for various slope classes in the Andit Tid catchment

Discussion

Both the location of run-off source areas and the effectiveness of a soil and water conservation practice depend on the dominating run-off processes in the watershed. Whether watershed run-off processes are ecologically (plant) or topographically controlled is an important consideration when selecting appropriate practices. Inherent in the assumption of ecologically based run-off is the concept of soil infiltration excess type of overland flow in which run-off occurs when rainfall intensity exceeds the infiltration capacity of the soil. Thus, for ecologically based models, improving plant cover and soil health will, in general, reduce overland flow and increase infiltration and interflow. On the other hand, topographically based run-off processes are, in general, based on the principle that if the soil saturates either above a slow permeable layer or a groundwater table, run-off occurs. In this case, changing plant cover will have little effect on run-off unless the conductivity of the most restricting layer is altered. These areas, which saturate easily, are called run-off source areas. They indicate where soil and water conservation practices would be most effective because most of the erosion originates in these areas. Understanding hydrological processes of a basin as diverse as the BNB is an essential prerequisite to understand the water resources and to ultimately design water management strategies for water access and improve water use in agriculture and other sectors. The results of various studies as briefly demonstrated above provide a wide range of tools and methods of analysis to explain such process. These finding will assist water resources and agricultural planners, designers and managers with a tool to better manage water resources in Ethiopian Highlands and to potentially mitigate impacts on water availability in downstream countries.

These results from the SCRP watersheds serve as the basis for the adoption of the models discussed next.

Adoption of models to the Blue Nile

Watershed management depends on the correct identification of when and where run-off and pollutants are generated. Often, models are utilized to drive management decisions, and focus resources where they are most needed. However, as discussed above, the hydrology and, by extension, biogeochemical processes in basins such as the Blue Nile, dominated by monsoonal conditions, often do not behave in a similar manner as watersheds elsewhere in the world. As a result, the models utilized to assess hydrology often do not correctly characterize the processes, or require excessive calibration and/or simplifying assumptions. Thus, watershed models that are capable of capturing these complex processes in a dynamic manner can be used to provide an enhanced understanding of the relationship between hydrological processes, erosion/sedimentation and management options.

There are many models that can continuously simulate streamflow, erosion/sedimentation or nutrient loss from a watershed. However, most were developed in temperate climates and were never intended to be applied in monsoonal regions, such as Ethiopia, with an extended dry period. In monsoonal climates, a given rainfall volume at the onset of the monsoon produces a drastically different run-off volume than the same rainfall volume at the end of the monsoon (Lui *et al.*, 2008).

Many of the commonly used watershed models employ some form of the Soil Conservation Service curve number (CN) to predict run-off, which links run-off response to soils, land use and five-day antecedent rainfall (e.g. antecedent moisture condition, AMC), and not the cumulative seasonal rainfall volume. The SWAT model is a basin-scale model where run-off is based on land use and soil type (Arnold *et al.*, 1998), and not on topography; therefore, run-off and sediment transport on the landscape are only correctly predicted for soil infiltration excess type of overland flow and not when saturation excess of overland flow from variable source areas (VSA) dominates. Thus critical sediment source areas might not be explicitly recognized as unique source areas. SWAT determines an appropriate CN for each simulated day by using this CN–AMC distribution in conjunction with daily soil moisture values determined by the model. This daily CN is then used to determine a theoretical storage capacity, S, of the watershed for each day. While a theoretical storage capacity is assigned and adjusted for antecedent moisture for each land use/soil combination, the storage is not used to directly determine the amount of water allowed to enter the soil profile. Since this storage is a function of the land's infiltration properties, as quantified by the CN–AMC, SWAT indirectly assumes that only infiltration excess processes govern run-off generation. Prior to any water infiltrating, the exact portion of the rainfall that will run-off is calculated via these infiltration properties. This determination of run-off volume before soil water volume is an inappropriate approach for all but the most intense rain events, particularly in monsoonal climates where rainfall is commonly of both low intensity and long duration and saturation processes generally govern run-off production. Several studies in the BNB or nearby watersheds have suggested that saturation excess processes control overland flow generation (Liu *et al.*, 2008; Collick *et al.*, 2009; Ashagre, 2009; Engda, 2009; Tebebu, 2009; Tebebu *et al.*, 2010; White *et al.*, 2011) and that infiltration excess run-off is rare (Liu *et al.*, 2008; Engda, 2009).

Many have attempted to modify the CN to better work in monsoonal climates, by proposing various temporally based values and initial abstractions. For instance, Bryant *et al.* (2006) suggest that a watershed's initial abstraction should vary as a function of storm size. While this is a valid argument, the introduction of an additional variable reduces the appeal of the one-parameter CN model. Kim and Lee (2008) found that SWAT was more accurate when CN values were averaged across each day of simulation, rather than using a CN that described moisture conditions only at the start of each day. White *et al.* (2009) showed that SWAT model results improved when the CN was changed seasonally to account for watershed storage variation due to plant growth and dormancy. Wang *et al.* (2008) improved SWAT results by using a different relationship between antecedent conditions and watershed storage. While these variable CN methods improve run-off predictions, they are not easily generalized for use outside of the watershed as they are tested mainly because the CN method is a statistical relationship and is not physically based.

In many regions, surface run-off is produced by only a small portion of a watershed that expands with an increasing amount of rainfall. This concept is often referred to as a variable source area (VSA), a phenomenon actually envisioned by the original developers of the CN method (Hawkins, 1979), but never implemented in the original CN method as used by the Natural Resource Conservation Service (NRCS). Since the method's inception, numerous attempts have been made to justify its use in modelling VSA-dominated watersheds. These adjustments range from simply assigning different CNs for wet and dry portions to correspond with VSAs (Sheridan and Shirmohammadi, 1986; White *et al.*, 2009), to full reinterpretations of the original CN method (Hawkins, 1979; Steenhuis *et al.*, 1995; Schneiderman *et al.*, 2007; Easton *et al.*, 2008).

To determine what portion of a watershed is producing surface run-off for a given precipitation event, the reinterpretation of the CN method presented by Steenhuis *et al.* (1995) and incorporated into SWAT by Easton *et al.* (2008) assumes that rainfall infiltrates when the soil is unsaturated or runs off when the soil is saturated. It has been shown that this saturated contributing area of a watershed can be accurately modelled spatially by linking this reinterpretation of the CN method with a topographic index (TI), similar to those used by the topographically driven TOPMODEL (Beven and Kirkby, 1979; Lyon *et al.*, 2004). This linked

CN–TI method has since been used in multiple models of watersheds in the north-eastern US, including the Generalized Watershed Loading Function (GWLF; Schneiderman *et al.*, 2007) and SWAT (Easton *et al.*, 2008). While the reconceptualized CN model is applicable in temperate US climates, it is limited by the fact that it imposes a distribution of storages throughout the watershed that need to fill up before run-off occurs. While this limitation does not seem to affect results in temperate climates, it results in poor model results in monsoonal climates.

SWAT–VSA, the CN–TI adjusted version of SWAT (Easton *et al.*, 2008), returned hydrological simulations as accurate as the original CN method; however, the spatial predictions of run-off-producing areas and, as a result, the predicted phosphorus export were much more accurate. While SWAT–VSA is an improvement upon the original method in watersheds, where topography drives flows, ultimately, it still relies upon the CN to model run-off processes and, therefore, it is limited when applied to the monsoonal Ethiopian Highlands. Water balance models are relatively simple to implement and have been used frequently in the BNB (Johnson and Curtis, 1994; Conway, 1997; Ayenew and Gebreegziabher, 2006; Liu *et al.*, 2008; Kim and Kaluarachchi, 2008; Collick *et al.*, 2009; Steenhuis *et al.*, 2009). Despite their simplicity and improved watershed outlet predictions they fail to predict the spatial location of the run-off generating areas. Collick *et al.* (2009) and, to some degree, Steenhuis *et al.* (2009) present semi-lumped conceptualizations of run-off-producing areas in water balance models. SWAT, a semi-distributed model can predict these run-off source areas in greater detail, assuming the run-off processes are correctly modelled.

Based on the finding discussed above, a modified version of the commonly used SWAT model (White *et al.*, 2011; Easton *et al.*, 2010) is developed and tested. This model is designed to more effectively model hydrological processes in monsoonal climates such as in Ethiopia. This new version of SWAT, including water balance (SWAT-WB), calculates run-off volumes based on the available storage capacity of a soil and distributes the storages across the watershed using a soil topographic wetness index (Easton *et al.*, 2008), and can lead to more accurate simulation of where run-off occurs in watersheds dominated by saturation-excess processes (White *et al.*, 2011). White *et al.* (2011) compared the performance of SWAT-WB and the standard SWAT model in the Gumera watershed in the Lake Tana Basin, Ethiopia, and found that even following an unconstrained calibration of the CN, the SWAT model results were 17–23 per cent worse than the SWAT-WB model results.

Application of models to the watershed, sub-basin and basin scales

Tenaw (2008) used a standard SWAT model for Ethiopian Highlands to analyse the rainfall-run-off process at various scales in the upper BNB. Gelaw (2008) analysed the Ribb watershed using Geographic Information System and analysed the meteorological and data for characterizing the flooding regime and extents of damage in the watershed.

At the sub-basin level, Saliha *et al.* (2011) compared artificial neural network and the distributed hydrological model WaSiM-ETH (where WaSiM is Water balance Simulation Model) for predicting daily run-off over five small to medium-sized sub-catchments in the BNB. Daily rainfall and temperature time series in the input layer and daily run-off time series in the output layer of a recurrent neural net with hidden layer feedback architecture were formed. As most of the watersheds in the basin are ungauaged, Saliha *et al.* (2011) used a Kohonen neural network and WaSiM-ETH to estimate flow in the ungauged basin. Kohonen neural network was used to delineate hydrologically homogeneous regions and WaSiM-ETH was used to generate daily flows. Twenty-five sub-catchments of the BNB in Ethiopia were grouped into five hydrologically homogenous groups. WaSiM was calibrated using automatic nonlinear parameter estimation (PEST) method coupled with shuffled complex evolution (SCE) algorithm and validated using an independent time series. In the coupled programme, the Kohonen neural network assigns the ungauged catchment into one of the five hydrologically homogeneous groups. Each homogeneous group has its own set of optimized WaSiM-ETH parameters, derived from simultaneous calibration and validation of gauged rivers in the respective homogeneous group. The coupled programme transfers the optimized WaSiM parameters from the homogeneous group (which the ungauged river belongs to) to the ungauged river, and WaSiM calculates the daily flow for this ungauged river.

The two approaches discussed above, developed by Easton *et al.* (2010) and Saliha *et al.* (2011), provided a means of estimating run-off in the BNB across a range of scales and locations. Readers are referred to Saliha *et al.* (2011) for detail models and results on these and the Kohonen neural network and WaSiM-ETH for a detailed discussion. What follows are examples of applications of the SWAT and SWAT-WB at multiple scales in the BNB.

At the watershed level (Gumera), results of a standard SWAT model (Tenaw, 2008) and modified SWAT-WB model (Easton *et al.*, 2010; White *et al.*, 2011) are compared. Tenaw (2008) initialized the standard SWAT model for the Gumera watershed, which provided good calibration results at the monthly time step with a Nash–Sutcliffe efficiency, E_{NS} (Nash and Sutcliffe, 1970), of 0.76, correlation coefficient, R^2, of 0.87, and mean deviation, D, of 3.29 per cent (Figure 6.6). Validation results also show good agreement between measured and simulated values, with E_{NS} of 0.72, R^2 of 0.82 and D of –5.4 per cent.

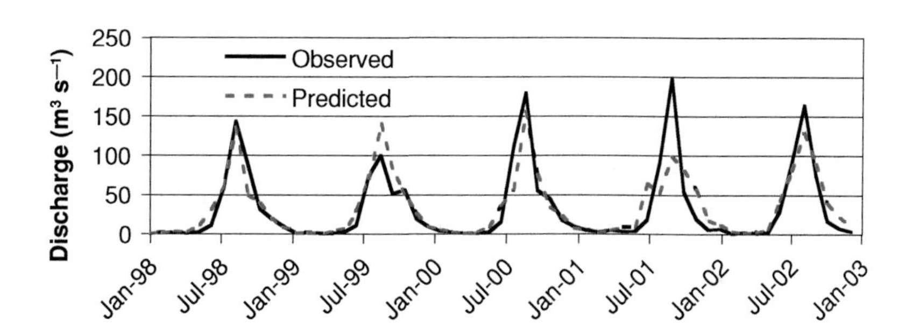

Figure 6.6 Calibration results of average monthly observed and predicted flow at the Gumera gauge using the SWAT

Source: Tenaw, 2008

At the sub-basin level, Habte *et al.* (2007) assessed the applicability of distributed WaSiM-ETH in estimating daily run-off from 15 sub-catchments in the Abbay River Basin. Input data in the form of daily rainfall and temperature data from 38 meteorological stations were used to drive model simulations. In a study by Saliha *et al.* (2011), the artificial neural network and distributed hydrological model (WaSiM-ETH) were compared for predicting daily run-off over five small to medium-sized sub-catchments in the Blue-Nile River Basin. Daily rainfall and temperature time series in the input layer and daily run-off time series in the output layer of recurrent neural net with hidden layer feedback architecture were formed.

The use of neural networks in modelling a complex rainfall–run-off relationship can be an efficient way of modelling the run-off process in situations where explicit knowledge of the internal hydrological process is not available. Indeed, most of the watersheds in the basin are ungauged, and thus there is little easily available data to run standard watershed models. Saliha *et al.* (2011) used a Self-Organizing Map (SOM) or Kohonen neural network (KNN) and WaSiM-ETH to estimate flow in the ungauged basin. The SOM groupings were used to delineate hydrologically homogeneous regions and WaSiM-ETH was then used to generate daily flows. The 26 sub-catchments of the BNB in Ethiopia were grouped into five hydrologically homogeneous groups, and WaSiM-ETH was then calibrated using the PEST method coupled with the SCE algorithm in neighbouring basins with available data. The results were then validated against an independent time series of flow (Habte *et al.*, 2007). Member catchments in the same homogenous group were split into calibration catchments and validation catchments. Each homogeneous group has a set of optimized WaSiM-ETH parameters, derived from simultaneous calibration and validation of gauged rivers in the respective homogeneous groups. Figure 6.7 shows a general framework of the couple model. In the coupled trained SOM and calibrated WaSiM-ETH programme, the trained SOM will assign the ungauged catchment into one of the five hydrologically homogenous groups based on the pre-defined catchment characteristics (e.g. red broken line in Figure 6.7). The coupled programme then transfers the whole set of optimized WaSiM-ETH parameters from the homogeneous group (to which the ungauged river belongs) to the ungauged river, and WaSiM-ETH calculates the daily flow for this ungauged river.

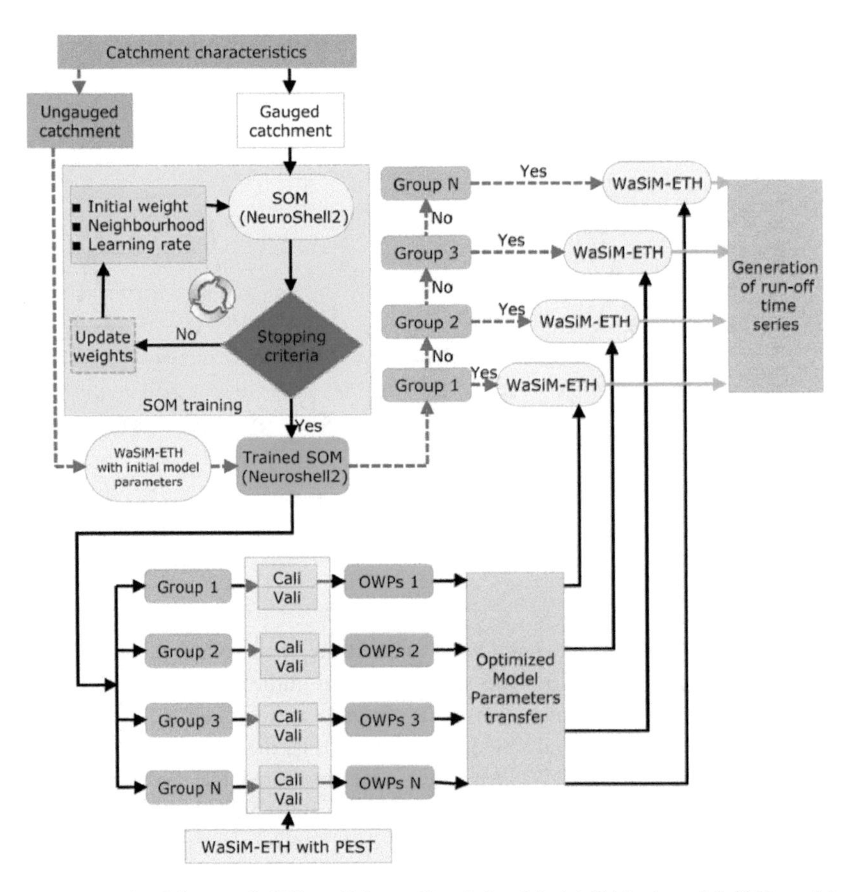

Figure 6.7 Framework of the coupled Water Balance Simulation Model–Ethiopia and Self-Organizing Map models

Source: Saliha *et al.*, 2011

Soil and Water Assessment Tool–Water Balance model

The SWAT-WB model is applied to the Ethiopian portion of the BNB that drains via the main stem of the river at El Diem on the border with Sudan (the Rahad and Dinder sub-basins that drain the north-east region of Ethiopia were not considered; Figures 6.8 and 6.9). Results show that incorporating a redefinition of how hydrological response units (HRUs) are delineated combined with a water balance to predict run-off can improve our analysis of when and where run-off and erosion occur in a watershed. The SWAT-WB model is initialized for eight sub-basins ranging in size from 1.3 to 174,000 km^2. The model is calibrated for flow using a priori

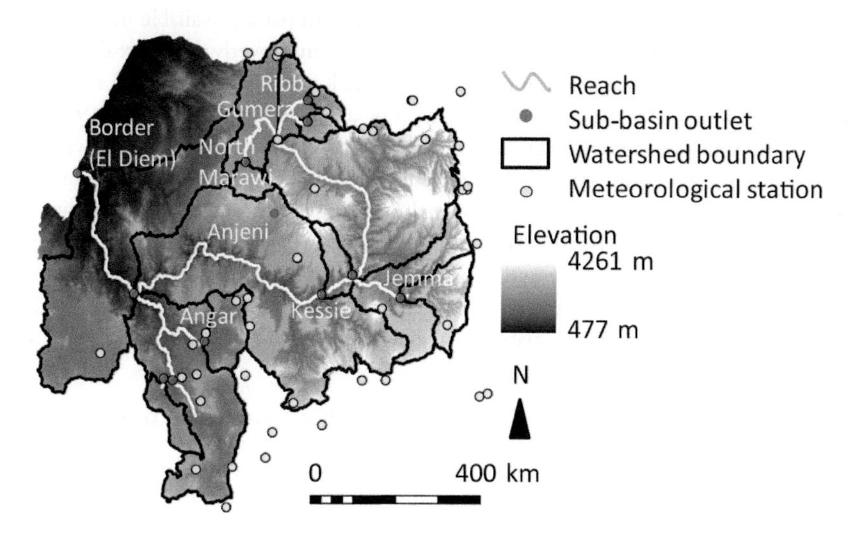

Figure 6.8 Digital elevation model reaches, sub-basins and sub-basin outlets initialized in the Blue Nile Basin SWAT model. Also displayed is the distribution of meteorological stations used in the model

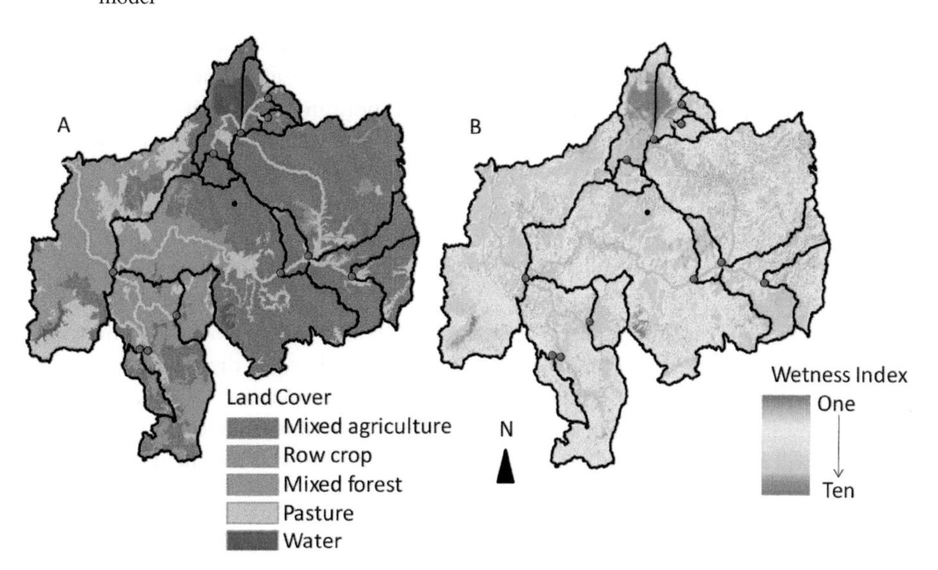

Figure 6.9 Land use/land cover (a) in the Blue Nile Basin (ENTRO) and (b) the Wetness Index used in the Blue Nile SWAT model

topographic information and validated with an independent time series of flows. The tested methodology captures the observed hydrological processes quite well across multiple scales, while significantly reducing the calibration data requirements. The reduced data requirements for model initialization have implications for model applicability to other data-scarce regions. Finally, a discussion of the implications of watershed management with respect to the model results is presented.

Summarized Soil and Water Assessment Tool model description

The SWAT model is a river basin model created to run with readily available input data so that general initialization of the modelling system does not require overly complex data-gathering, or calibration. SWAT was originally intended to model long-term run-off and nutrient losses from rural watersheds, particularly those dominated by agriculture (Arnold *et al.*, 1998). SWAT requires data on soils, land use/management information and elevation to drive flows and direct sub-basin routing. While these data may be spatially explicit, SWAT lumps the parameters into HRUs, effectively ignoring the underlying spatial distribution. Traditionally, HRUs are defined by the coincidence of soil type and land use. Simulations require meteorological input data including precipitation, temperature and solar radiation. Model input data and parameters were parsed using the ARCSWAT 9.2 interface. The interface combines SWAT with the ARCGIS platform to assimilate the soil input map, digital elevation model (DEM) and land use coverage.

Soil and Water Assessment Tool–Water Balance saturation excess model

The modified SWAT model uses a water balance in place of the CN for each HRU to predict run-off losses. Based on this water balance, run-off, interflow and infiltration volumes are calculated. While these assumptions simplify the processes that govern water movement through porous media (in particular, partly saturated regions), for a daily model, water balance models have been shown to better capture the observed responses in numerous African watersheds (Guswa *et al.*, 2002). For Ethiopia, water balance models outperform models that are developed in temperate regions (Liu *et al.*, 2008; Collick *et al.*, 2009; Steenhuis *et al.*, 2009; White *et al.*, 2011). For the complete model description see Easton *et al.*, 2010 and White *et al.*, 2011. In its most basic form, the water balance defines a threshold moisture content over which the soil profile can neither store nor infiltrate more precipitation; thus additional water becomes either run-off or interflow ($q_{E,i}$):

$$q_{E,i} = \begin{cases} (\theta_s - \Theta_{i,t})d_i + P_t - Et_t & \text{for: } P_t > (\theta_s - \Theta_{i,t})d_i - Et_t \\ 0 & \text{for: } P_t \leq (\theta_s - \Theta_{i,}t)d_i - Et_t \end{cases} \tag{6.1}$$

where θ_s (cm^3 cm^{-3}) is the soil moisture content above which storm run-off is generated, $\theta_{I,t}$ (cm^3 cm^{-3}) is the current soil moisture content, d_i (mm) is the depth of the soil profile, P_t (mm) is the precipitation and Et_t (mm) is the evapotranspiration. In SWAT, there is no lateral routing of interflow among watershed units, and thus no means to distribute watershed moisture; thus Equation 6.1 will result in the same excess moisture volume everywhere in the watershed given similar soil profiles.

To account for the differences in run-off generation in different areas of the basin, the following threshold function for storm run-off that varies across the watershed as a function of topography is used (Easton *et al.*, 2010):

$$\tau_i + (\rho_i \theta s - \theta_{i,t}) \tag{6.2}$$

where, ρ_i is a number between 0 and 1 that reduces θ_s to account for water that should drain down slope, and is a function of the topography (as defined by a topographic wetness index, λ; e.g. Beven and Kirkby, 1979). Here it is assumed that the distribution of ρ_i values is inversely proportional to the soil topographic index (λ_i) averaged across each wetness index class or HRU and that the lowest λ, (λ_o) corresponds to the highest λ_i (λ_o):

$$\rho_i = \frac{\lambda_o}{\lambda_i} \tag{6.3}$$

Easton *et al.* (2010) showed that using the baseflow index reliably constrained the distribution of these ρ_i values. Note that Equation 6.2 applies only to the first soil layer. Once the soil profile has been adequately filled, Equation 6.2 can be used to write an expression for the depth of run-off, $q_{R,i}$, (mm) from a wetness index, i:

$$q_{R,i} = \begin{cases} P_t - \tau_i d_i & \text{for } P_t > \tau_i d_i \\ 0 & \text{for } P_t \leq \tau_i d_i \end{cases} \tag{6.4}$$

While the approach outlined above captures the spatial patterns of VSAs and the distribution of run-off and infiltrating fractions in the watersheds, Easton *et al.* (2010) noted there is a need to maintain more water in the wettest wetness index classes for evapotranspiration, and proposed adjusting of the available water content (*AWC*) of the soil layers below the first soil layer (recall that the top soil layer is used to establish our run-off threshold; Equation 6.2) so that higher topographic wetness index classes retain water longer (i.e. have *AWC* adjusted higher), and the lower classes dry faster (i.e. *AWC* is adjusted lower by normalizing by the mean ρ_i value, similar to Easton *et al.*, 2008, for example).

Note that, since this model generates run-off when the soil is above saturation, total rainfall determines the amount of run-off. When results are presented on a daily basis rainfall intensity is assumed to be inconsequential. It is possible that under high-intensity storms (e.g. storms with rainfall intensities greater than the infiltration capacity of the soil) the model might under-predict the amount of run-off generated, but this is the exception rather than the rule (Liu *et al.*, 2008; Engda, 2009).

Model calibration

The water balance methodology requires very little direct calibration, as most parameters can be determined a priori. Soil storage was calculated as the product of soil porosity and soil depth from the soils data. Soil storage values were distributed via the λ described above, and the effective depth coefficient (ρ_i, varies from 0 to 1) was adjusted along a gradient in λ values as in Equation 6.3. Here it is assumed that the distribution of λ_i values is inversely proportional to λ_i (averaged across each wetness index class or HRU) and that the lowest λ, (λ_o) corresponds to the highest λ_i (λ_o). In this manner, the ρ_i distribution requires information on the topography (and perhaps on soil). If a streamflow record is available baseflow separation can be employed to further parameterize the model.

In constraining or 'calibrating' ρ_i, it is recognized that, since the ρ-value controls how much precipitation is routed as run-off, it also controls how much precipitation water can enter the soil for a given wetness index class. Thus, a larger fraction of the precipitation that falls on an

Table 6.2 Effective depth coefficients (ρ_i) for each wetness index class and watershed in the Blue Nile Basin model from Equation 6.3. The ΠB is determined from baseflow separated run-off of the streamflow hydrograph and distributed via the topographic wetness index, λ

Wetness index class	ρ_i *(Border)*	ρ_i *(Kessie)*	ρ_i *(Jemma)*	ρ_i *(Angar)*	ρ_i *(Gumera)*	ρ_i *(Ribb)*	ρ_i *(N. Marawi)*	ρ_i *(Anjeni)*
10 (most saturated)	0.22	0.20	0.16	0.15	0.26	0.24	0.24	0.15
9	0.58	0.51	0.24	0.22	0.31	0.41	0.43	0.25
8	0.75	0.68	0.31	0.26	0.40	0.51	0.53	0.30
7	0.87	0.78	0.35	0.30	0.47	0.59	0.62	0.32
6	0.97	0.87	0.37	0.34	0.61	0.66	0.69	0.36
5	1.00	0.94	0.43	0.38	0.75	0.72	0.75	0.44
4	1.00	1.00	0.57	0.42	0.89	0.80	0.83	0.46
3	1.00	1.00	0.64	0.47	1.00	0.88	0.91	0.57
2	1.00	1.00	0.74	0.52	1.00	0.99	1.00	0.86
1 (least saturated)	1.00	1.00	1.00	0.63	1.00	1.00	1.00	1.00
*Π_B	**0.84**	**0.80**	**0.48**	**0.37**	**0.67**	**0.68**	**0.70**	**0.47**

Note: *Π_B partitions moisture in the above saturation to run-off and infiltration

area with a large ρ_i will potentially recharge the groundwater than in an area with a small ρ_i. As a first approximation, then, assume ρ_i can be equated with the ratio of groundwater recharge, $q_{B,i}$ to total excess precipitation, $q_{E,i}$ (i.e. precipitation falling on wetness class *i* that eventually reaches the watershed outlet). Baseflow is determined directly from the digital signal filter baseflow separation technique of several years of daily streamflow hydrographs (Hewlett and Hibbert, 1967; Arnold *et al.*, 1995; for greater detail see Easton *et al.*, 2010).

The primary difference between the CN-based SWAT and the water-balance-based SWAT is that run-off is explicitly attributable to source areas according to a wetness index distribution, rather than by land use and soil infiltration properties as in original SWAT (Easton *et al.*, 2008). Soil properties that control saturation-excess run-off generation (saturated conductivity, soil depth) affect run-off distribution in SWAT-WB since they are included in the wetness index via Equation 6.4. Flow calibration was validated against an independent time series that consisted of at least one half of the observed data. To ensure good calibration, the calibrated result maximized the coefficient of determination (r^2) and the Nash–Sutcliffe efficiency (E_{NS}; Nash and Sutcliffe, 1970). Table 6.2 summarizes the calibrated ρ_i values for each wetness index class while Table 6.3 summarizes the calibration statistics. Since flow data at some of the available gauge locations were available at the monthly time step (Angar, Kessie, Jemma) and daily at others (Anjeni, Gumera, Ribb, North Marawi, El Diem; Figure 6.10), the model was run for both time steps, and the results presented accordingly.

Results

Run-off from saturated areas and subsurface flow from the watershed were summed at the watershed outlet to predict streamflow. The graphical comparison of the modelled and measured daily streamflow at the El Diem station at the Sudan border (e.g. integrating all sub-basins above) is shown in Figure 6.10. The model was able to capture the dynamics of the basin response well (E_{NS} = 0.87, r^2 = 0.92; Table 6.3; Figure 6.10). Both baseflow and storm flow were correctly predicted with a slight over-prediction of peak flows and a slight under-prediction of low flows (Table 6.3); however, all statistical evaluation criteria indicated the model predicted well. In fact, all calibrated sub-basins predicted streamflow at the outlet reasonably well (e.g. Table 6.3). Model predictions showed good accuracy (E_{NS} ranged from 0.53 to 0.92) with measured data across all sites except at Kessie, where the water budget could not be closed; however, the timing of flow was well captured. The error at Kessie appears to be due to underestimated precipitation at the nearby gauges, as measured flow was nearly 15 per cent higher than precipitation – evapotranspiration (P – E). Nevertheless, the prediction is within 25 per cent of the measured data. Observed normalized discharge (Table 6.3) across the sub-basins shows a large gradient, from 210 mm at Jemma to 563 mm at Anjeni. For the basin as a whole, approximately 25 per cent of precipitation exits at El Diem of the BNB.

Table 6.3 Calibrated sub-basins (Figure 6.10), drainage area, model fit statistics (coefficient of determination, r^2 and Nash–Sutcliffe Efficiency, E_{NS}), and observed and predicted flows

Sub-basin	*Area (km^2)*	*r^2*	*E_{NS}*	*Observed mean annual discharge (million m^3)*	*Observed normalized discharge (mm yr^{-1})*	*Predicted direct run-off (mmyr^{-1})*	*Predicted ground-water (mmyr^{-1})[3]*
Anjeni[1]	1.3	0.76	0.84	0.40	563	44	453
Gumera[1]	1286	0.83	0.81	501	390	22	316
Ribb[1]	1295	0.74	0.77	495	382	25	306
North Marawi[1]	1658	0.78	0.75	646	390	17	274
Jemma[2]	5429	0.91	0.92	1142	210	19	177
Angar[2]	4674	0.87	0.79	1779	381	34	341
Kessie[2]	65,385	0.73	0.53	19,237	294	19	259
Border (El Diem)[1]	174,000	0.92	0.87	56,021	322	13	272

Notes: [1] Statistics are calculated on daily time step
[2] Statistics are calculated on monthly time step
[3] Includes both baseflow and interflow

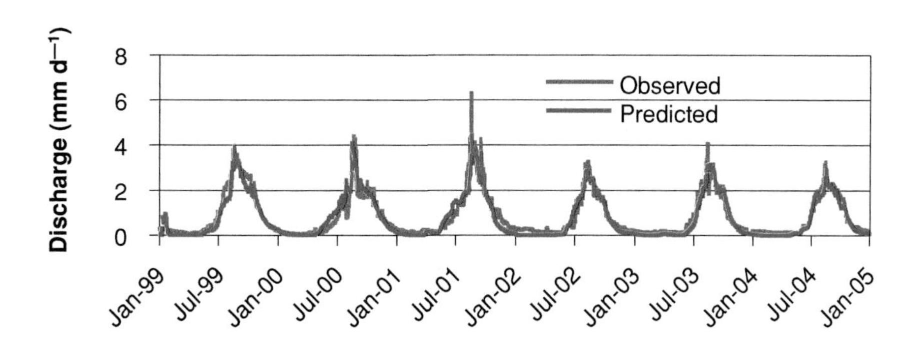

Figure 6.10 Daily observed and predicted discharge at the Sudan border

Table 6.2 shows the adjusted ρ_i parameter values (e.g. Equation 6.3) for the various sub-basins in the BNB; these values are scalable, and can be determined from topographical information (i.e. the ρ_i values vary by sub-basin, but the distribution is similar).

The SWAT-WB model was able to accurately reproduce the various watershed responses across the range of scales. Notice, for instance, that the hydrographs at the Sudan border (174,000 km^2; Figure 6.12), Gumera (1200km^2; Figure 6.11) and Anjeni (1.13km^2; Figure 6.12) reasonably capture the observed dynamics (i.e. both the rising and receding limbs and the peak flows are well represented). There was a slight tendency for the model to bottom out during baseflow, probably due to overestimated ET, but the error is relatively minor. More importantly, the model captures peak flows, which are critical to correctly predict to assess sediment transport and erosion.

Run-off and streamflow are highly variable both temporally (over the course of a year; Figure 6.10) and spatially (across the Ethiopian Blue Nile Basin; Table 6.3). Daily watershed outlet discharge during the monsoonal season at Gumera is four to eight times larger than at the Sudan border (after normalizing flow by the contributing area; Figures 6.10 and 6.11). Anjeni, the smallest watershed had the largest normalized discharge, often over 20 mm d^{-1} during the rainy season (Figure 6.12). Discharges (in million m^3 y^{-1}) intuitively increase with drainage area, but precipitation also has a large impact on overall sub-basin discharge. Both Jemma and Angar are of approximately the same size (Jemma is actually slightly bigger), yet discharge from Angar is nearly 40 per cent higher, a result of the higher precipitation in the southwestern region of the basin. Temporally, outlet discharges typically peak in August for the small and medium-sized basins and slightly later for Kessie and the Sudan border, a result of the lag time for lateral flows to travel the greater distances. Due to the monsoonal nature of the basin, there is a very low level of baseflow in all tributaries and, in fact, some dry up completely during the dry season, which the model reliably predicts, which is important when considering the impacts of intervention measures to augment flow.

Run-off losses predicted by the model varied across the basin as well, and were generally well corroborated by run-off estimates from baseflow separation of the streamflow hydrograph. Predicted run-off losses (averaged across the entire sub-basin) varied from as low as 13 mm y^{-1} for the BNB as a whole sub-basin to as high as 44 mm y^{-1} in Anjeni. Of course, small areas of

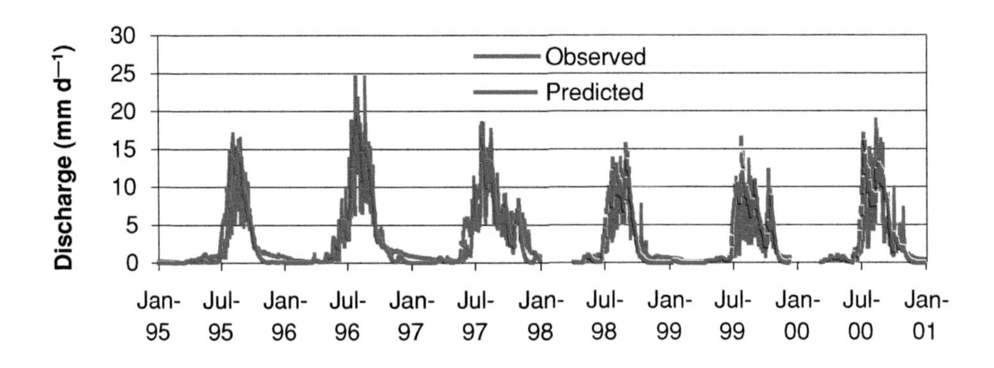

Figure 6.11 Daily observed and predicted discharge from the Gumera sub-basin. See Table 6.3 for model performance for the Ribb and North Marawi sub-basins

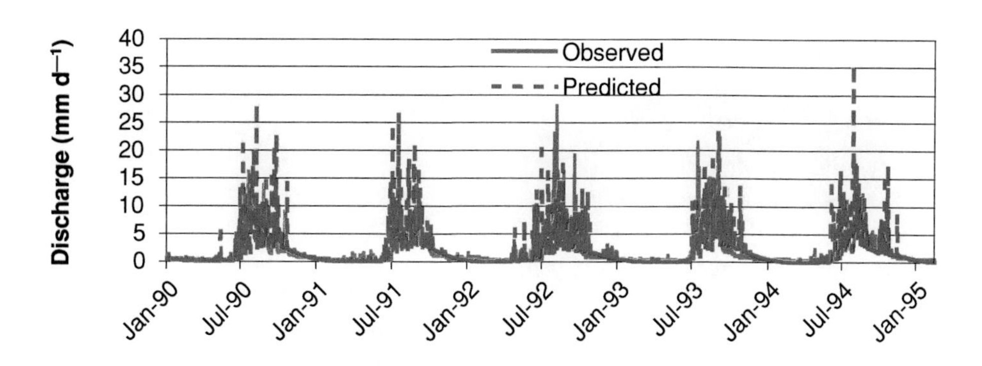

Figure 6.12 Daily observed and predicted discharge from the Anjeni micro-watershed

the individual sub-basins produce significantly higher run-off losses and others significantly less. These differences are well reflected in the average baseflow coefficient (Π_B) for the sub-basins (Table 6.2). Notice that Π_B for Anjeni (smallest watershed, highest run-off losses) is significantly lower than for Gumera and the Sudan border (Table 6.2). A lower Π_B reflects less average available storage in the watershed (i.e. more rainfall ends up as run-off). This Π_B value is determined from the baseflow separation of the streamflow hydrograph (Hewlett and Hibbert, 1967), and can thus be considered a measured parameter. It is also interesting to note how the distribution of the individual ρ_i differs between basins. For instance, there are more classes (areas) in Anjeni and Angar that are prone to saturate, and would thus have lower available storage, and create more run-off. This is relatively clear in looking at the streamflow hydrographs (Figures 6.10–6.13) where the smaller watersheds tend to generate substantially more surface run-off. Conversely, as basin size increases (Kessie, Sudan border) the saturated fraction of the watershed decreases, and more of the rainfall infiltrates, resulting in greater baseflow, as reflected in the higher Π_B, or, in terms of run-off, the smaller upland watersheds have higher run-off losses than the larger basins. This is not unexpected, as the magnitude of the subsurface flow paths have been shown to increase with the size of the watershed, because as watershed size increases more and more deep flow paths become activated in transport (Steenhuis *et al.*, 2009).

The ability to predict the spatial distribution of run-off source areas has important implications for watershed intervention, where information on the location and extent of source areas is critical to effectively managing the landscape. For instance, the inset of Figure 6.13 shows the predicted spatial distribution of average run-off losses for the Gumera watershed for an October 1997 event. As is evident from Figure 6.13, run-off losses vary quite dramatically across the landscape; some HRUs are expected to produce no run-off, while others have produced more than 90 mm of run-off. When averaged spatially at the outlet, run-off losses were 22 mm (Table 6.3). Other sub-basins responded in a similar manner. These results are consistent with data collected in the Anjeni SCRP watershed (SCRP, 2000; Ashagre, 2009), which show that run-off losses roughly correlate with topography.

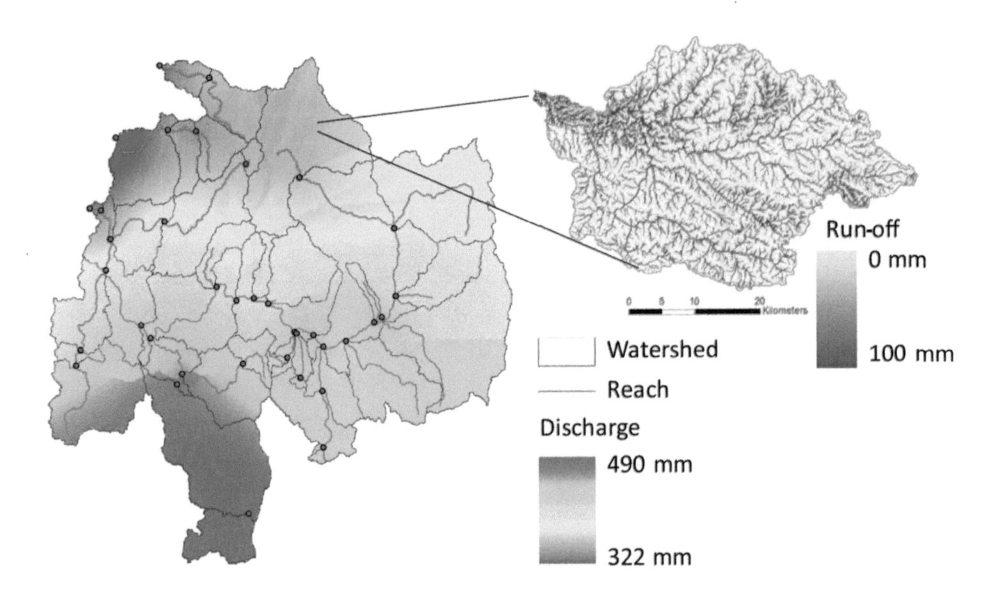

Figure 6.13 Predicted average yearly spatial distribution of discharge in the BNB (main) and predicted run-off distribution in the Gumera sub-watershed for an October 1997 event (inset)

Discussion

Flows in the Blue Nile Basin in Ethiopia show large variability across scales and locations. Sediment and water yields from areas of the basin range more than an order of magnitude (a more in-depth discussion of sediment is given in Chapter 7). The use of the modified SWAT-WB model that more correctly predicts the spatial location of run-off source areas is a critical step in improving the ability to manage landscapes, such as the Blue Nile, to provide clean water supplies, enhance agricultural productivity and reduce the loss of valuable topsoil. Obviously, the hydrological routines in many of the large-scale watershed models do not incorporate the appropriate mechanistic processes to reliably predict when and where run-off occurs, at least at the scale needed to manage complex landscapes. For instance, the standard SWAT model predicts run-off to occur more or less equally across the various land covers (e.g. croplands produce approximately equal run-off losses and pastureland produces approximately equal erosive losses, etc.) provided they have similar soils and land management practices throughout the basin. The modified version of SWAT used here recognizes that different areas of a basin (or landscape) produce different run-off losses and thus different sediment losses. However, all crops or pasture within a wetness index class in the modified SWAT produce the same run-off or erosion losses.

Water balance models are consistent with the saturation excess run-off process because the run-off is related to the available watershed storage capacity and the amount of precipitation. The implementation of water balances into run-off calculations in the BNB is not a novel concept and others have shown that water balance type models often perform better than more complicated models in Ethiopian-type landscapes (Johnson and Curtis, 1994; Conway, 1997; Liu *et al.*, 2008). However, these water balance models are typically run on a monthly or yearly time steps because the models are generally not capable of separating base-, inter- and surface run-off flow. To truly model erosion and sediment transport (of great interest in the BNB), large events must be captured by the model and daily simulations are required to do so. Thus SWAT-WB not only maintains a water balance but also calculates the interflow and the baseflow component, and gives a reasonable prediction of peak flows. SWAT-WB is therefore more likely to be capable of predicting erosion source areas and sediment transport than either SWAT-CN or water budget models with monthly time steps. Indeed, Tebebu *et al.* (2010) found gully formation and erosion in the Ethiopian Highlands to be related to water table levels and saturation dynamics, which SWAT-WB reliably predicts.

Conclusions

A modified version of the SWAT model appropriate for monsoonal climates is presented as a tool to quantify the hydrological and sediment fluxes in the BNB, Ethiopia. The model requires very little direct calibration to obtain good hydrological predictions. All parameters needed to initialize the model to predict run-off are obtained from baseflow separation of the hydrograph (Π_B), and from topographical information derived from a DEM and soils data (λ). The reduced parameterization/calibration effort is valuable in environments such as Ethiopia where limited data are available to build and test complicated biogeochemical models.

The model quantified the relative contributions from the various areas of the BNB with relatively good accuracy, particularly at a daily time step. The analysis showed that not all sub-basins contribute flow or run-off equally. In fact, there is large variation in average flow and run-off across the watershed. Additionally, within any one watershed the model indicates that there are areas that produce significantly more run-off and areas that produce almost no run-off, which, of course, has implications for the management of these areas. This model is helpful to identify areas of a basin that are susceptible to erosive or other contaminant losses, due to high run-off production. These areas should be targeted for management intervention to improve water quality.

References

Arnold, J. G., Allen, P. M., Muttiah, R. and Bernhardt G. (1995) Automated base-flow separation and recession analysis techniques, *Ground Water*, 33, 1010–1018.

Arnold, J. G., Srinivasan, R., Muttiah, R. S. and Williams, J. R. (1998) Large area hydrologic modelling and assessment part I: model development, *Journal of the American Water Resources Association*, 34, 1, 73–89.

Arseno, Y. and Tamrat, I. (2005) Ethiopia and the eastern Nile Basin, *Aquatic Sciences*, 16, 15–27.

Ashagre, B. B. (2009) Formulation of best management option for a watershed using SWAT (Anjeni watershed, Blue Nile Basin, Ethiopia), MPS thesis, Cornell University, Ithaca, NY.

Bekele, S. and Horlacher, H. H. B. (2000) Development and application of 2-parameter monthly water balance model in limited data situation: the case of Abaya-Chamo Basin, Ethiopia, *Zede (Ethiopian Journal of Engineers and Architects)*, 17, 56–79.

Awulachew, S. B., Yilma, A. D., Luelseged, M., Loiskandl, W., Ayana, M. and Alamirew, T. (2007) *Water Resources and Irrigation Development in Ethiopia*, Working paper 123, International Water Management Institute, Colombo, Sri Lanka.

Ayenew, T. and Gebreegziabher, Y. (2006) Application of a spreadsheet hydrological model for computing long-term water balance of Lake Awassa, Ethiopia, *Hydrological Sciences*, 51, 3, 418–431.

Bayabil, H. K. (2009) Are runoff processes ecologically or topographically driven in the (sub) humid Ethiopian Highlands? The case of the Maybar watershed, MPS thesis, Cornell University, Ithaca, NY.

Beven, K. J. and Kirkby, M. J. (1979) Towards a simple physically-based variable contributing model of catchment hydrology, *Hydrological Science Bulletin*, 24 1, 43–69.

Bosshart, U. (1997) *Measurement of River Discharge for the SCRP Research Catchments: Gauging Station Profiles*, Soil Conservation Reserve Program, Research reports 31, University of Bern, Bern, Switzerland.

Bryant, R. D., Gburek, W. J., Veith, T. L. and Hively, W. D. (2006) Perspectives on the potential for hydropedology to improve watershed modeling of phosphorus loss, *Geoderma*, 131, 299–307.

Collick, A. S., Easton, Z. M., Adgo, E., Awulachew, S. B., Gete, Z. and Steenhuis, T. S. (2009) Application of a physically-based water balance model on four watersheds throughout the upper Nile basin in Ethiopia, *Hydrological Processes*, 23, 3718–372.

Conway, D. (1997) A water balance model of the upper Blue Nile in Ethiopia, *Hydrological Sciences*, 42, 2, 265–286.

Conway, D. and Hulme, M. (1993) Recent fluctuations in precipitation and runoff over the Nile sub-basins and their impact on main Nile discharge, *Climatic Change*, 25, 127–151.

Derib, S. D. (2005) Rainfall-runoff processes at a hill-slope watershed: case of simple models evaluation at Kori-Sheleko catchments of Wollo, Ethiopia, MSc thesis, Wageningen University, The Netherlands.

Easton, Z. M., Fuka, D. R., Walter, M. T., Cowan, D. M. Schneiderman, E. M. and Steenhuis, T. S. (2008) Re-conceptualizing the Soil and Water Assessment Tool (SWAT) model to predict runoff from variable source areas, *Journal of Hydrology*, 348, 3–4, 279–291.

Easton, Z. M., Fuka, D. R., White, E. D., Collick, A. S., Ashagre, B. B., McCartney, M., Awulachew, S. B., Ahmed, A. A. and Steenhuis, T. S. (2010) A multibasin SWAT model analysis of runoff and sedimentation in the Blue Nile, Ethiopia, *Hydrology and Earth System Sciences*, 14, 1827–1841.

Engda, T. A. (2009) Modeling rainfall, runoff and soil loss relationships in the northeastern Highlands of Ethiopia, Andit Tid watershed, MPS thesis, Cornell University, Ithaca, NY.

Gelaw, S. (2008) Causes and impacts of flooding in Ribb river catchment, MSc thesis, School of Graduate Studies Department of Geography and Environmental Studies, Addis Ababa University, Addis Ababa, Ethiopia.

Grunwald, S. and Norton, L. D. (2000) Calibration and validation of a non-point source pollution model, *Agricultural Water Management*, 45, 17–39.

Guswa, A. J., Celia, M. A. and Rodriguez-Iturbe, I. (2002) Models of soil dynamics in ecohydrology: a comparative study, *Water Resources Research*, 38, 9, 1166–1181.

Habte, A. S., Cullmann, J. and Horlacher, H. B. (2007) Application of WaSiM distributed water balance simulation model to the Abbay River Basin, FWU, *Water Resources Publications*, 6, 1613–1045.

Hawkins, R. H. (1979) Runoff curve numbers from partial area watersheds, *Journal of Irrigation, Drainage and Engineering–ASCE*, 105, 4, 375–389.

Hewlett, J. D. and Hibbert, A. R. (1967) Factors affecting the response of small watersheds to precipitation in humid area, in *Proceedings of International Symposium on Forest Hydrology*, W. E. Sopper and H. W. Lull (eds), pp275–290, Pergamon Press, Oxford, UK.

Horton, R. E. (1940) An approach toward a physical interpretation of infiltration capacity, *Soil Science Society of America Proceedings*, 4, 399–418.

Hu, C. H., Guo, S. L., Xiong, L. H. and Peng, D. Z. (2005) A modified Xinanjiang model and its application in northern China, *Nordic Hydrology*, 3, 175–192.

Hurni, H. (1984) *The Third Progress Report*, Soil Conservation Reserve Program, vol 4, University of Bern and the United Nations University, Ministry of Agriculture, Addis Ababa, Ethiopia.

Ibrahim, A. M. (1984) The Nile – Description, hydrology, control and utilization, *Hydrobiologia*, 110, 1–13.

Johnson, P. A. and Curtis, P. D. (1994) Water balance of Blue Nile river basin in Ethiopia, *Journal of Irrigation, Drainage and Engineering–ASCE*, 120, 3, 573–590.

Kim, N. W. and Lee, J. (2008) Temporally weighted average curve number method for daily runoff simulation, *Hydrological Processes*, 22, 4936–4948.

Kim, U. and Kaluarachchi, J. J. (2008) Application of parameter estimation and regionalization methodologies to ungauged basins of the upper Blue Nile river basin, Ethiopia, *Journal of Hydrology*, 362, 39–56.

Lange, J., Greenbaum, N., Husary, S., Ghanem, M., Leibundgut, C. and Schick, A. P. (2003) Runoff generation from successive simulated rainfalls on a rocky, semi-arid, Mediterranean hillslope, *Hydrological Processes*, 17, 279–296.

Liu, B. M., Collick, A. S., Zeleke, G., Adgo, E., Easton, Z. M. and Steenhuis, T. S. (2008) Rainfall-discharge relationships for a monsoonal climate in the Ethiopian highlands, *Hydrological Processes*, 22, 7, 1059–1067.

Lyon, S. W., Walter, M. T., Gerard-Marchant, P. and Steenhuis, T. S. (2004) Using a topographic index to distribute variable source area runoff predicted with the SCS curve-number equation, *Hydrological Processes*, 18, 2757–2771.

McHugh, O.V. (2006) Integrated water resources assessment and management in a drought-prone watershed in the Ethiopian highlands, PhD thesis, Department of Biological and Environmental Engineering, Cornell University, Ithaca, NY.

Merz, J., Dangol, P. N., Dhakal, M. P., Dongol, B. S., Nakarmi, G. and Weingartner, R. (2006) Rainfall-runoff events in a middle mountain catchment of Nepal, *Journal of Hydrology*, 331, 3–4, 446–458.

MoWR (Ministry of Water Resources) (2002) *Ethiopian Water Sector Strategy*, MoWR, Addis Ababa, Ethiopia.

Nash, J. E. and Sutcliffe, J.V. (1970) River flow forecasting through conceptual models, Part I a discussion of principles, *Journal of Hydrology*, 10, 282–290.

Nyssen, J., Clymans, W., Descheemaeker, K., Poesen, J., Vandecasteele, I., Vanmaercke, M., Zenebe, A., Camp, M.V., Mitiku, H., Nigussie, H., Moeyersons, H., Martens, K., Tesfamichael, G., Deckers, J. and Walraevens, K. (2010) Impact of soil and water conservation measures on catchment hydrological response – a case in north Ethiopia, *Hydrological Processes*, 24, 13, 1880–1895.

Saliha, A. H., Awulachew, S. B., Cullmann, J. and Horlacher, H. B. (2011) Estimation of flow in ungauged catchments by coupling a hydrological model and neural networks: case study, *Hydrological Research*, 42, 5, 386–400.

Schneiderman, E. M., Steenhuis, T. S., Thongs, D. J., Easton, Z. M., Zion, M. S., Neal, A. L., Mendoza, G. F. and Walter, M. T. (2007) Incorporating variable source area hydrology into a curve-number-based watershed model, *Hydrological Processes*, 21, 3420–3430.

SCRP (Soil Conservation Reserve Program) (2000) *Area of Anjeni, Gojam, Ethiopia: Long-Term Monitoring of the Agricultural Environment 1984–1994*, 2000 Soil Erosion and Conservation Database, Soil Conservation Reserve Program, Centre for Development and Environment, Berne, Switzerland in association with the Ministry of Agriculture, Addis Ababa, Ethiopia.

Sheridan, J. M. and Shirmohammadi, A. (1986) Application of curve number procedure on coastal plain watersheds, in *Proceedings from the 1986 Winter Meeting of the American Society of Agricultural Engineers,* paper no. 862505, American Society of Agricultural Engineers, Chicago, IL.

Steenhuis, T. S., Collick, A. S., Easton, Z. M., Leggesse, E. S., Bayabil, H. K., White, E. D., Awulachew, S. B., Adgo, E. and Ahmed, A. A. (2009) Predicting discharge and erosion for the Abay (Blue Nile) with a simple model, *Hydrological Processes*, 23, 3728–3737.

Steenhuis, T. S., Winchell, M., Rossing, J., Zollweg, J. A. and Walter, M. F. (1995) SCS runoff equation revisited for variable-source runoff areas, *Journal of Irrigation, Drainage and Engineering–ASCE*, 121, 3, 234–238.

Sutcliffe, J.V. and Parks, Y. P. (1999) *The Hydrology of the Nile*, Special Publication 5, IAHS, Wallingford, UK, p180.

Tebebu, T. Y. (2009) Surface and subsurface flow effects on permanent gully formation and upland erosion near Lake Tana in the northern Highlands of Ethiopia, MPS thesis, Cornell University, Ithaca, NY.

Tebebu, T.Y., Abiy, A. Z., Dahlke, H. E., Easton, Z. M., Zegeye, A. D., Tilahun, S. A., Collick, A. S., Kidnau, S., Moges, S., Dadgari, F. and Steenhuis, T. S. (2010) Surface and subsurface flow effects on permanent gully formation and upland erosion near Lake Tana in the northern Highlands of Ethiopia, *Hydrological Earth System Sciences*, 14, 2207–2217.

Tenaw, A. (2008) SWAT based run off and sediment yield modeling (a case study of Gumera watershed in Lake Tana sub basin), Ethiopia, MSc thesis, Arba Minch University, Ethiopia.

Wang, X., Shang, S., Yang, W. and Melesse, A. M. (2008) Simulation of an agricultural watershed using an improved curve number method in SWAT, *Transactions of the ASAE*, 51, 4, 1323–1339.

White, E. D., Easton, Z. M., Fuka, D. R., Collick, A. S., Adgo, E., McCartney, M., Awulachew, S. B., Selassie, Y. G. and Steenhuis, T. S. (2011) Development and application of a physically based landscape water balance in the SWAT model, *Hydrological Processes*, 25, 15–25.

White, E. D., Feyereisen, G. W., Veith, T. L. and Bosch, D. D. (2009) Improving daily water yield estimates in the Little River watershed: SWAT adjustments, *Transactions of the ASAE*, 52, 1, 69–79.

7

The Nile Basin sediment loss and degradation, with emphasis on the Blue Nile

Tammo S. Steenhuis, Zachary M. Easton, Seleshi B. Awulachew, Abdalla A. Ahmed, Kamaleddin E. Bashar, Enyew Adgo, Yihenew G. Selassie and Seifu A. Tilahun

Key messages

- Run-off and erosion are spatially distributed in the landscape. Contrary to the prevailing consensus, the steep slopes in well-established agricultural watersheds in humid climates are usually not the main sources of sediment.
- Most run-off and sediments originate from both the severely degraded areas with shallow topsoils and points near a river when they are becoming saturated, around the middle of the rainy phase of the monsoon. Degraded areas that are bare deliver greater amounts of sediment than the saturated areas that are vegetated.
- Two simulation models (SWAT-WB and the water balance type model), adapted to the Ethiopian Highlands, widely ranging in complexity, were able to simulate the available sediment concentrations equally well. This illustrates the point that conceptual correctness is more important than complexity in simulating watersheds.
- Gully formation is an important source of sediment in the Blue Nile Basin. Sediment concentration can be up to an equivalent of 400 t ha^{-1} in the watershed. Although gullies are formed on the hillsides, the largest gullies of up to 5 m depth and 10 m width are formed in the periodically saturated and relatively flat lands near the river.
- On average, the annual sediment loss in the Blue Nile Basin at the border with Sudan is 7 t ha^{-1} and is equivalent 0.5 mm of soil over the entire basin. Although this seems to be relatively small, it is an enormous amount of sediment for the Rosaries reservoir and, consequently, its capacity to store water has been decreased significantly since 1966 when it was completed.

Introduction

High population pressure, poor land-use planning, over-dependency on agriculture as a source of livelihoods and extreme dependence on natural resources are inducing deforestation, overgrazing, expansion of agriculture to marginal lands and steep slopes, declining agricultural productivity and degradation of the environment. Poor agricultural and other practices affect run-off characteristics resulting in increased erosion and siltation and reduced water quality in the BNB (Awulachew *et al.*, 2008). FAO (1986) estimates an annual loss of over 1.9 billion tonnes of soil from the Ethiopian Highlands. Only approximately 122 million tonnes reach the Ethiopia border (Ahmed and Ismail, 2008). Erosion from the land surface occurs in the form of sheet erosion, rill and inter-rill erosion, or gully erosion, part of which is delivered to rivers. This, together with deposition of erosion from in-stream beds and banks of rivers, constitutes the sediment load in the river (Awulachew *et al.*, 2008). According to Hydrosult *et al.* (2006), the Ethiopian plateau is the main source of the sediment in the Blue Nile system. The main area of sheet erosion is within the Ethiopian Highlands. Some sheet erosion occurs within Sudan, mainly on and around the rock hills, which have become devoid of vegetative cover. Most of this is deposited on the foot slope and does not enter the drainage system. Those streams reaching the river during the rainy season can carry high sediment concentrations. The eroded and transported sediment ultimately reaches Sudan, and can cause significant loss of reservoir volume and transmittance capacity in irrigation canals. In fact, some sediment from the Highlands is transmitted to the Aswan High Dam as suspended sediment. As a result, we have focused on the study of erosion, sedimentation and understanding the impacts of intervention measures in the BNB (including from the Ethiopian Highlands to the Roseires reservoir in Sudan).

Modelling of the processes governing erosion and sedimentation can further help our understanding of the basin-wide issues in terms of the critical factors controlling erosion and associated sediment transport. However, sediment modelling on a daily or weekly basis in Ethiopia has generally not been very successful, because the underlying hydrological models have not predicted run-off well (e.g. the Agricultural Non-Point Source Pollution, or AGNPS, model; Haregeweyn and Yohannes, 2003; Mohammed *et al.*, 2004), Soil and Water Assessment Tool (SWAT; Setegn *et al.*, 2008) and Water Erosion Prediction Project (WEPP; Zeleke, 2000). Various modelling approaches have been attempted with a limited degree of success because of an ineffective ability to link erosion and sediment transport to the correct hydrological process. We present an in-depth analysis of the various landscape sediment sources to better understand the erosion–sediment transport relationship of the BNB watershed. Unfortunately, there is a general lack of sediment data, particularly time series of sediment concentrations in the various reaches of the basin. We first explore which data are available from routine measurements and in experimental watersheds followed by a description of how we have modelled these data using the hydrology models presented in Chapter 6; finally, we discuss the implication of our findings on structural and non-structural practices.

Available sediment concentration data

Similar to the rainfall run-off studies discussed in Chapter 6, primary and secondary data are collected routinely at sub-basin level on tributary rivers and the main stream of the Blue Nile; additional data on quantities of sediment can be obtained at the Ethiopia Sudan border (El Diem station) where the Roseires reservoir has been trapping sediment for over 40 years. Finally, data on sediment concentration are available from experimental watershed stations of the Soil Conservation Reserve Program (SCRP; Herweg, 1996), which have long records (such as from Anjeni, Maybar and Andit Tid), while shorter records are available from the Debre-Mawi and Koga watersheds.

Routine collection of sediment data

An examination of the sediment stations available from the Ministry of Water Resources (MoWR) in Ethiopia shows that there are altogether 45 stations in the Abbay Basin. However, most of these have only very sporadic measurements and most are related to periods during which stage, discharge relationships were developed or revisions to such relationships were made. A consolidated list of stations with data records for the Abbay is provided by Awulachew *et al.* (2008).

Preliminary analyses of the routinely measured data show that sediment peaks during the rainy season, particularly in July. However, almost no sediment is measured in the streams in the dry season. The annual sediment concentration (sediment weight per volume of water) measured in mg l^{-1} shows the highest sediment concentrations from June to September, generally peaking in July, while rainfall and run-off peaks occur in August.

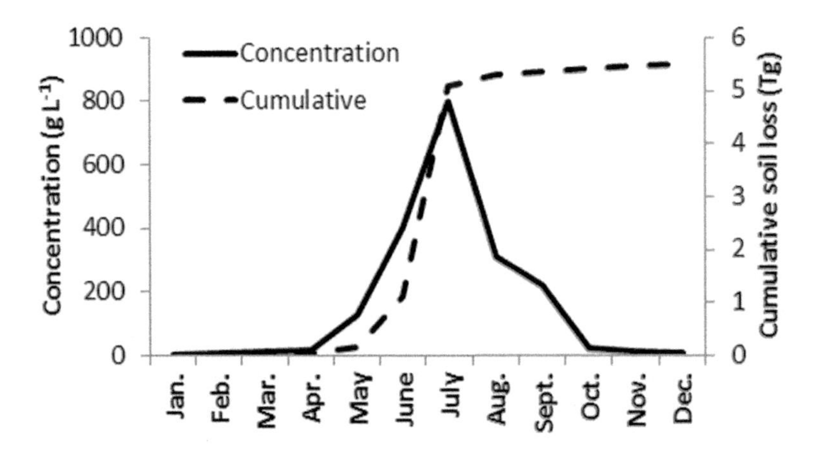

Figure 7.1 Typical monthly sediment concentrations, cumulative sediment load over time at Ribb at Addis Zemen station, a tributary of Lake Tana and the Blue Nile

Figure 7.1 illustrates a typical sediment concentration time series for the Ribb River at Addis Zemen in the Lake Tana watershed. The river is a medium-sized watershed tributary, with a drainage area of about 1600 km². An important implication is that the sediment rating curve established on flow volume or river stage alone cannot provide accurate estimation of sediment yield. Rainfall and run-off are the driving factors for the onset of the erosion process. However, timing of rainfall, land use and land cover have a major influence on the erosion process (Ahmed and Ismail, 2008).

Data derived from trapping of sediments in reservoirs

Accumulation of sediment in a reservoir can be used to indicate the severity of the land degradation, erosion, sediment transport and minimum yield at that particular point. The Roseires reservoir is of particular interest as it acts as the sediment sink for the entire Ethiopian Highlands. The annual amount of sediment delivered by the Blue Nile at the entrance of the

Roseires reservoir is, on average, 122 million tonnes per year (t yr^{-1}). The bed load is less than 10 per cent. The coarser sand is deposited in the upper portion of the Blue Nile near the Ethiopia–Sudan border, while the lighter sediment is carried by the flow downstream. The suspended sediment load distribution is 30 per cent clay, 40 per cent silt and 30 per cent fire sand (Ahmed and Ismail, 2008).

Bashar and Khalifa (2009) used bathymetric surveys conducted in 1976, 1981, 1985, 1992, 2005 and 2007 to assess the rate of sedimentation in the Roseires reservoir. In order to calculate the amount of sediment deposited, the design storage capacity in 1966 for the different reservoir levels was taken as a baseline. The reservoir storage capacity as a function of the reservoir for the years that the bathymetric surveys were taken is depicted in Figure 7.2. In the 41 years of operation (1966–2007), the maximum storage capacity at 481 m decreased from 3330 million to 1920 million m^3. This represents a loss of storage capacity of 1410 million m^3, and thus only 60 per cent of the initial storage is still available after 41 years. The reservoir is currently filled with sediment up to the 470 m level (Figure 7.2). The total amount of sediment delivered over the 41 years to the Roseires reservoir is approximately 5000 million tonnes, or 3000 million m^3. This represents a decrease of twice the storage capacity in the reservoir, since the reservoir operation rule maintains that only the end of the rainy season flow is stored, which is when the sediment concentration is smallest (Figure 7.1).

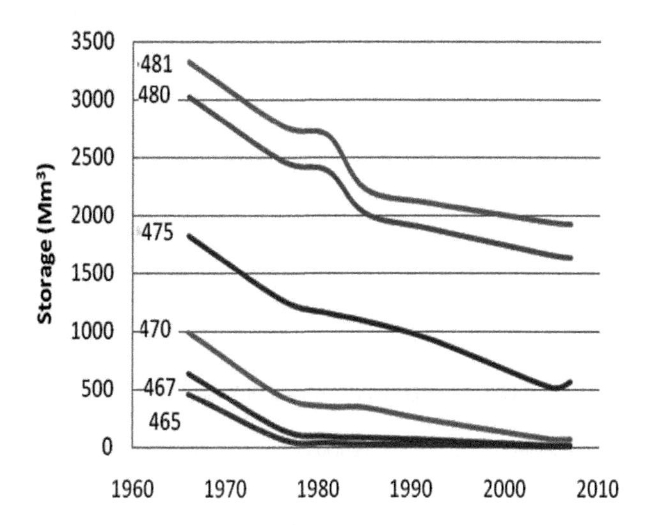

Figure 7.2 Variation of storage with time at various reservoir levels (m) in the Roseires reservoir

Source: Bashar and Khalifa, 2009

Upland and gully erosion in micro watersheds

Since the establishment of the micro-watersheds by the SCRP in 1981, high-resolution data on climate, hydrology and suspended sediment, from both rivers and test plots, have been collected. Hence, an expansive database has been established that has served as a data source for understanding and quantifying erosion processes and for validating models. The Anjeni, Andit Tid and Maybar watersheds are located near or in the BNB and have over 10,000 combined observations.

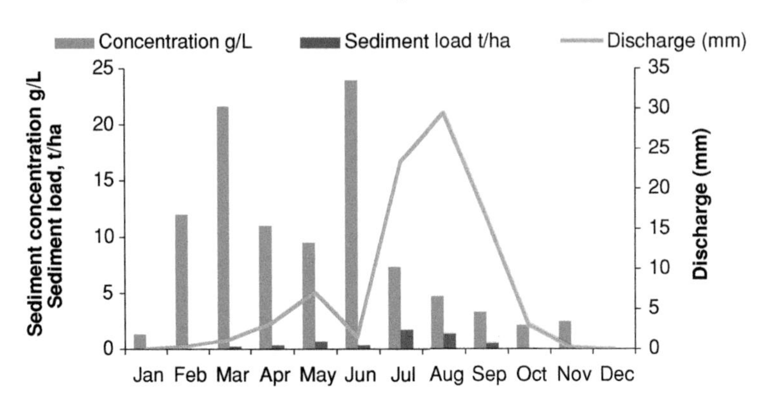

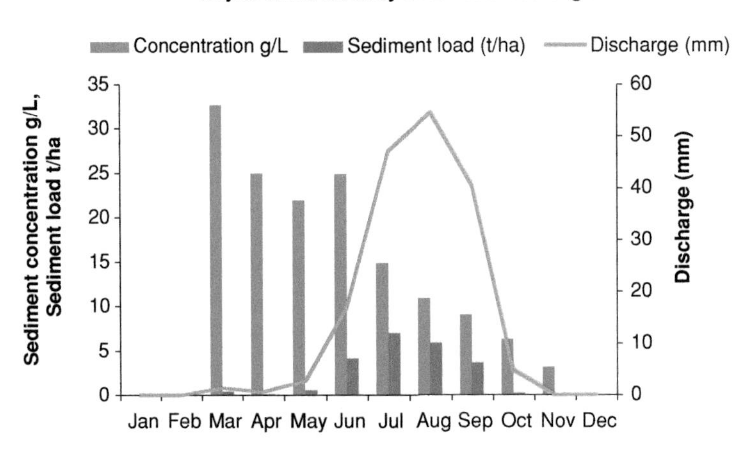

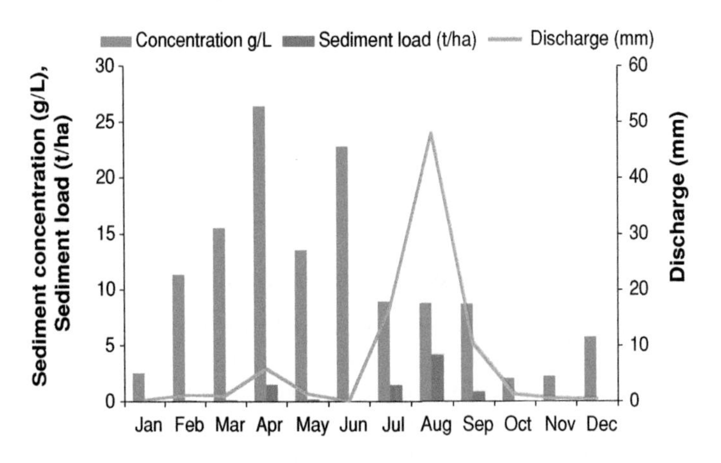

Figure 7.3 Mean monthly concentration of sediment in the SCRP watersheds

Both sediment concentration and discharge data are available for each measurement with a resolution of 10 minutes during run-off events. Based on these measurements, annual sediment yields during run-off events were 5.4, 22.5 and 8.8 t $ha^{-1} \cdot yr^{-1}$ for Andit Tid, Anjeni and Maybar, respectively (Guzman, 2010). During the main rainy season, there was a decrease in average sediment concentration over the course of the season (Figure 7.4). This decrease was less noticeable at the Maybar site than at the other two sites, possibly caused by the more variable year-to-year fluctuations in precipitation and discharge for the Maybar watershed (Hurni *et al.*, 2005). Unfortunately, a simple sediment rating curve could not be developed for these watersheds either. The maximum correlation coefficient did not exceed 0.22 for any watershed when all discharge–sediment data were used. These small watersheds offer an ideal opportunity to investigate the reason for the non-uniqueness in the sediment rating curve. This is best illustrated in Anjeni where the average concentrations are calculated over daily periods. Two storms are depicted, one in the beginning of the short rainy season (24 April 1992; Figure 7.4a) and the other, later in the rainy season (19 July 1992; Figure 7.4b; Tilahun *et al.*, 2012).

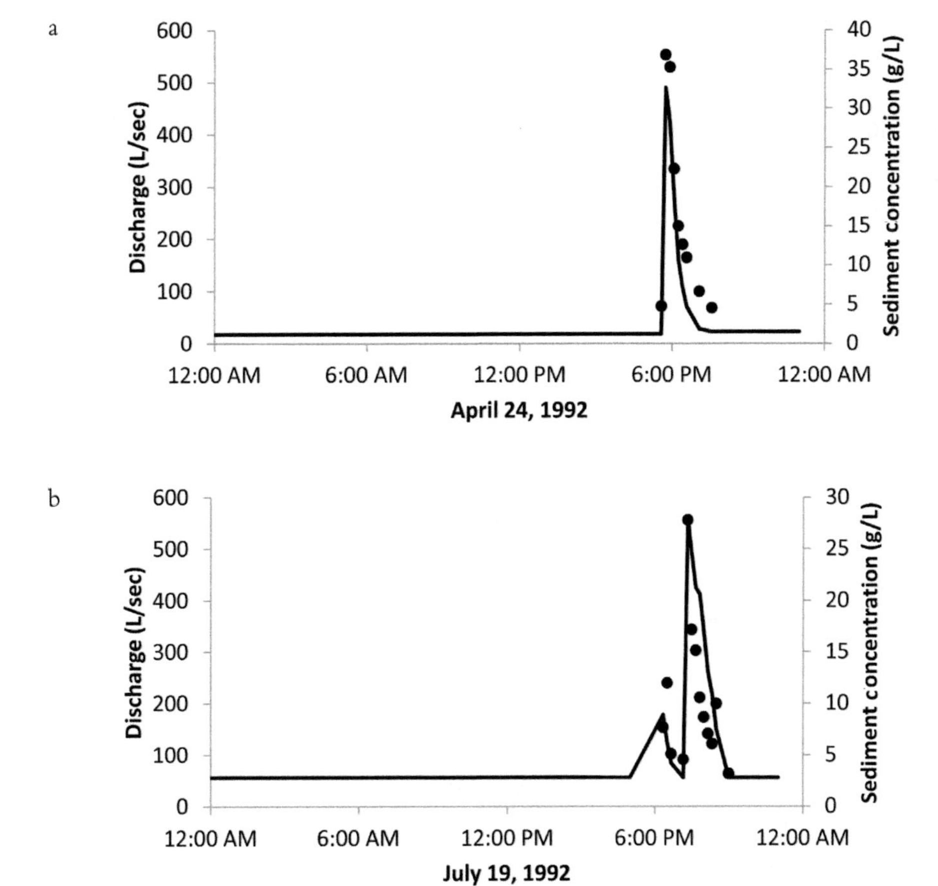

Figure 7.4 Measured discharge (solid line) and sediment concentration (closed circles) for the Anjeni watershed on (a) 24 April 1992 and (b) 19 July 1992

Source: Tilahun *et al.*, 2012

The surface run-off for both events is similar with a peak run-off of 400–500 l s^{-1}. The duration of the run-off event was approximately 2 hours. The peak sediment concentrations were nearly the same, around 30–35 g l^{-1}. Baseflow discharge is low during the beginning of the rainy season (around 10 l sec^{-1} for April or equivalent to 0.8 mm day^{-1} over the whole watershed). Baseflow increases during the rainy season and is approximately 50 l sec^{-1} (equivalent to 4.0 mm day^{-1}) in July. Despite the similar surface run-off characteristics the total flows for April and July were 2400 and 6500 m^3, respectively. The averages of the daily sediment concentration can be obtained by dividing the load by the total flow, resulting in a concentration of 11.3 g l^{-1} for the April storm and 4.4 g l^{-1} for the July storm. In essence, the baseflow dilutes the peak storm concentration when simulated on a daily basis later in the rainy season. Thus, since the ratio of the baseflow to surface run-off is increasing during the wet season the temporally averaged concentration is decreasing. Figure 7.5 shows this clearly for the Anjeni watershed. Therefore, it is important to incorporate the contribution of baseflow in the prediction of sediment concentrations.

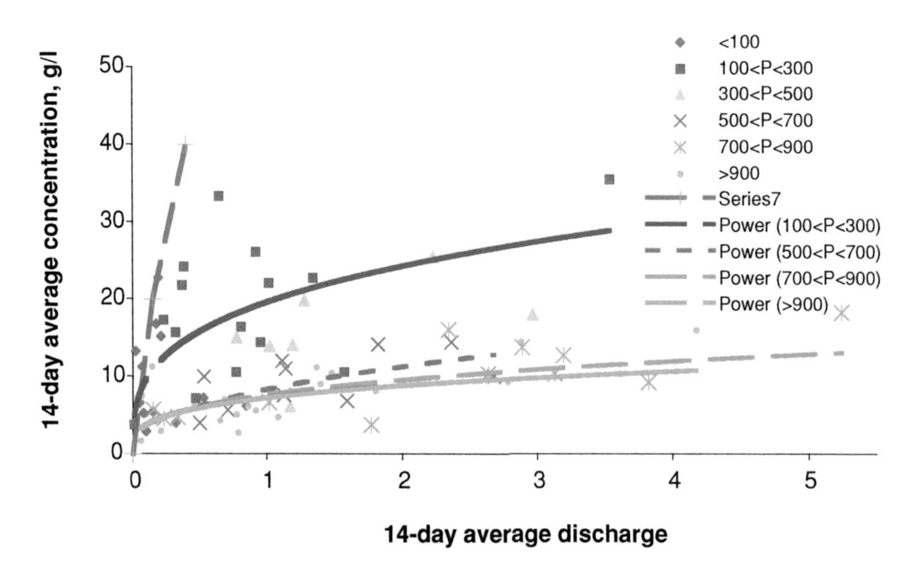

Figure 7.5 Stratified biweekly storm concentration versus discharge for Anjeni

Note: Symbols indicate the amount of cumulative effective precipitation, P, since the beginning of the wet period: diamonds, P<100 mm; squares, 100<P<300 mm; triangles, 300<P<500 mm; crosses, 500<P<700 mm; stars, 700<P<900 mm; circles, P>900 mm

Gully erosion

Gully formation and upland erosion were studied in the Debre-Mawi watershed south of Lake Tana by Abiy (2009), Tebebu *et al.* (2010) and Zegeye *et al.* (2010). We selected one of the gullies (Figure 7.6a, b) with a contributing area of 17.4 ha. According to farmers' interviews through the AGERTIM (Assessment of Gully Erosion Rates through Interviews and Measurements method; Nyssen *et al.*, 2006), the gully erosion started in the early 1980s, which corresponds to the time when the watershed was first settled and the indigenous vegetation on the hillsides was converted gradually to agricultural land. Almost all farmers agreed on the incision location of the current gully and confirm that the locations of the two gully incisions were related to three springs on the hill slope. Erosion rates for the main stem and two branches are given in Table 7.1. Walking along the gully with a Garmin GPS with an accuracy of 2 m, gully boundaries were determined before the rainy season in 2008 (indicated as 2007 measurement) and after the rainy season on 1 October (the 2008 measurement). The increase in the main stem erosion rate (gully C) from an average of 13.2 to 402 t ha^{-1} yr^{-1} from 1980 to 2007 is due to the recently enlarged and deepened gully at the lower end (Figure 7.6b). Although not shown in the table, our measurements showed that from 2005 to 2007, the gully system increased from 0.65 to 1.0 ha, a 54 per cent increase in area. In 2008, it increased by 43 per cent to cover 1.43 ha from the year before. This is a significant amount of loss of land in a 17.4 ha watershed. The increase in rate of expansion of gully formation 20–30 years after its initial development is in agreement with the finding of Nyssen *et al.*, 2008 in the May Zegzeg catchment, near Hagere Selam in the Tigray Highlands, at which the gully formation follows an S shape pattern. Although it is slow in the beginning and end, there is a rapid gully formation phase after 20–30 years after initialization of the gully and then erosion rates decrease again.

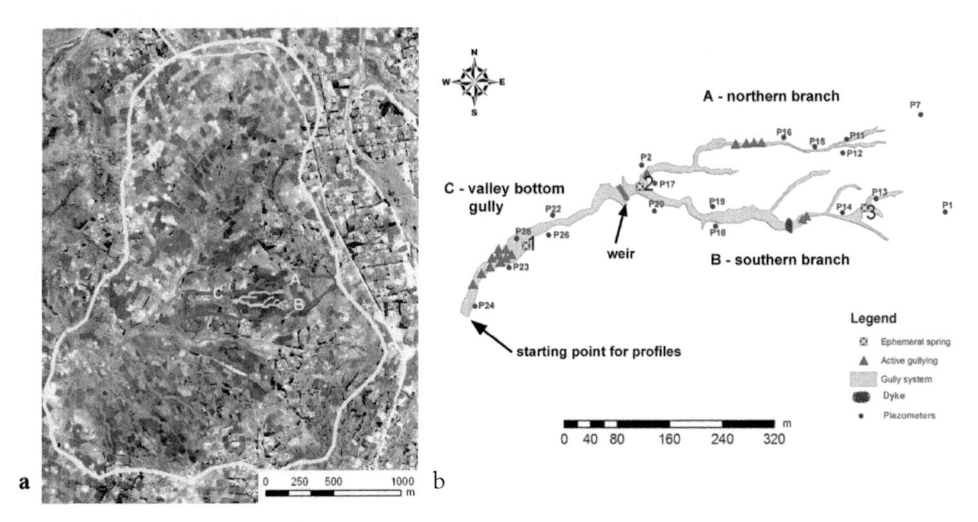

Figure 7.6 Map of the Debra-Mawi watershed (a) with the gully area outlined in red and (b) the Debre-Mawi gully extent generated by hand-held GPS tracking; active erosion areas are indicated by triangles. Ephemeral springs and piezometer locations are shown as well

Table 7.1 Erosion losses for gullies A, B and C. Erosion rates calculated from the gullies are then distributed uniformly over the contributing area

Gully location	*Soil loss*		
	1980–2007 (t ha^{-1} yr^{-1})	*2007–2008 (t ha^{-1} yr^{-1})*	*2007–2008 (cm yr^{-1})*
Branches (gullies A and B)*	17.5	128	1
Main stem (gully C)	13.2	402	3
Total	30.7	530	4

Note: *The calculated erosion rates for gullies A and B were nearly identical, and are thus presented in aggregates

In order to investigate the cause of the gully erosion, 24 piezometers were installed at depths up to 4 m in the gully bottom, as well as in the gully's contributing area (Figure 7.6b; Tebetu *et al.*, 2010). The gully is very active in a few areas as indicated by the red triangles in Figure 7.6b. Active widening of the gully occurs when the water table is above the gully bottom. This is best illustrated in the large gully near the valley bottom (Figure 7.6b). The depth of the gully (Figure 7.7a) and the corresponding widths (Figure 7.7b) are depicted before the 2007 and after the 2008 rainy seasons for the area of the valley bottom as a function of the distance from the point where this gully joins the main branch. The average water table depths for adjacent piezometers (from bottom to top, P24, P23, P22, P26 and P17) are shown as well, and indicate that the valley bottom is saturated, while further uphill the water table is below the gully bottom. During the 2008 rainy season the gully advanced up the hill, past the 187 m point (Figure 7.7a) and increased up to 20 m in the top width (Figure 7.7b). In this region, the water table was near the surface and approximately 4 m above the gully bottom (Figure 7.7a). This means that under static conditions the pore water pressure near the gully advance point is 4 m, which might be sufficient to cause failure of the gully wall.

The piezometers P24 and P26 at 244 and 272 m indicate that the water table is at the surface (Figures 7.6 and 7.7), but that the gully is not incised as yet (Figure 7.7a). The area is flatter and, in the past, the sediment had accumulated here. It is likely that over the next few years the head wall will rapidly go uphill in these saturated soils. At the 323 and 372 m points the water table is below the bottom of the 4 m long P17 piezometer, and thus below the bottom of the gully. Here the gully is stable despite its 3 m depth.

Upland erosion

The second watershed was used to study upland erosion (rill and inter-rill erosion) processes in cultivated fields. The location of the upland site relative to the gully site is given in Figure 7.1. Soils consisted of clay and clay loam, and land use/land cover was similar to the gully site.

For determining rill erosion, 15 cultivated fields were selected in the contributing area, representing a cumulative area of 3.6 ha. These fields were classified into three slope positions: upslope (slope length of 100 m), mid-slope (slope length of 250 m) and toe-slope (slope length of 100 m). A series of cross-slope transects were established with an average distance of 10 m between two transects, positioned one above another to minimize interference between transects. During the rainy season, each field was visited immediately after the rainfall events in July and August, when the peak rainfall occurred. During these visits the length, width and depth of the rills were measured along two successive transects. The length of a rill was measured from its upslope starting point down to where the eroded soil was deposited. Widths were measured at several points along a rill and averaged over the rill length (Herweg, 1996). From these measurements, different magnitudes of rill erosion were determined, including rill volumes, rates of erosion, density of rills, area impacted by the rills, and the percentage of area covered by the rills in relation to the total area of surveyed fields (Herweg, 1996; Bewket and Sterk, 2005). The average upland erosion of the 15 agricultural fields is 27 t ha^{-1} over the 2008 rainy season in the Debre-Mawi watershed (Zegeye *et al.*, 2010). The lower watersheds had significantly greater soil erosion and greater area covered by rills than either the middle watershed or the upper watershed, which had the least of both (Table 7.2). It is hypothesized that this is related to the greater amount of run-off produced on the lower slopes (Bayabil *et al.*, 2010) causing the greater volume of rills. In addition, the Teff plots had the greatest density of rills, possibly caused by the repeated cultivation of the field and compaction of the soil by livestock traffic before sowing, and possibly because of the reduced ground cover from the later planting date for Teff

a

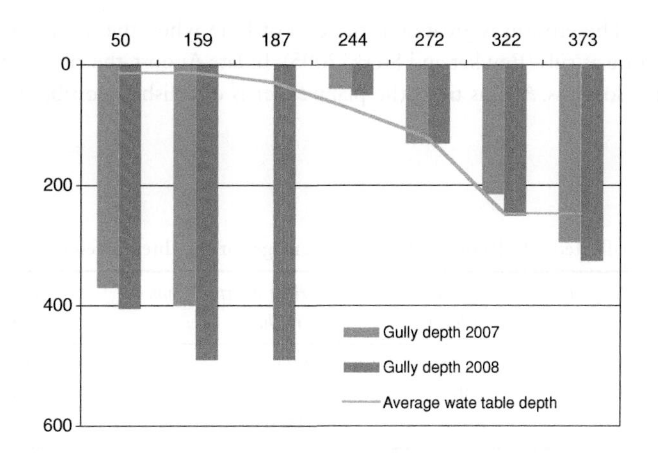

b

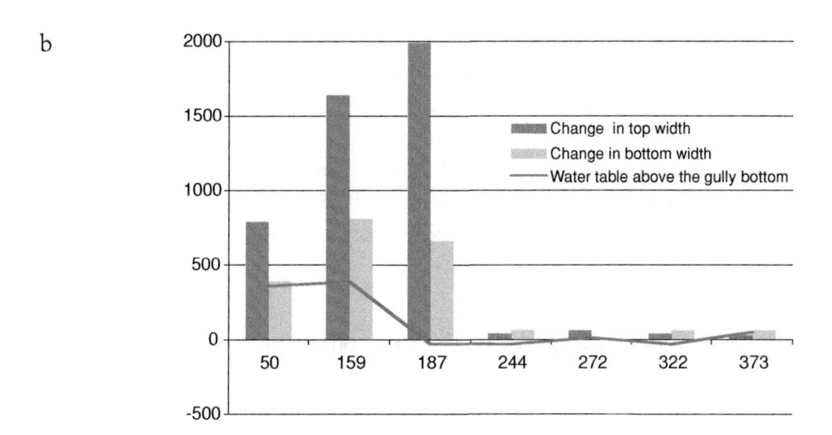

c

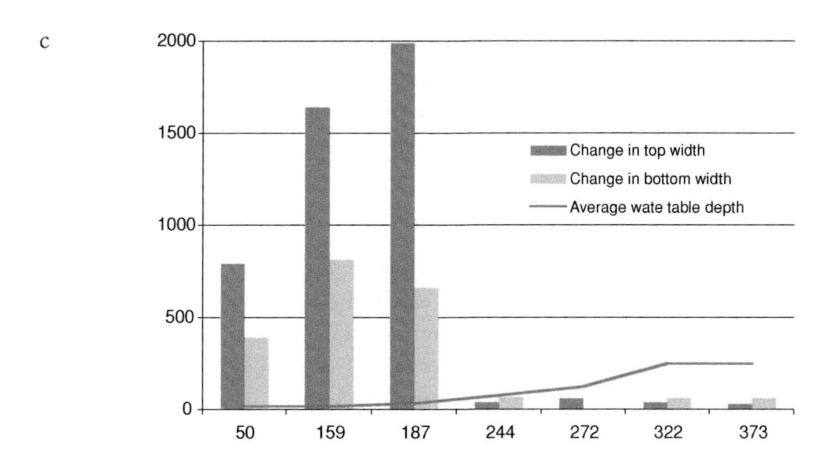

Figure 7.7 (a) Average water table and gully depths (m) before and after the 2008 rainy season for the main stem (gully C) using the soil surface as a reference elevation point, and (b) change in top and bottom widths (m) of the gully and average water table depth (m) above the gully bottom

(Zegeye *et al.*, 2010). The erosion is greatest at the end of June when the soil is loose and dry, making it easy to erode as rills (Bewket and Sterk, 2005). In late August, the rills degrade giving an apparent negative soil loss. At this time, the plant cover is established possibly reducing soil losses.

Table 7.2 Soil loss, area affected, rill density and slope percentage for the three different slope positions

Slope position	*Slope (%)*	*Soil loss ($t\ ha^{-1}$)*	*Area of actual damage ($m^2\ ha^{-1}$)*	*Rill density ($m\ ha^{-1}$)*
Down slope	14a	34a*	884a	4469a
Mid-slope	10b	23b	662b	2860b
Upslope	9b	8c	256c	1029c

Note: *Means followed by different letters (a,b,c) within columns are significantly different at α=0.05

A comparison of the gully and upland erosion rates in the Debre-Mawi watershed indicates that the soil loss rate of the gully system is approximately 20 times higher than the erosion rates for the rill and inter-rill systems. While significantly lower than gully erosion, rill erosion is still nearly four times greater than the generally accepted soil loss rate for the region and thus cannot be ignored in terms of agricultural productivity and soil fertility. However, if reservoir siltation and water quality of Lake Tana and the Blue Nile constitute the primary impetus for soil conservation, gully erosion has far greater consequences.

Simulating erosion losses in the Blue Nile Basin

Schematization of the Blue Nile Basin for sediment modelling

To better understand the issues and processes controlling sediment, the BNB was divided into a set of nested catchments from micro-watershed to basin level that include micro- watershed level, watershed and small dam level, sub-basins and major lakes, basin outlet and a large reservoir (Awulachew *et al.*, 2008). In this section, we discuss some of the methods and models to predict erosion and sediment loads. We will start with the simple Universal Soil Loss Equation (USLE) and end with the more complicated SWAT-WB model.

Universal Soil Loss Equation

The simplest method to predict the erosion rates is using USLE, originally developed empirically based on a 72.6 m long plot for the United States east of the 100th meridian. It has been adapted to Ethiopian conditions using the data of long-term upslope erosion data of the SCRP sites of Mitiku *et al.* (2006). More recently, Kaltenrieder (2007) adapted USLE for Ethiopian conditions to predict annual soil losses at the field scale. The erosion data set collected by Zegeye *et al.* (2010) offers an ideal opportunity to check the modified USLE. The predicted erosion rate was calculated assuming that 25 per cent of the erosion was splash erosion. The predicted and observed erosion rates for the individual plots in the Debre-Mawi watershed (described above) are shown in Figure 7.8. Although USLE seems to predict the general magnitude of the plot-scale erosion well, it does not include erosion due to concentrated flow

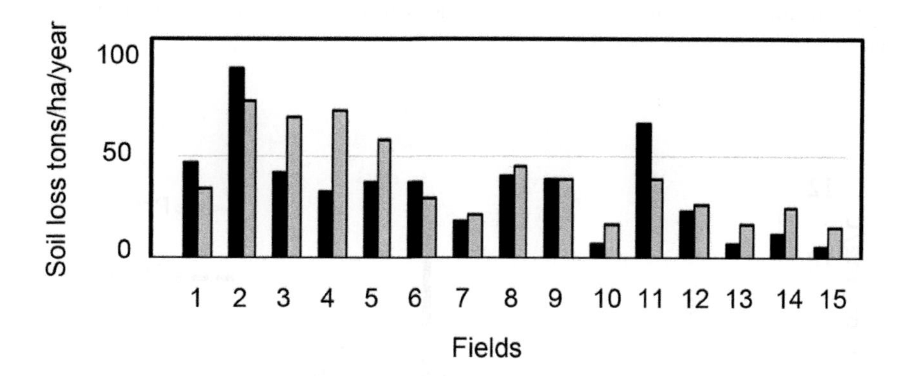

Figure 7.8 Comparison of modified USLE for Ethiopia and observed soil losses in the Debra-Mawi watershed. Observed soil loss is indicated by black bars and predicted loss by grey bars

channels and gullies (Capra *et al.*, 2005), which is another reason for not using USLE predictions without field verification.

Simple erosion model

We can use the simple model presented in Chapter 6 for predicting sediment concentrations and loads by assuming that baseflow and interflow are sediment-free and only surface run-off mobilizes sediment. If we assume that the velocity of the water (Hairsine and Rose, 1992) is linearly related to the concentration in the water, it is possible to predict the concentration of the sediment (Tilahun *et al.*, 2012):

$$C_t = \frac{a_s A_s Q_{s_t}^{1.4} = a_d A_d Q_{d_t}^{1.4}}{A_s Q_{s_t} + A_d Q_{d_t} + A_h \left(Q_{b_t} + Q_{I_t}\right)}$$

where C_t is the sediment concentration, A is the fraction in the watershed area that is saturated (s), degraded (d) or hillsides (h); the constant a represents the sediment and watershed characteristics for the saturated area (s) and degraded areas (d); and Q is the discharge per unit area at time t from either the saturated area (s) or degraded area (d) as overland flow or as subsurface flow from the hillside as baseflow (B) and as interflow (I).

Thus there are only two calibration parameters, one for each source area, that determine contribution to the sediment load of the source area at the outlet of the watershed. Although it is recognized that by incorporating more calibration parameters, such as plant cover, or soil type for the different areas, we might obtain a better agreement between observed and predicted sediment yield, the current methods seem to provide a reasonably accurate prediction of sediment yield, as shown in Figure 7.9. Note that in Table 7.3, the coefficient, a, for the degraded areas is significantly larger than the saturated areas, and thus the degraded areas produce the majority of the sediment load. The agreement between observed and predicted sediment loads deteriorates rapidly if we increase the sediment concentration in the interflow from zero. Thus the simple model clearly demonstrates (Table 7.3) that most of the sediments originate from the degraded surface areas. Practically, these areas can be recognized easily in the landscape during the growing season as the areas with little or no vegetation and not often farmed.

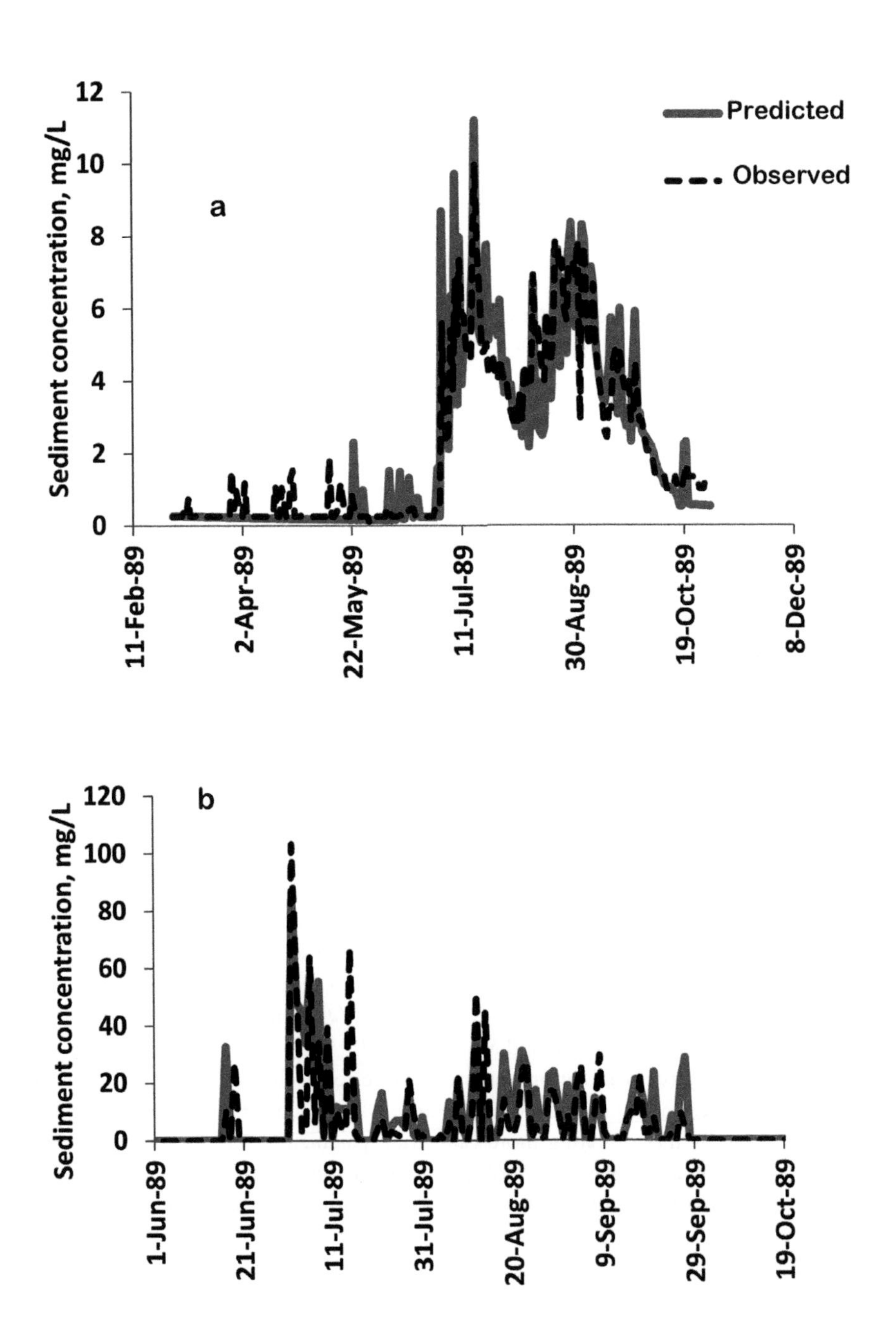

Figure 7.9 Predicted and observed (a) streamflow and (b) sediment concentration for the Anjeni watershed

Table 7.3 Model input parameters for the Anjeni watershed

Components	*Description*	*Parameters*	*Unit*	*Calibrated values*
Hydrology	Saturated area	Area A_s	%	2
		S_{max} in A_s	mm	70
	Degraded area	Area A_d	%	14
		S_{max} in A_d	mm	10
	Hillside	Area A_h	%	50
		S_{max} in A_h	mm	100
	Subsurface flow parameters	BS_{max}	mm	100
		$t_{1/2}$	days	70
		$\tau^{\star}$	days	10
Sediment	Saturated area	a_s	g $(mm\ day^{-1})^{-0.4}$	l^{-1} 1.14
	Degraded area	a_d	g l^{-1} $(mm\ day^{-1})^{-0.4}$	4.70

Notes: A is fractional area for components of saturated area (s), degraded area (d) and hillside (h); S_{max} is maximum water storage capacity; $t_{1/2}$ is the time it takes in days to reduce the volume of the baseflow of the reservoir by a factor of 2 under no recharge condition; BSmax is maximum baseflow storage of linear reservoir; $\tau^{\star}$ is the duration of the period after a single rainstorm until interflow ceases; the constant a represents the sediment and watershed characteristics for the saturated area (s) and degraded areas (d) for obtaining the sediment concentration in the runoff

Soil and Water Assessment Tool

The Soil and Water Assessment Tool–Water Balance (SWAT-WB) model introduced in Chapter 6 allows us to study sediment losses for watersheds ranging from the micro-watershed (Anjeni) to the entire Ethiopian section of the Blue Nile (Easton *et al.*, 2010). Landscape erosion in SWAT is computed using the Modified Universal Soil Loss Equation (MUSLE), which determines sediment yield based on the amount of surface run-off. The SWAT-WB model improves the ability to correctly predict the spatial distribution of run-off by redefining the hydrologic response units (HRUs) by taking the topography into account and results in wetness classes that respond similarly to rainfall events. Thus by both predicting the run-off distribution correctly and using MUSLE the erosion from surface run-off producing areas is incorporated into the distributed landscape erosion predictions.

The robust SCRP data sets were used to calibrate and parameterize the SWAT-WB model (Easton *et al.*, 2010). The discharge in SWAT-WB model in Chapter 6 was calibrated using a priori topographic information and validated with an independent time series of discharge at various scales.

In Anjeni, the sediment hydrograph (Figure 7.10 and Table 7.4) has mimicked the flashy nature of the streamflow hydrograph. The fitting statistics were good for daily predictions (Table 7.4). For parameterization we assumed, in accordance with the SCRP watershed observation, that terraces have been utilized by approximately 25 per cent of the steeply sloped agricultural land to reduce erosion (Werner, 1986). To include this management practice, slope and slope length were reduced and the overland Mannings-n values were specified as a function of slope steepness (Easton *et al.*, 2010). Finally, Anjeni has a large gully providing approximately 25 per cent of the sediment (Ashagre, 2009). Since SWAT is incapable of realistically modelling gully erosion, the soil erodibility factor (USLE_K) in the MUSLE (Williams, 1975) was increased by 25 per cent to reflect this.

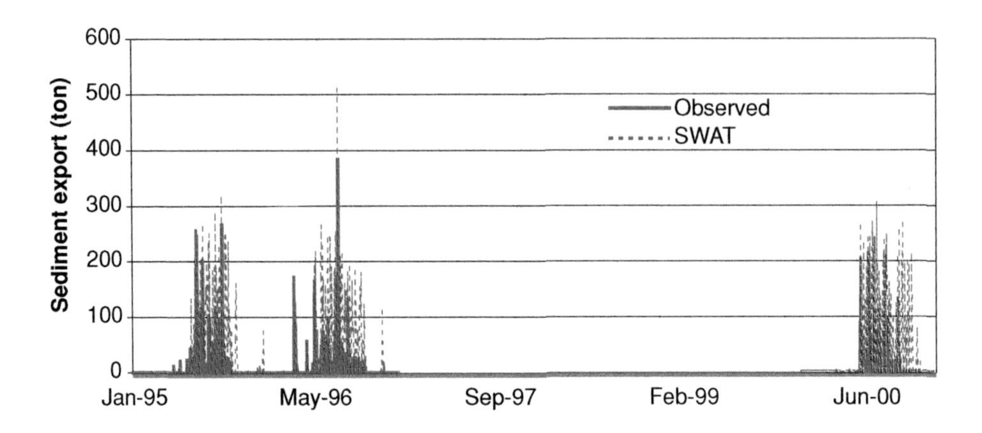

Figure 7.10 Measured and Soil and Water Assessment Tool–Water Balance predicted sediment export from the Anjeni micro-watershed

Table 7.4 shows the statistics for the measured and predicted sediment export for the two other locations for which we had data available, El Diem at the border with Sudan and the Ribb. In these simulations, the most sensitive parameters controlling erosion in the watershed were those used for calculating the maximum amount of sediment that can be entrained during channel routing (Easton *et al.*, 2010). The daily Nash–Sutcliff Efficiency (NSE) factor for the simulation period for the watersheds was approximately 0.7, indicating acceptable model performance. Nearly 128 million t yr^{-1} were delivered during the 2 years of measurements (Ahmed, 2003), with a measured daily average during the rainy season of 1.22 million tonnes. The model predicted 121 million tonnes over the 2 years, with a daily average of 1.16 million tonnes during the rainy season. The average sediment concentration at El Diem was 3.8 g l^{-1}, while the model predicted a slightly higher concentration of 4.1 g l^{-1}. The higher concentration was somewhat counterbalanced by the slightly under-predicted flow. Despite this, model performance appears to be adequate. El Diem sediment export was much less flashy than that in the Anjeni watershed (compare Figures 7.10 and 7.11). While the total sediment export intuitively increases with basin size, the normalized sediment export (in t km^2) was inversely proportional to the basin size (Table 7.4). This is a direct result of the difference in the baseflow coefficients (Π_B) among the basins of various sizes (e.g. 0.47 for Anjeni to 0.84 for the border at El Diem) and is similar to what is predicted with the simple spreadsheet erosion model.

Table 7.4 Model fit statistics (coefficient of determination, r^2 and NSE), and daily sediment export for the Anjeni, Ribb, and border (El Diem) sub-basins during the rainy season

Sub-basin	r^2	*NSE*	*Measured sediment export (t d^{-1})*	*Modelled sediment export (t d^{-1})*	*Modelled sediment export (t·km^2 d^{-1})*
Anjeni	0.80	0.74	239	227	201.2
Ribb*	0.74	0.71	30.6×10^3	29.5×10^3	22.7
Border (El Diem)	0.67	0.64	1.23×10^6	1.23×10^6	7.1

Note: *Consists of four measurements

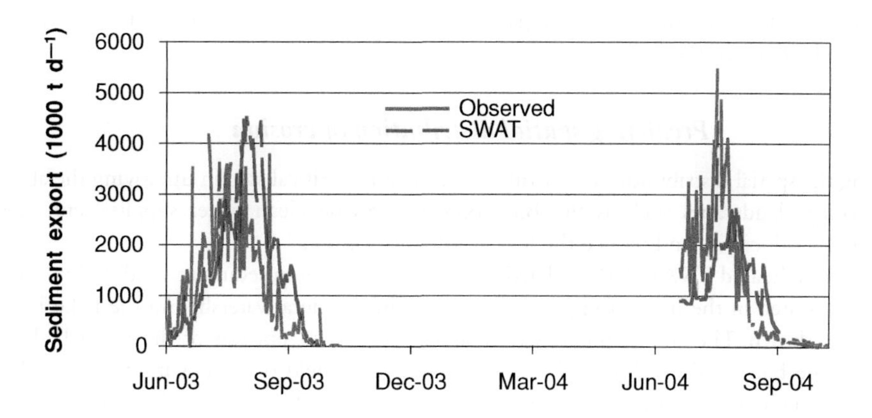

Figure 7.11 Observed and Soil and Water Assessment Tool–Water Balance modelled sediment export at the Sudan/Ethiopia border

Table 7.5 Annual predicted sediment yield for each wetness index class and for the pasture, crop and forest land covers. Wetness index one produces the lowest run-off and wetness class and ten the most

Land cover	*Wetness index class sediment yield (t ha⁻¹ yr⁻¹)*									
	One	*Two*	*Three*	*Four*	*Five*	*Six*	*Seven*	*Eight*	*Nine*	*Ten*
Pasture	1.2	3.6	3.4	3.6	3.9	5.6	8.8	10.1	12.5	14.3
Crop	2.1	2.3	3.4	3.5	4.6	5.9	10.7	9.9	14.2	15.6
Forest	0.3	0.5	0.9	1.5	1.7	1.6	2.8	3.1	3.7	4.1

Interestingly, the SAWT-WB model predicted that landscape-based erosion forms agricultural areas, particularly tilled fields in the lower slope positions which dominated sediment delivery to the river reaches during the early part of the growing season (approximately mid-end August), after which landscape-based erosion was predicted to decrease. The reduction in landscape-borne sediment reflects the growth stages of plants in the highlands, which in mid-late August are reasonably mature, or at least have developed a canopy and root system that effectively reduce rill and sheet erosion (Zegeye *et al.*, 2010). The reduction of sediment load can also be caused by a stable rill network (resulting in very little erosion losses) that is established once the fields in the watershed are not ploughed anymore. The plant cover is a good proxy for this phenomenon since the fields with plants are not ploughed. After the upland erosion stops, the sediment export from the various sub-basins is controlled by channel erosion and re-entrainment/resuspension of landscape sediment deposited in the river reaches in the early part of the growing season. This sediment is subsequently mobilized during the higher flows that typically peak after the sediment peak is observed (e.g. the sediment peak occurs approximately 2 weeks in July, before the flow peaks in August). This, of course, has implications for reservoir management in downstream countries in that much of the high sediment flow can pass through the reservoir during the rising limb, and the relatively cleaner flows stored during the receding limb. Nevertheless, the sheer volume of sediment exported from the Ethiopian Highlands threatens many downstream structures regardless of their operation, and clearly impacts agricultural productivity in the highlands.

Predicting spatial distribution of erosion

Predicting the spatial distribution of run-off source areas is a critical step in improving the ability to manage landscapes such as the Blue Nile to provide clean water supplies, enhance agricultural productivity and reduce the loss of valuable topsoil.

Using the validated SWAT-WB model, the predicted gradient in sediment yield within sub-basins is illustrated in the inset of Figure 7.12, where the Gumera watershed in the Lake Tana sub-basin is shown. The model predicts only a relatively small portion of the watershed to contribute the bulk of the sediment (75% of the sediment yield originates from 10% of the area) while much of the area contributes low sediment yield. The areas with high sediment yield are generally predicted to occur at the bottom of steep agricultural slopes, where sub-surface flow accumulates, and the stability of the slope is reduced from tillage and/or excessive livestock traffic. Indeed, Table 7.5 shows that these areas (higher wetness index classes or areas with higher λ values) inevitably produce substantially higher sediment yields than other areas as the latter produce higher run-off losses as well. This seems to agree with what has been observed in the basin (e.g. Tebebu *et al.*, 2010), and points towards the need to develop management strategies that incorporate the landscape position into the decision-making process. Interestingly, both pastureland and cropland in the higher wetness classes had approximately equivalent sediment losses, while the forests in these same areas had substantially lower erosive losses, likely due to the more consistent ground cover and better root system.

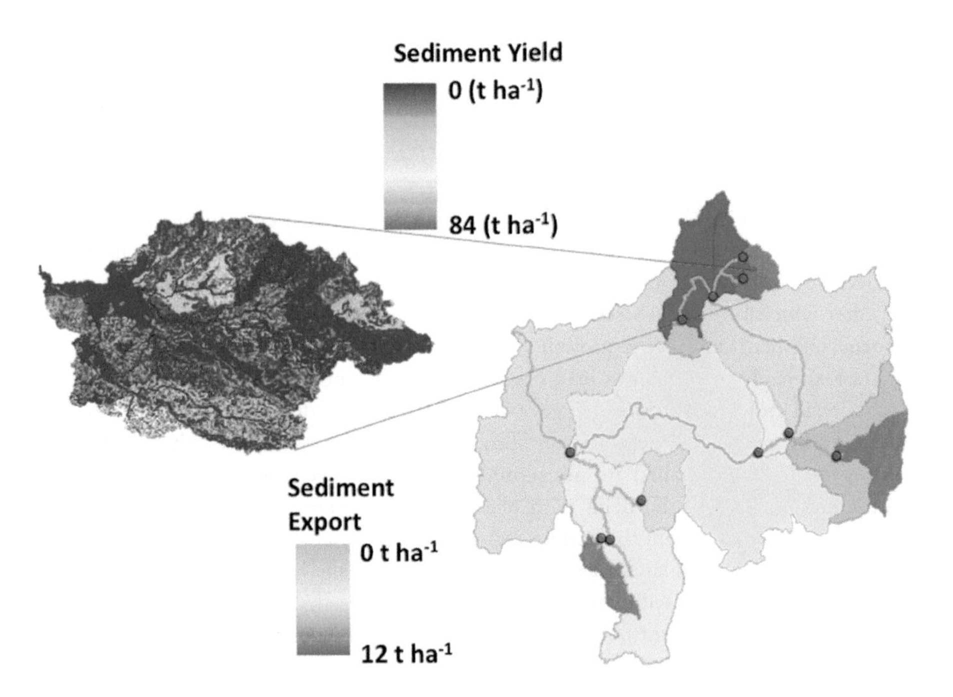

Figure 7.12 Sediment export (t ha^{-1} yr^{-1}) in the sub-basins predicted by the SWAT-WB model (main figure) and sediment yield by hydrologic response unit (HRU) for the Gumera sub-basins (inset)

Spatial distribution of sediment in the Gumera watershed (east of Lake Tana) simulated by Betrie *et al.* (2009) with the original infiltration-excess-based SWAT-CN model is shown in Figure 7.13. A comparison with the saturation-excess-based, SWAT-WB model (inset in Figure 7.12) shows that the amounts of sediment and its distribution are different. Although the sediment leaving the watershed might not be that different, the location where soil and water conservation practices have the most effect is quite different.

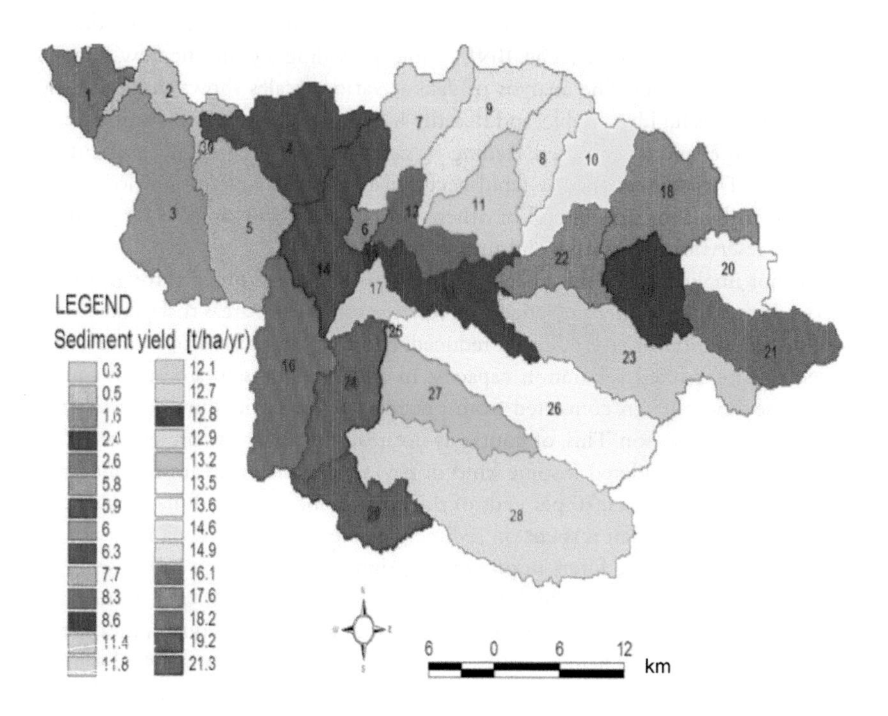

Figure 7.13 Spatial distribution of average annual sediment yield by sub-watershed (t ha^{-1} yr^{-1}) simulated using SWAT

Note: 1–29 are sub-watershed numbers in the Gumera watershed

Obviously, in addition, the erosion routines (USLE, RUSLE, MUSLE, sediment rating curves) in many of the large-scale watershed models are crude at best, and do not incorporate the appropriate mechanistic processes to reliably predict when and where erosion occurs, at least at the scale needed to manage complex landscapes. For instance, the MUSLE routine in SWAT does not predict gully erosion, which is a large component of the sediment budget in the Blue Nile. To correctly capture the integrated watershed-wide export of sediment, the original SWAT model predicts erosion to occur more or less equally across the various land covers (e.g. cropland and pastureland produce approximately equal erosive losses provided they have similar soils and land management practices throughout the basin). The modified version of SWAT used here recognizes that different areas of a basin (or landscape) produce differing run-off losses and thus differing sediment losses (Table 7.5). However, all crops or pastures within a wetness index class in the modified SWAT produce the same erosive losses, rill or sheet erosion (as predicted by MUSLE), but not the same gully erosion. Thus, rill and sheet erosion are likely over-predicted to obtain the correct sediment export from the basin.

Concluding remarks

Erosion, sediment transport and sedimentation of reservoirs are critical problems in the BNB. The current levels of erosion are causing irreversible levels of soil degradation and loss of livelihoods and are already resulting in significant costs in canal and reservoir dredging or in heightening plans of reservoirs. The BNB, while providing significant flow, also contributes substantial sediment loads. The analysis of data at various scales shows that the seasonal sediment distribution is highly variable, and that the highest sediment concentration occurs in July, when most of the land is cultivated, leading to significant loss of soil and nutrients from agricultural fields. The consequence is rapid accumulation (of sediment) and loss of capacity of small reservoirs built for agricultural or other water supplies and rapid filling of the dead storage of large reservoirs and natural and man-made lakes.

The major implication of this chapter is that erosion is distributed through the watershed. By incorporating management practices to reduce erosion from areas that generate most of the run-off, sedimentation in rivers can be reduced. Most of the erosion occurs in the areas with degraded soils or limited infiltration capacity. In addition, the saturated areas can potentially contribute sediment when converted from grazing land to agricultural land where crops are grown after the wet season. This, of course, is not realistic under the present economic conditions but could be considered if some kind of payment by downstream beneficiaries is made. According to unofficial data, 70 per cent of the cost of operation and maintenance (O&M) in the Blue Nile part of Sudan is spent on sediment-related canal maintenance.

The utility of vegetative filters in providing a significant reduction in the sediment load to the upper Blue Nile has been demonstrated in a study by Tenaw and Awulachew (2009). Application of the vegetative filter and other soil and water conservation interventions throughout the basin could help to reverse land degradation and improve the livelihoods of the people upstream, and at the same time reduce the cost of O&M of hydraulic infrastructure and sedimentation damage downstream. In order to target the critical areas requiring interventions, more fieldwork and model validation are required on the exact locations of the high erosion-risk areas.

References

Abiy, A. Z. (2009) Geological controls in the formations and expansions of gullies over hillslope hydrological processes in the Highlands of Ethiopia, northern Blue Nile region, MPS thesis, Cornell University, Ithaca, NY.

Ahmed, A. A. (2003) *Towards Improvement of Irrigation Systems Management*, AMCOW Conference, Addis Ababa, Ethiopia.

Ahmed, A. A. and Ismail, U. H. A. E. (2008) *Sediment in the Nile River System*, UNESCO/IHP/International Sediment Initiative, www.irtces.org/isi/isi_document/Sediment%20in%20the%20Nile%20River%20System.pdf, accessed 27 October 2011.

Ashagre, B. B. (2009) SWAT to identify watershed management options: Anjeni watershed, Blue Nile basin, Ethiopia, MPS thesis, Cornell University, Ithaca, NY.

Awulachew, S. B., McCartney, M., Steenhuis, T. S. and Ahmed, A. A. (2008) *A Review of Hydrology, Sediment and Water Resource Use in the Blue Nile Basin,* Working Paper 131, IWMI, Colombo, Sri Lanka.

Bashar, K. E and Khalifa, E. A. (2009) Sediment accumulation in the Roseires reservoir, in *Improved Water and Land Management in the Ethiopian Highlands: Its Impact on Downstream Stakeholders Dependent on the Blue*

Nile, Intermediate Results Dissemination Workshop held at the International Livestock Research Institute (ILRI), Addis Ababa, Ethiopia, 5–6 February, International Water Management Institute, Colombo, Sri Lanka, http://publications.iwmi.org/pdf/H042510.pdf, accessed 17 August 2011.

Bayabil, H. K., Tilahun, S., Collick, A. S., Yitaferu, B. and Steenhuis, T. S. (2010) Are run-off processes ecologically or topographically driven in the (sub)humid Ethiopian highlands? The case of the Maybar watershed, *Ecohydrology*, 3, 4, 457–466.

Betrie, G. D., Mohamed, Y. A., van Griensven, A., Popescu, I. and Mynett, A. (2009) Modeling of soil erosion and sediment transport in the Blue Nile Basin using the open model interface approach, in *Improved Water and Land Management in the Ethiopian Highlands: Its Impact on Downstream Stakeholders Dependent on the Blue Nile,* Intermediate Results Dissemination Workshop held at the International Livestock Research Institute (ILRI), Addis Ababa, Ethiopia, 5–6 February, International Water Management Institute, Colombo, Sri Lanka, http://publications.iwmi.org/pdf/H042513.pdf, accessed 17 August 2011.

Bewket, W. and Sterk, G. (2005) Dynamics in land cover and its effect on stream flow in the Chemoga watershed, Blue Nile basin, Ethiopia, *Hydrological Process,* 19, 445–458.

Capra, A., Mazzara, L. M. and Scicolone, B. (2005) Application of EGEM model to predict ephemeral gully erosion in Sicily, Italy, *Catena,* 59, 133–146.

Easton, Z. M., Fuka, D. R., White, E. D., Collick, A. S., McCartney, M., Awulachew, S. B., Ahmed, A. A. and Steenhuis, T. S. (2010) A multi basin SWAT model analysis of run-off and sedimentation in the Blue Nile, Ethiopia, *Hydrology and Earth System Sciences*, 14, 1827–1841.

FAO (Food and Agriculture Organization of the United Nations) (1986) *Highlands Reclamation Study: Ethiopia*, Final Report, vols I and II, FAO, Rome, Italy.

Guzman, C. (2010) Suspended sediment concentration and discharge relationships in the Ethiopian Highlands, MS thesis, Department of Biological and Environmental Engineering Cornell University, Ithaca, NY.

Hairsine, P. B. and Rose, C. W. (1992) Modeling water erosion due to overland flow using physical principles 1. Sheet flow, *Water Resources Research*, 28, 1, 237–243.

Haregeweyn, N. and Yohannes, F. (2003) Testing and evaluation of the agricultural non-point source pollution model (AGNPS) on Augucho catchment, western Hararghe, Ethiopia, *Agriculture Ecosystems and Environment*, 99, 1–3, 201–212.

Herweg, K. (1996) *Field Manual for Assessment of Current Erosion Damage,* Soil Conservation Research Programme (SCRP), Ethiopia and Centre for Development and Environment (CDE), University of Berne, Berne, Switzerland.

Hurni, H., Kebede, T. and Gete, Z. (2005) The implications of changes in population, land use, and land management for surface run-off in the upper Nile basin area of Ethiopia, *Mountain Research and Development*, 25, 2, 147–154.

Hydrosult Inc., Tecsult, DHV and their Associates Nile Consult, Comatex Nilotica and T and A Consulting (2006) *Trans-Boundary Analysis: Abay-Blue Nile Sub-basin*, Nile Basin Initiative-Eastern Nile Technical Regional Organization (NBI-ENTRO), Addis Ababa, Ethiopia.

Kaltenrieder, J. (2007) *Adaptation and Validation of the Universal Soil Loss Equation (USLE) for the Ethiopian-Eritrean Highlands*, Diplomarbeit der Philosophisch-Naturwissenschaftlichen Fakultät der Universität Bern, Centre for Development and Environment Geographisches www.cde.unibe.ch/CDE/pdf/Kaltenrieder%20Juliette_USLE%20MSc%20Thesis.pdf, accessed 17 August 2011.

Mitiku, H., Herweg, K. and Stillhardt, B. (2006) *Sustainable Land Management – a New Approach to Soil and Water Conservation in Ethiopia*, Land Resource Management and Environmental Protection Department, Mekelle University, Mekelle, Ethiopia, Center for Development and Environment (CDE), University of Bern and Swiss National Center of Competence in Research (NCCR) North-South, Bern, Switzerland.

Mohamed, Y. A., Bastiaanssen, W. G. M. and Savenije, H. H. G. (2004) Spatial variability of evaporation and moisture storage in the swamps of the upper Nile studied by remote sensing techniques, *Journal of Hydrology*, 289, 1–4, 145–164.

Nyssen, J., Poesen, J., Moeyerson, J. and Mitiku Haile Deckers, J. (2008) Dynamics of soil erosion rates and controlling factors in the Northern Ethiopian Highlands – towards a sediment budget, *Earth Surface Processes and Landforms*, 33, 5, 695–711.

Nyssen, J., Poesen, J., Veyret-Picot, M., Moeyersons, J., Mitiku Haile Deckers, J., Dewit, J., Naudts, J., Kassa Teka and Govers, G. (2006) Assessment of gully erosion rates through interviews and measurements: a case study from Northern Ethiopia, *Earth Surface Processes and Landforms*, 31, 2, 167–185.

Setegn, S. G., Srinivasan, R. and Dargahi, B. (2008) Hydrological modelling in the Lake Tana Basin, Ethiopia using SWAT model, *The Open Hydrology Journal*, 2, 24–40.

Tebebu, T.Y., Abiy, A. Z., Zegeye, A. D., Dahlke, H. E., Easton, Z. M., Tilahun, S. A., Collick, A. S., Kidnau, S., Moges, S., Dadgari, F. and Steenhuis, T. S. (2010) Surface and subsurface flow effect on permanent gully formation and upland erosion near Lake Tana in the Northern Highlands of Ethiopia, *Hydrology and Earth System Sciences*, 7, 5235–5265.

Tenaw, M. and Awulachew, S. B. (2009) *Soil and Water Assessment Tool (SWAT)-Based Runoff and Sediment Yield Modeling: A Case of Gumera Watershed in Lake Tana Subbasin*, http://home.agrarian.org:8080/ethiopia%20work%20frank/vertisols/ethiopian%20%20soils/H042511.pdf, accessed 26 April 2012.

Tilahun, S. A., Guzman, C. D., Zegeye, A. D., Sime, A., Collick, A. C., Rimmer, A. and Steenhuis, T. S. (2012) An efficient semi-distributed hillslope erosion model for the sub-humid Ethiopian Highlands, *Hydrological Earth System Sciences Discussions*, 9, 2121–2155.

Werner, C. (1986) *Soil Conservation Experiments in the Anjeni Area*, Gojam Research Unit (Ethiopia), Institute of Geography, University of Bern, Bern, Switzerland.

Williams, J. R. (1975) *Sediment – Yield Prediction with Universal Equation Using Runoff Energy Factor*, Proceedings of the Sediment-Yield Workshop, USDA Sedimentation Laboratory, Oxford, MS.

Zegeye, A. D., Steenhuis, T. S., Blake, R. W., Kidnau, S., Collick, A. S. and Dadgari, F. (2010) Assessment of upland erosion processes and farmer perception of land conservation in Debre-Mewi watershed, near Lake Tana, Ethiopia, presented at Ecohydrology for water ecosystems and society in Ethiopia, International Symposium, 18–20 November, Addis Ababa, Ethiopia.

Zeleke, G. (2000) *Landscape Dynamics and Soil Erosion Process Modeling in the North-Western Ethiopian Highlands*, African Studies Series A 16, Geographica Bernensia, Berne, Switzerland.

8
Nile Basin farming systems and productivity

Poolad Karimi, David Molden, An Notenbaert and Don Peden

Key messages

- Farming systems in the Nile are highly variable in terms of size, distribution and characteristics. The most prevailing system in the Nile Basin is the pastoral system, followed by mixed crop–livestock and agro-pastoral systems, covering 45, 36 and 19 per cent, respectively, of the land area.
- While productivity in irrigated agriculture in the Nile Delta and Valley is high, productivity is low in the rest of the basin with rain-fed agriculture being the prevailing agricultural system.
- The average water productivity in the Nile Basin is US\$0.045 m^{-3}, ranging from US\$0.177 m^{-3} in the Nile Delta's irrigated farms to US\$0.007 m^{-3} in the rain-fed dry regions of Sudan.
- Water productivity variations in the basin closely follow land productivity variations; thus land productivity gains result in water productivity gains.
- While improved scheme management is key to improving productivity in low productive irrigated agriculture in Sudan (i.e. in Gezira), interventions like supplemental irrigation, rainwater harvesting and application of soil water conservation techniques can increase productivity in many rain-fed areas that receive favourable rainfall throughout the year, including Ethiopian Highlands and the great lake areas.

Introduction

Agriculture is a major livelihood strategy in the Nile Basin, sustaining tens of millions of people. It provides occupations for more than 75 per cent of the total labour force and contributes to one-third of the GDP in the basin. Enhancing agriculture could directly contribute to poverty alleviation in the region as most of the poor live in agricultural areas, and are therefore largely reliant on agriculture as their primary (and often only) source of income and living. Increased agricultural production can also be effective to reduce the cost of living for both rural and urban poor through reduced food prices (OECD, 2006).

Basin-wide agricultural development and management of water resources on which production depends require an appropriate understanding of the environmental characteristics, farmers' socio-economic assets, and the spatial and temporal variability of resources. Exposure to risk, institutional and policy environments, and conventional livelihood strategies all vary over space and time. Hence, it is difficult to design intervention options that properly address all these different circumstances (Notenbaert, 2009). Therefore, agricultural development should take a farming systems approach aimed at delivering suites of institutional, technological and policy strategies that are well targeted to heterogeneous landscapes and diverse biophysical and socioeconomic contexts where agricultural production occurs (Pender, 2006).

One major constraint that agricultural development faces in the Nile Basin is water scarcity, in terms of both physical water scarcity and economic water scarcity. In areas with physical water scarcity – arid and semi-arid areas – the agriculture sector competes for water with domestic and industrial sectors, and it is likely that water allocation for agriculture will decrease as the population grows (Ahmad *et al.*, 2009). In areas with economic water scarcity, investments in water storage and control systems will increase water availability; nonetheless, polices are needed to ensure that water is used wisely (de Fraiture *et al.*, 2010). This requires agricultural development strategies to aim for more productive use of water and to maximize the profit gained from the water consumed.

This chapter describes major Nile farming systems that are sometimes referred to as agricultural production systems. It introduces the concept of agricultural water productivity (WP) and provides an overview of crop WP across the Nile Basin (livestock WP is addressed in Chapter 9). Then we will briefly present several case studies on agricultural production from across the Nile Basin.

Farming systems classifications for the Nile Basin

A farming system can be defined as a group of farms with similar structure, production and livelihood strategies, such that individual farms are likely to share relatively similar production functions (Dixon *et al.*, 2001). The advantage of classifying farming systems is that, as a group of farms and adjacent landscapes, each operates in a relatively homogeneous environment compared with other basin farming systems. This provides a useful scheme for the description and analysis of crop and livestock development opportunities and constraints (Otte and Chilonda, 2002). A farming systems approach facilitates spatial targeting of development interventions including those related to water management and offers a spatial framework for designing and implementing proactive, more focused and sustainable development and agricultural policies.

Farming systems classification for this study was performed based on a classification described by Seré and Steinfeld (1996). For the purpose of distinguishing the degree of agricultural intensification and industrialization, and inclusion of spatial variability of dominant crops in mixed farming systems, we integrated global crop data layers from the Spatial Allocation Model (SPAM) data set (You *et al.*, 2009) with the Seré and Steinfeld classification. Crops were assigned to four crop types: cereals, legumes, root crops, and tree crops (Table 8.1). In some cases, one specific crop group dominates the landscape by covering at least 60 per cent of the land area. In other cases, cropping patterns are more diverse with two or more crops combined covering at least 60 per cent of the land area. The combination of both layers enabled the creation of a new hierarchical systems classification that gives a clearer indication of the main crop types grown. Pastoral, agro-pastoral, urban and peri-urban areas were also differentiated. For the purpose of this chapter, we excluded any indication of agro-ecology because of the trade-off between clarity, readability and the variety of criteria included.

Table 8.1 Crop group classification for mapping Nile Basin farming systems

Broad farming system classes		*Major Nile crops*
Where >60% of production	*Where <60% of production*	
Cereals	Cereals+	Maize, millet, sorghum, rice, barley, wheat, teff
Legumes	Legumes+	Bean, cowpea, soybean, groundnut
Root crops	Root crops+	Cassava, (sweet) potato, yam
Tree crops	Tree crops+	Coffee, cotton, oil palm, banana

Note: Forage crops and sugar cane were excluded. The + symbol indicates that the crop is mixed with other commodities

The resultant modification of the Seré and Steinfeld (1996) farming systems classification for the Nile is shown in Figure 8.1, which includes two levels. The first retains division of the land area into grazing-based farming systems, mixed rain-fed crop-livestock systems, and mixed irrigated crop-livestock systems. Although conceptualizing irrigated areas as mixed crop-livestock systems is counterintuitive, Africa's highest livestock densities are associated with irrigation (Chapter 9). The second level splits mixed crop-livestock systems into eight sub-criteria based on type of crop (cereals, tree crops, root crops, and legumes) and the degree of dominance of each crop type. For example, in Figure 8.1, 'cereals' implies that cereals make up at least 60 per cent of farm production whereas 'cereals+' indicates that cereals are most common but are mixed with other important commodities.

The degree of intensification in major farming systems in the Nile is shown in Figure 8.2. Agricultural potential and market access were two criteria that we used in order to assess intensification potential in the existing farming systems. Areas with high agricultural potential were defined as irrigated areas and areas with length of growing period of more than 180 days per year. Good market access was defined using the time required to travel to the nearest city with a population of 250,000 or more. We applied a threshold of 8 hours for travel. According to the results besides the Nile Delta, Nile Valley, and irrigated areas in Sudan, areas around Lake Victoria have high potential for agricultural development.

The Nile's farming systems vary greatly in size, distribution and characteristics. Mixed crop-livestock, agro-pastoral and pastoral systems occupy about 36, 19 and 45 per cent, respectively, of the land area (2.85 million km^2) of the basin excluding urban, peri-urban and other land uses. The mixed crop-livestock systems are composed of large-scale irrigation (28,000 km^2) and rain-fed cultivation and pasture (1.0 million km^2.). These farming systems are also home to a population of about 160 million, with 139 million living in the mixed rain-fed systems and with the large-scale irrigation systems having the highest densities of about 1681 persons per km^2.

Numerous biophysical constraints to farm production, particularly in densely populated areas, potentially limit agricultural production. About 50, 33, 28 and 9 per cent of the mixed irrigated, mixed rain-fed, agro-pastoral, and pastoral systems, respectively, are degraded. Aluminium toxicity, high leaching potential and low nutrient reserves are especially acute in mixed rain-fed systems while salinity and poor drainage are problematic in some irrigated areas.

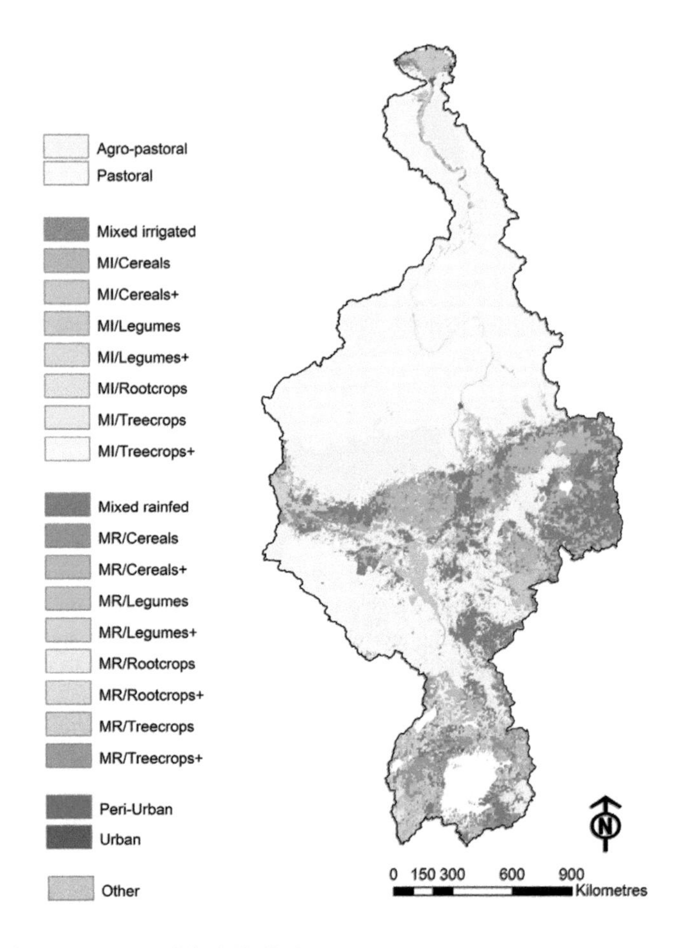

Figure 8.1 Farming system map of the Nile Basin

Agricultural productivity in the Nile Basin

Land productivity

Land productivity is the ratio of farm output per unit of land cultivated. Figure 8.3 shows land productivity of sorghum and maize. These two major Nile Basin crops serve as proxies for a wide range of water-dependent food crops. Sorghum and maize cover 20 per cent (8 million ha) and 10 per cent (4 million ha), respectively, of the cropped area in the basin. Well over 90 per cent is produced through rain-fed cultivation, particularly in the mixed rain-fed crop-livestock farming systems. The average land productivity of sorghum in the rain-fed system in the Nile is about 0.64 tonnes (t) ha^{-1}, ranging from 2 t·ha^{-1} in the southeastern part of the basin, Tanzania, where annual rainfall is about 1000 mm, to less than 0.2 t ha^{-1} in the dry regions of Sudan. Irrigated sorghum is cultivated in parts of Egypt and some Sudanese states namely White Nile, Sennar, Kassala, and Gadaref. The average land productivity of irrigated sorghum is about 3.1 t ha^{-1} and ranges from 6.3 t ha^{-1} in the Asyiut State in Egypt to 1.2 t ha^{-1}) in the Blue Nile State, Sudan. The average yield of rain-fed maize in the basin is near 1.3 t ha^{-1},

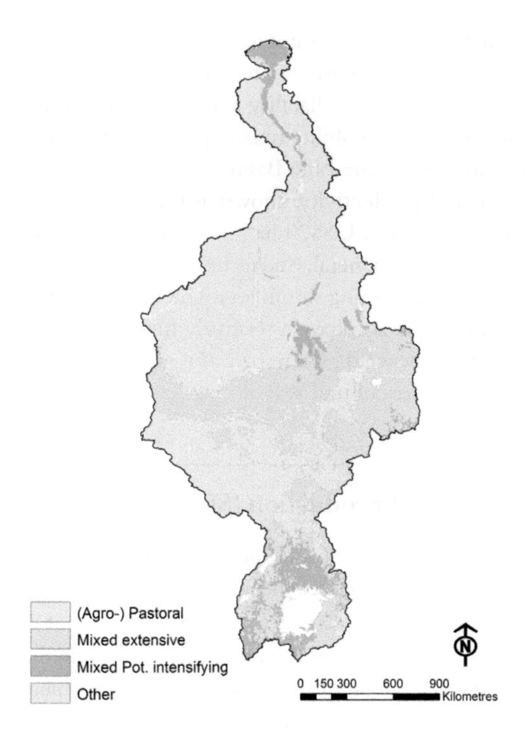

Figure 8.2 The degree of intensification in the Nile Basin

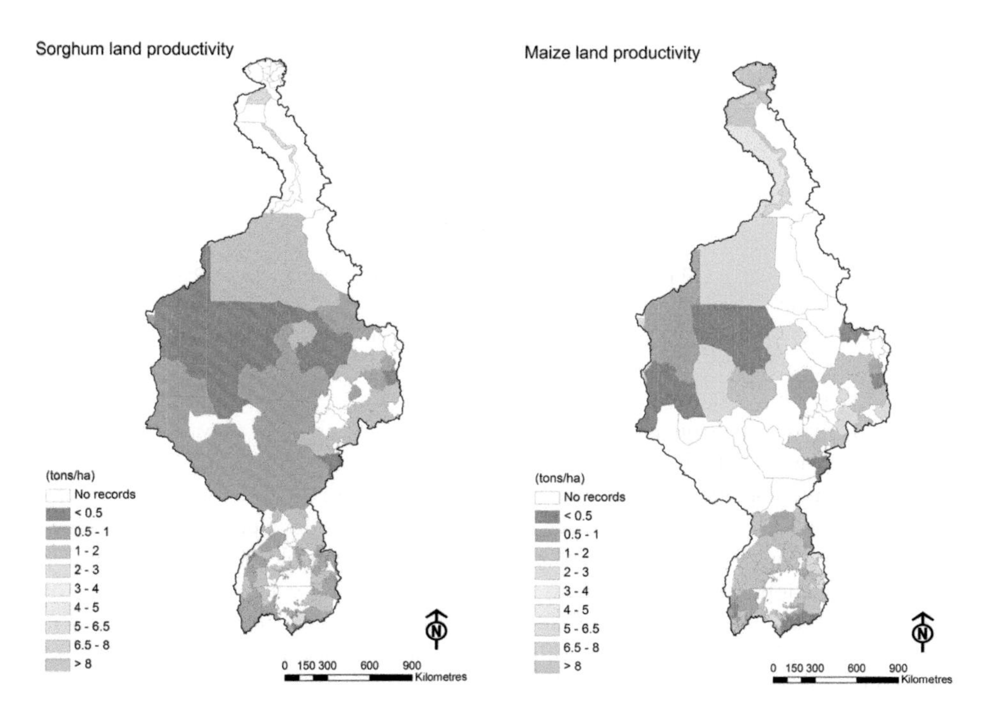

Figure 8.3 Land productivity of (a) Sorghum and (b) maize in the Nile Basin

ranging from 2.7 t ha^{-1} in East Wellega, Ethiopia, to less than 0.3 t ha^{-1} in southern Darfur, Sudan. Irrigated maize production averages 8.3 t ha^{-1} in Egypt. The huge gap between irrigated and rain-fed yields suggests that water availability and access are key constraints to maize and sorghum production. Similar spatial variability in land productivity characterizes about 70 crops commonly found in various parts of the Nile Basin.

The economic value of land productivity, known as the *standardized gross value of production* (SGVP), in the Nile Basin varies from US\$20 ha^{-1} in some Sudanese states to more than US\$ 1832 ha^{-1} in Egypt (Figure 8.4). In general, Sudan has the lowest land productivity except in states like Gezira where irrigated farming dominates. The densely populated highland areas of Ethiopia and the great lakes region also have a relatively high SGVP. Low land productivity in many areas suggests that significant yield gaps remain (Figure 8.4). One major factor contributing to gaps in crop yield is low agricultural WP.

Standardized gross value of production (SGVP)

Different pricing systems and local market fluctuations complicate efforts to estimate the total value of agricultural goods and services in large transboundary river basins. One way to overcome this challenge is the use of an index, the SGVP, which enables comparison of the economic value of mixtures of different crops regardless of the country or location where they are produced. This index converts values of different crops into equivalent values of a dominant crop and uses the international price of a dominant crop to evaluate the gross value of production. For the Nile River Basin, wheat was chosen as the base crop. About 70 other crops were pegged to the 'wheat standard' by assessing the price gaps between each of them and wheat in each country. The International price of wheat (US\$ t^{-1}) from 1990 to 2005 was used as the standard value against which other crops were pegged. For details, refer to Molden *et al.*, 1998.

Crop water productivity

Large gaps between actual and potential crop yields reflect the presence of socio-environmental conditions that limit production. In much of the Nile, lack of farmers' access to available water is the prime constraint to crop production. With increasing numbers of people and their growing demand for food, combined with little opportunity to access new water sources, great need exists to make more productive use of agricultural water.

WP is the ratio of benefits produced, such as yield, to the amount of water required to produce those benefits (Molden *et al.*, 2010). WP varies greatly among crop types and according to the specific conditions under which they are grown. WP can be estimated at scales ranging from pots, to fields, to the watershed, and to river basins. The typical unit of measurement for single crops is kg m^{-3} (e.g. Qureshi *et al.*, 2010). At larger scales WP estimates need to include multiple crops, and monetary units such as US dollars per cubic metre are used. The WP index serves as a useful indicator of the performance of rain-fed and irrigated farming in water-scarce areas. It can further help with planning water allocation among different uses while ensuring water availability for agro-ecosystem functioning (Loeve *et al.*, 2004; Molden *et al.*, 2007).

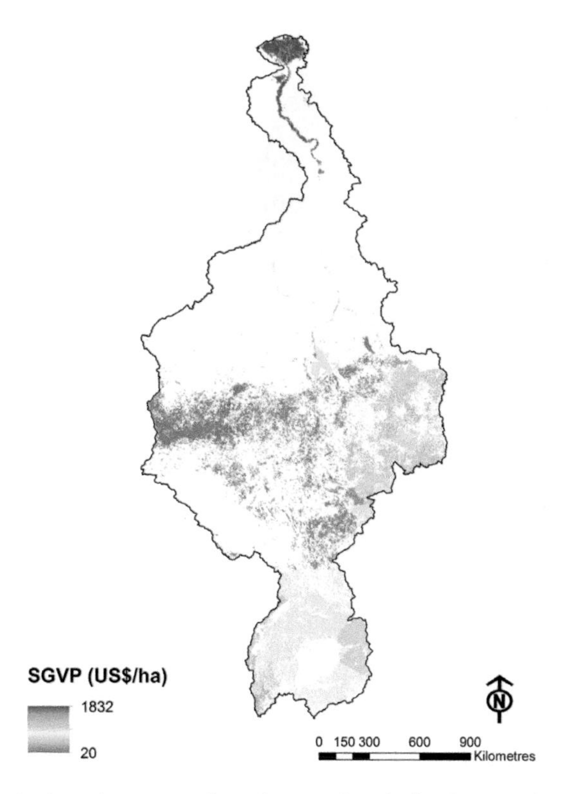

Figure 8.4 Economic land productivity in the Nile Basin (standardized gross value of production per hectare)

Mapping WP across the Nile Basin enables understanding of spatial distribution of effectiveness of water use. To assess consumptive water use of crops in the Nile Basin we used actual evapotranspiration (ETa) data produced by WaterWatch. Variation in the ETa across the basin is high. It ranges from 8 mm yr^{-1} in the desert to nearly 2460 mm yr^{-1} from free water surfaces at the Lake Nasser (Figure 8.5). Except for the Nile Delta, irrigated agriculture covers a very small fraction of the land in the Nile Basin. Therefore, ETa is chiefly a result of natural processes and is driven by the availability of water. The pattern and variation in the ETa map, thus, can represent the general water availability pattern, although areas along the river and the delta are exceptions to this rule. From this point of view, the map depicts that water availability is relatively high in the southern part of the basin and, as we move to north, water becomes scarce and vegetation becomes possible only close to the river.

SGVP and ETa were calculated to estimate crop WP across the Nile Basin (Figure 8.6), which is US$0.045 m^{-3}, and the minimum, maximum, and standard deviation of WP are US$0.007, US$0.177 and US$0.039 m^{-3}, respectively. As in land productivity, WP shows a huge variation across the basin.

Based on WP, spatial distribution of the basin can be divided into three zones: the *high productivity zone*, the *average productivity zone* and the *low productivity zone*.

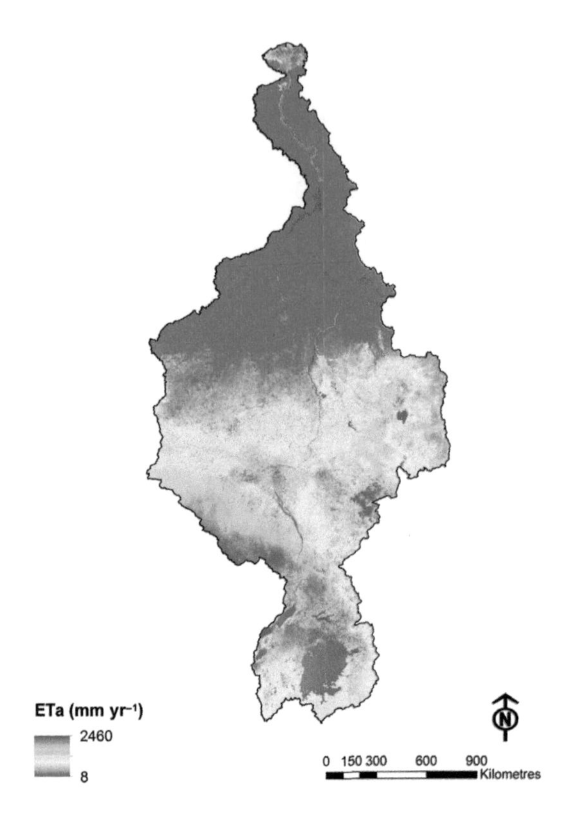

Figure 8.5 Actual evapotranspiration (ETa) in the Nile Basin in 2007

High productivity zone

The high productivity zone includes the delta and irrigated areas along the Nile River in the northern part of the basin. This zone is characterized by intensive irrigation, high yields and high-value crops. These characteristics collaboratively contribute to the high level of the WP attained and are in fact correlated. Access to irrigation results in higher yields; higher yield results in higher incomes; and higher incomes result in higher investment in farm inputs by farmers. Furthermore, access to irrigation and higher income make it possible for farmers to afford growing high-value crops that often have higher risk and require better water management. Further improvement in already high lands and WP might be possible using a higher rate of fertilizer application or adaption of new technologies but the environmental and economic cost might prove to be too high to make it a feasible option for future plans. However, interventions like supporting cropping rotations that produce higher economic returns and promoting aquaculture mixed with crops might be viable options for investment to gain more benefits from water and eventually increase overall productivity of water.

Average productivity zone

The average productivity zone consists of two major areas, one in the eastern part (Ethiopia mainly) and the other in the southern part (areas around the Lake Victoria). Despite the fact that

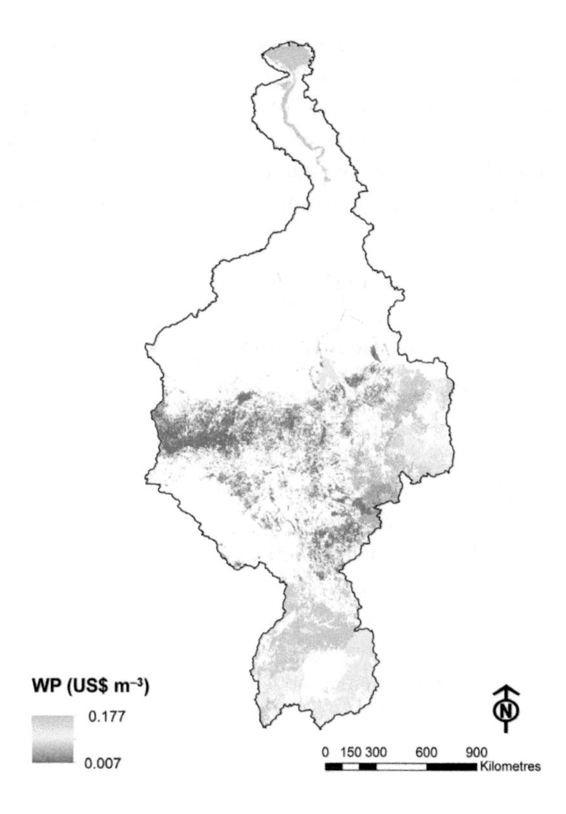

Figure 8.6 Crop water productivity in the Nile Basin

most of the areas in this zone receive relatively good amounts of rainfall, the predominantly rain-fed agriculture has rather low yields and, therefore, relatively low WP. This indicates poor farm water management practices and farmer's financial inability to invest in on-farm inputs like fertilizer, good-quality seeds, etc. The fact that rainfall is sufficient to grow crops in this zone opens a wide prospect for improvement in this region. Two parallel strategies that could be applied are, first, improving farm water management and, second, promoting irrigated agriculture. Common methods to enhance farm water management are supplemental irrigation (wherever possible), rainwater harvesting and application of soil water conservation techniques. These methods have proved to be effective in many parts of the world and helped to gain significantly more yields. Promoting irrigated agriculture, however, requires investment in water control and storage infrastructure. The main obstacle for irrigated agriculture in this zone is accessibility to water rather than its availability. For example, in Ethiopia, due to lack of storage infrastructure the majority of generated run-off leaves the country without being utilized. Controlling these flows and diverting the water to farms can drastically improve both land and water productivity.

Low productivity zone

The low productivity zone covers the central and western part of the basin. Agriculture in this zone is rain-fed and it receives a low amount of rainfall. In most areas rainfall amounts received cannot meet the crop water demands and therefore crops suffer from high water stress. As a result, yields are extremely low. In this zone improving water and land productivity is contingent upon expanding irrigated agriculture. A good example that shows how irrigation can bring in improvements is the Gezira scheme in Sudan. This scheme is located in the same zone (geographically) but irrigation has resulted in significantly higher WP in the scheme compared to its surrounding rain-fed areas. However, due to poor water management, WP in the Gezira scheme is much lower than in irrigated areas in northern parts of the basin (i.e. in the delta).

Irrigated agriculture

The Gezira scheme, Sudan

The Gezira scheme is one of the largest irrigation schemes in the world. It is located between the Blue and White Nile in the south of Khartoum (Figure 8.7). The area has an arid and hot climate with low annual rainfall, nearly 400 mm yr^{-1} in the southern part to 200 mm yr^{-1} in the northern part near Khartoum. The area of the scheme is about 880,000 ha, and represents more than 50 per cent of irrigated agriculture in Sudan. It produces about two-thirds of Sudan's cotton exports, and considerable volumes of food crops and livestock for export and domestic consumption, thereby generating and saving significant foreign exchange. The scheme is of crucial importance for Sudan's national food security and generates livelihoods for the 2.7 million inhabitants of the command area of the scheme (Seleshi *et al.*, 2010). The Sennar Dam, located at the southern end of the scheme, supplies water to Gezira through a network of irrigation canals of about 150,000 km (Plusquellec, 1990).

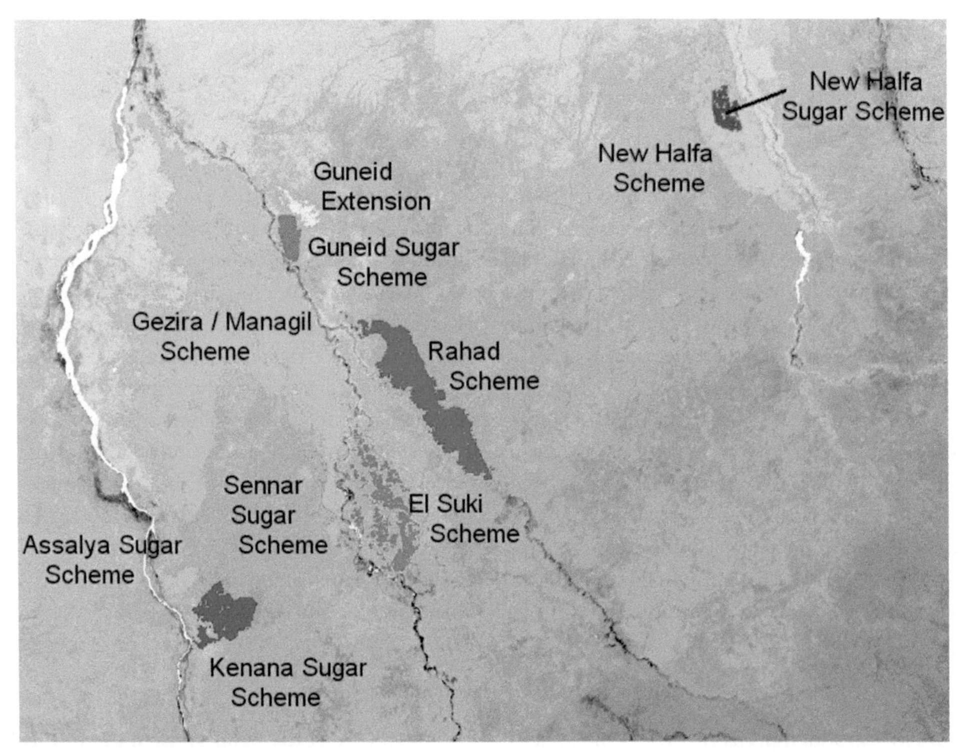

Figure 8.7 Major irrigation schemes in Sudan

Source: WaterWatch, 2009

The main crops in Gezira are cotton, sorghum, groundnut, wheat and vegetables. Yields and cropping intensities are rather low and unstable, irrigation management is poor, and operation and maintenance are organized in a highly centralized way, as is cotton production and marketing (Laki, 1993; Al-Feel and Al-Bashir, 2012; Mahir and Abdelaziz, 2011; Yasir *et al.*, 2011). Cotton was a mandatory crop for farmers, and was financed and marketed by the government before introducing liberalization of choice of crop in 1981. After adoption of the liberalization policy in the agriculture sector, farmers started to grow other crops, such as sorghum, wheat and groundnut. As a result, the cotton area and production decreased (Gamal, 2009). However, despite the financial benefits of growing multiple crops for farmers, diversifying from cotton has implications on foreign exchange acquisitions by the government of Sudan (Guvele, 2001).

Figure 8.8a shows actual annual evapotranspiration in the Gezira scheme in 2007. Total water consumption in the scheme and its surrounding extensions is about 9.3 billion m^3 yr^{-1}, with an average ETa of 830 mm yr^{-1}. ETa shows a huge variation across the scheme, ranging from 150 to 1700 mm yr^{-1}, which shows water is poorly distributed. Evidently areas in the head end receive too much of water whereas areas in the tail end receive very little water. Therefore, ETa is generally considerably low in the northern part while some areas in the south have extremely high ETa for which a possible explanation could be the waterlogging issue.

Comparison of actual transpiration (Ta) and potential transpiration (Tp) is an indicator for assessing performance of crops. High Ta Tp^{-1} ratio indicates good performance, while a low ratio is a sign of low performance because biomass production and subsequently food production have a close to linear relation with crop transpiration (Howell, 1990). This ratio is, in fact, suggested to also have a proportional relation with the ratio of actual yields to potential yields (de Wit, 1958; Hanks, 1974). Figure 8.8b depicts Ta Tp^{-1} values in the scheme. As is evident from the figure, crop performance is generally very low. The average Ta Tp^{-1} ratio in Gezira is about 0.5, and ranges from 0.1 to 0.85. This high variation is mainly attributed to poor scheme management and extremely uneven water distribution. In effect, except for some areas near the head end, the rest of the scheme suffers from high water stress.

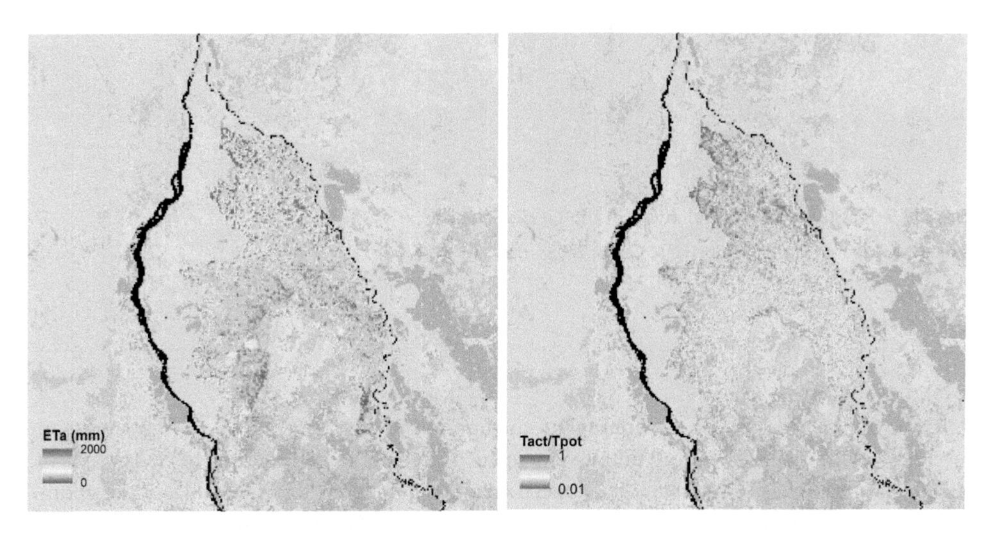

Figure 8.8 (a) Annual actual evapotranspiration (Eta) and (b) ratio of actual to potential transpiration (Ta Tp^{-1}) in the Gezira scheme in 2007

Source: Background image is Globe Land Cover (2008, http://postel.mediasfrance.org/en/PROJECTS/Preoperational-GMES/GLOBCOVER)

To gain an insight into WP variation in the scheme, it was estimated based on produced biomass and crops consumptive water use. The results, then, were presented in a relative term that offers a basis to compare it within the scheme. As illustrated in Figure 8.9, in general, WP in the Gezira scheme is uniformly low and the variation does not follow the same pattern as that in actual evapotranspiration and Ta Tp^{-1} ratio. There is no significant difference in WP in the head and tail ends of the scheme, although higher WP pixels, to some extent, are more prevalent in the tail ends than in the head ends. This shows that some areas in the head ends, despite having relatively higher yields (higher Ta Tp^{-1}) have low WP, the which indicates excessive evaporation as a result of poor water management.

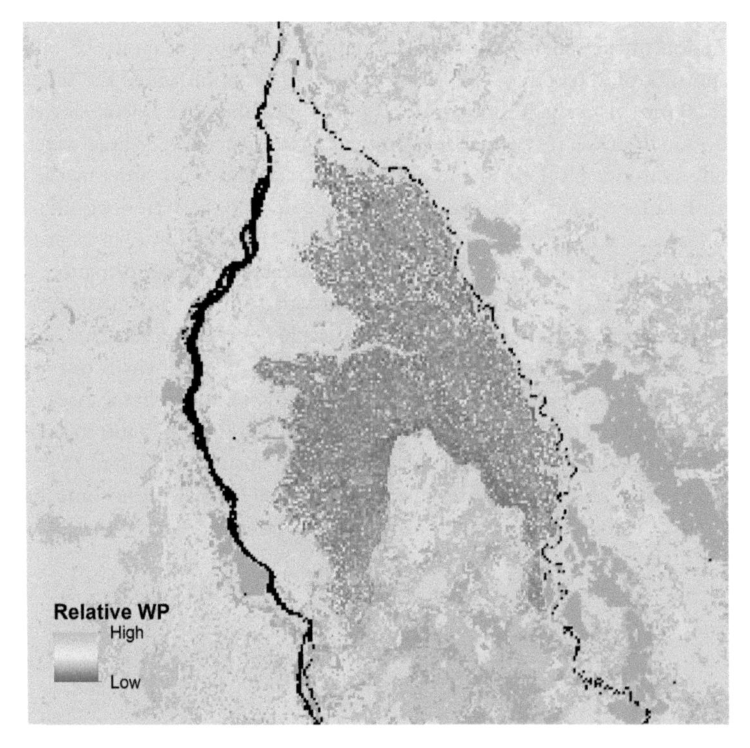

Figure 8.9 Relative water productivity in the Gezira scheme

Source: Background image is Globe Land Cover, 2008

Opportunities to increase agricultural production in most areas of Sudan are limited due to severe water shortage. Therefore, improvement in managing available water in Sudan and in already existing irrigation schemes is a crucial factor to cope with food demands of the country's growing population at present and in the future. In the Gezira scheme, low performance is a direct consequence of poor management rather than of problems with water availability as the water supply appears to be adequate across the basin regardless of the location (Yasir *et al.*, 2011). Hence, agricultural policies have to target improving the scheme management to enhance scheme performance that will subsequently increase WP.

Irrigated Egypt

The Nile is a lifeline for Egypt, its population and its almost entirely irrigated agriculture. Agricultural activities provide employment for 35 per cent of the labour force and contribute to 13.5 per cent of the country's GDP. The Nile River is the main source of water for Egypt, providing 55.5 billion of its 58.3 billion m^3 total actual water resources, out of which 85 per cent is committed to irrigating 3.42 million ha of cropped lands. The Nile Valley and Delta are the main agricultural areas of Egypt encompassing 85 per cent of the total irrigated area of 2.9 million ha (Figure 8.10a). The main cultivated crops are wheat, rice, clover and maize. Crop intensity is high and, in most of the areas, a double-cropping system is a common practice. Land productivity is also high in Egypt with the average yields of some crops in the country being among the highest in the world.

The Nile Delta covers two-thirds of the total irrigated agriculture (Stanley, 1996) and is the food basket of Egypt (Figure 8.10b). Figure 8.11a shows crops actual evapotranspiration in the Nile Delta in 2007. ETa in most areas across the delta is high with an average of 1200 mm yr^{-1}. Lower ETa at the areas close to the edge of the delta could be because these areas receive less water and have a lower crop intensity. As we move towards the centre, crop intensity grows, and so does ETa. Actual transpiration (Ta) is very close to potential, with an average Tp Tp^{-1} ratio of 0.85 (Figure 8.11b). This indicates overall high performance of irrigated agriculture in the Delta, which is also reflected in its high relative water productivity (Figure 8.11c).

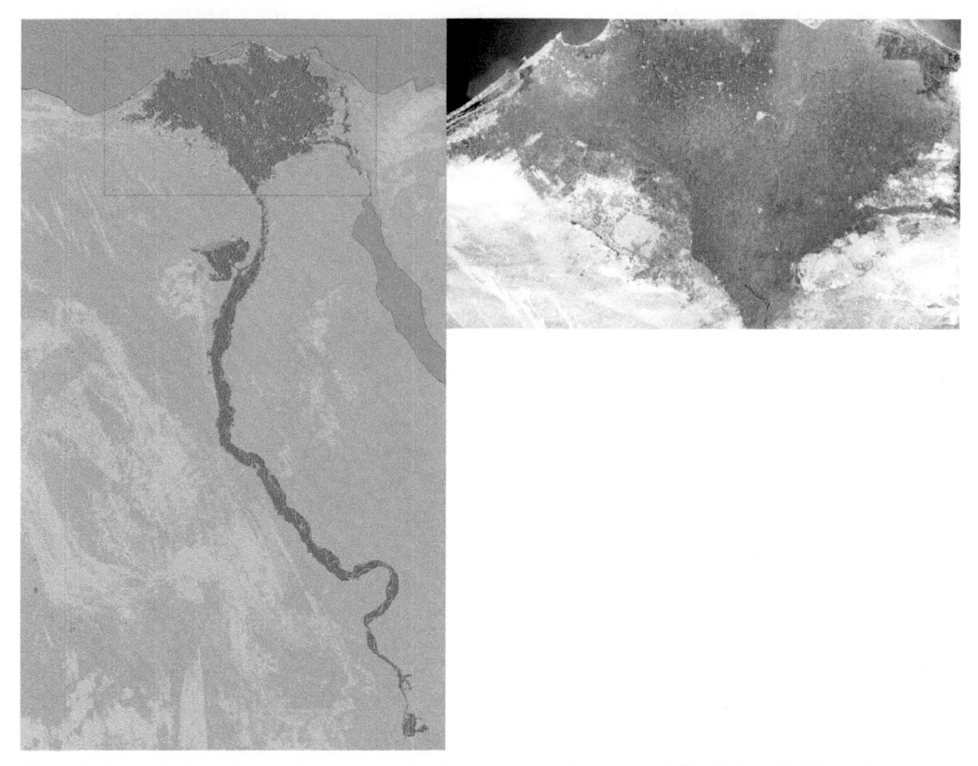

Figure 8.10 (a) Irrigated agriculture along the Nile river banks and the Nile Delta; (b) false colour composite image of the Nile Delta based on Landsat thematic mapper measurements

Note: Red colour characterizes vigorous crop growth

Source: WaterWatch, 2009

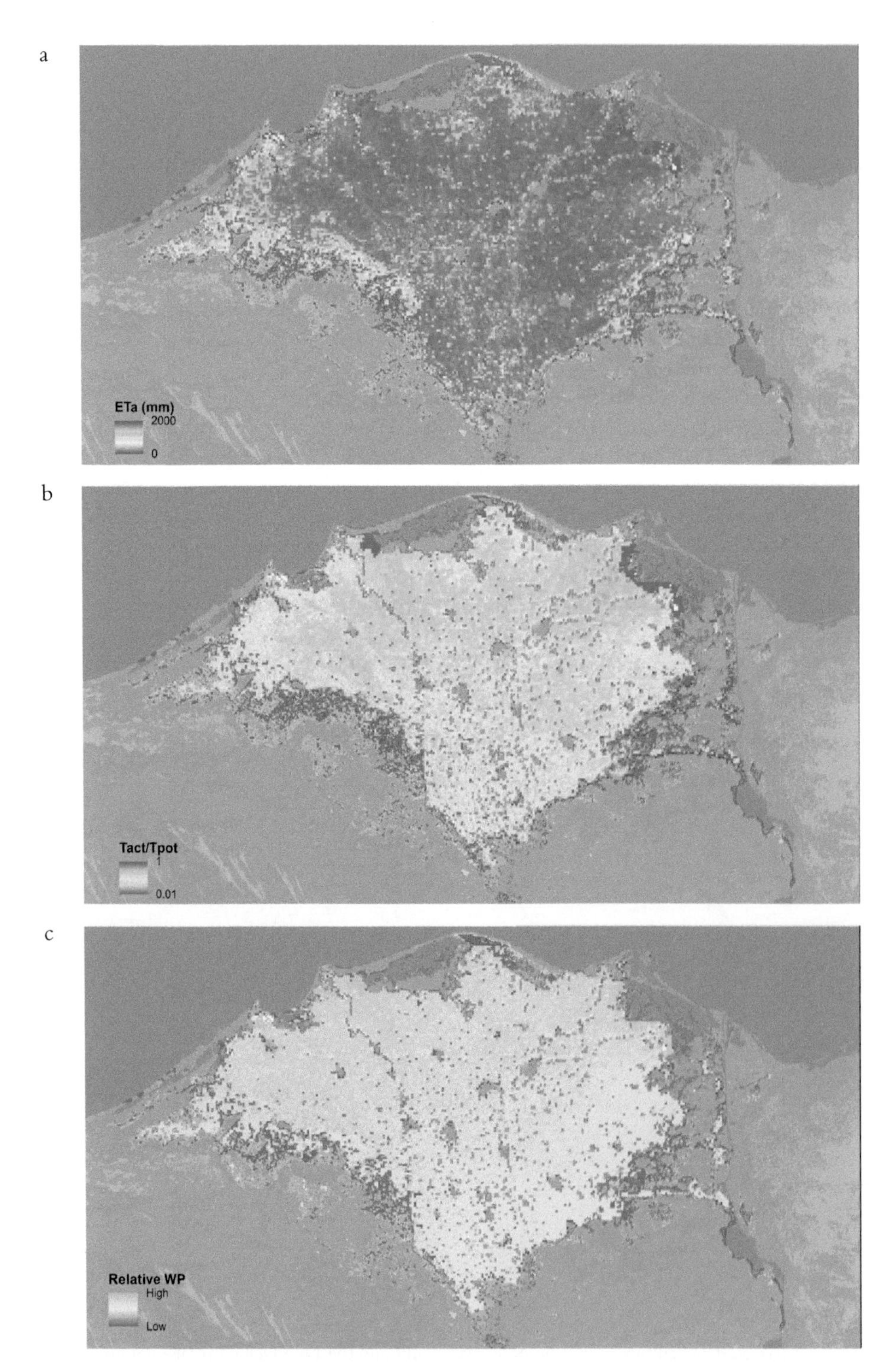

Figure 8.11 (a) Annual actual evapotranspiration (Eta) and (b) ratio of actual to potential transpiration (Ta Tp–1); (c) relative water productivity in the Nile Delta in 2007

Source: Background image is Globe Land Cover, 2008

Despite the current high performance of irrigated agriculture, coping with water scarcity remains a challenge for any recent and planned agricultural expansions in Egypt. Therefore, maximizing physical and economic crop WP plays a vital role in drawing future sustainable agricultural development. Increasing economic WP can be achieved through enhancing cropping patterns and promoting high-value crops. Institutional bodies like agricultural extension offices and water user associations should play a more active role to provide farmers with the necessary information about financially rewarding crop rotations and individual crops, and coordinate with the farmers to cultivate the most profitable crops for different seasons and areas.

Rain-fed agriculture

Rain-fed farming in the Nile Basin

Rain-fed farming, covering 33.2 Mha, is the dominant agricultural system in the Nile Basin. Over 70 per cent of the basin population depend on rain-fed agriculture (Seleshi *et al.*, 2010). Sudan, with 14.7 Mha, accounts for 45 per cent of the total rain-fed lands, followed by Uganda, Ethiopia, Tanzania, Kenya, Rwanda and Burundi (Figure 8.12). Low rainfall does not allow rain-fed farming in Egypt, and rain-fed areas of Eritrea that fall within the Nile boundary are almost negligible.

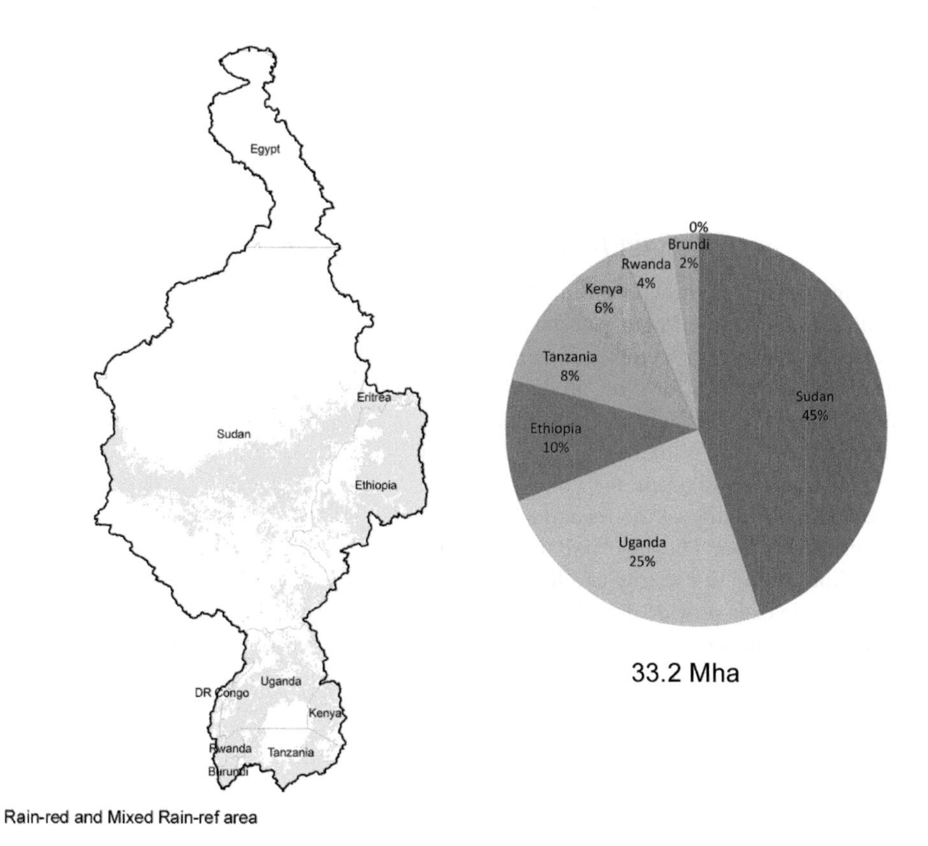

Figure 8.12 Distribution of rain-fed agriculture in the Nile Basin

The main rain-fed crop in the Nile Basin in terms of cultivated area is sorghum, followed by sesame, maize, pulses and millet, covering 7.39, 3.68, 3.35, 2.94 and 2.86 Mha, respectively (Table 8.2). Rain-fed agriculture in the Nile Basin is characterized by low yields with the majority of crops having an average yield of less than 1 t ha^{-1}. Different sets of reasons have been proposed for the low yields in rain-fed systems from natural causes such as poor soils and drought-prone rainfall regimes to distance from urban markets (Allan, 2009). However, the opportunity of favourable rainfall in many rain-fed areas of the basin provides a high potential for yields to increase by improved farm water management techniques such as rainwater harvesting.

Table 8.2 Rain-fed crops in the Nile Basin

	Crop area (ha)	*Yield (t ha^{-1})*
Sorghum	7,392,154	0.64
Sesame	3,688,529	0.35
Maize	3,354,597	1.43
Pulses	2,943,231	0.86
Millet	2,869,540	0.58
Groundnut	1,793,453	0.68
Sweet Potato	1,661,132	4.63
Banana	1,647,751	5.77
Other crops	7,877,708	–
Total	33,228,095	

Rain-fed farming in the Blue Nile

The farming systems of the upper Blue Nile region are categorized as mixed farming in the highland areas and pastoral/agro-pastoralism in the lowland areas. Mixed farming of cereal-based crops, teff, ensete, root crops, and coffee crops compose one system.

The major constraints for crop production are soil erosion, shortage and unreliability of rainfall, shortage of arable land, and weeds, disease and pests, which damage crops in the field; after harvest, there is also utilization of a low level of agricultural inputs (fertilizers, seed, organic matter) and shortage of oxen for cultivation. The magnitude of resource degradation in Ethiopia and the inability of the fragmented approaches to counter it are two key challenges reinforcing each other. The highland mixed farming systems are characterized by varying degrees of integration of the crop and livestock components. Crop residues often provide livestock feed, while oxen provide draught power, and cattle can provide manure for improvement of soil fertility. With increasing population pressure, there is increasing competition for land between crops and grazing, which often goes in favour of the crops. As grazing land is converted to cropland, the importance of crop residues as livestock feed also increases. There is a need for sustainable land management. Resource degradation is the most critical environmental problem in the Ethiopian Highlands (Woldeamlak, 2003).

Figure 8.13 shows crops ETa, gross value of production (GVP) and WP in the Ethiopian part of the Nile. Average crop water consumption is about 450 mm. GVP ranges from US$286 ha^{-1} in Zone 2 to US$823 ha^{-1} in Shaka, where high-value crops like coffee and fruit trees are

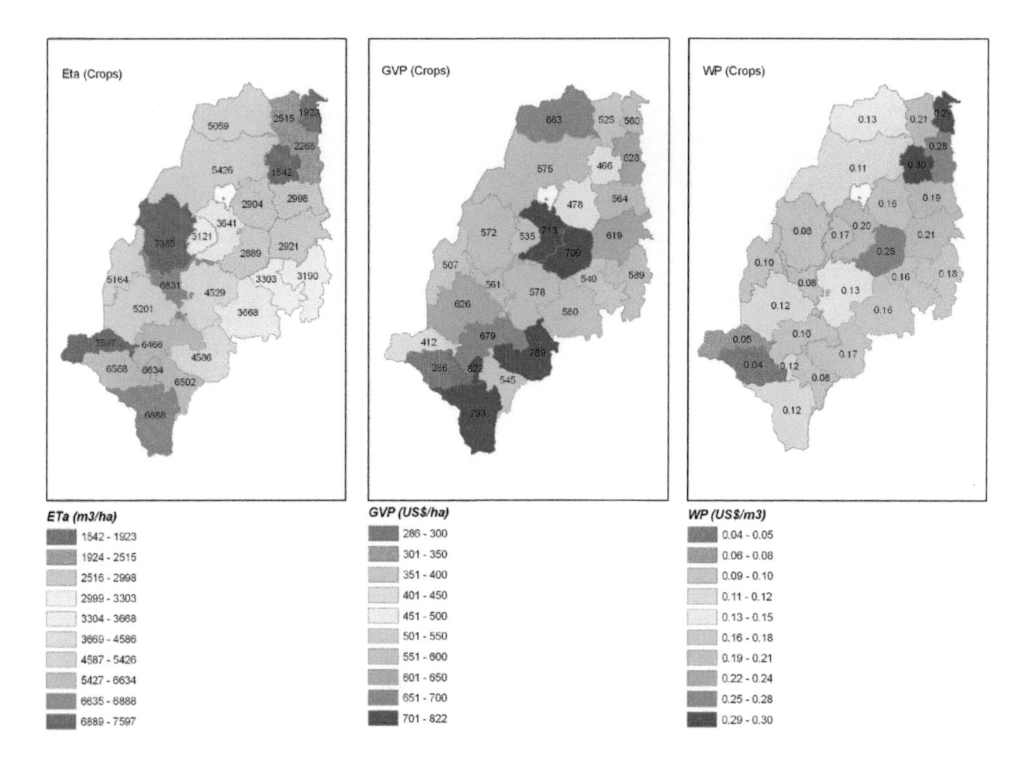

Figure 8.13 Evapotranspiration, gross value of production and water producitivity maps of the Ethiopian part of the Nile

cultivated. Average crop WP in the area is about US$0.16 ha^{-1}, ranging from US$0.04 to US$0.3 ha^{-1}. Zone 2 has the lowest WP, mainly due to low land productivity, despite high water availability in the region. Generally, WP increases toward east, due to cultivation of high-value crops and the existence of irrigated farms.

Overview of the Nile Basin fisheries and aquaculture

Fisheries

Fisheries and aquaculture are an important component of agricultural production and productivity in the Nile. Nile Basin fisheries are mainly freshwater lakes, rivers and marsh sources and human-derived aquaculture. Freshwater fisheries have a large potential to enhance income opportunities for many thousands of people and contribute towards food and nutritional security of millions in Kenya, southern Sudan, Tanzania and Uganda. Figure 8.14 summarizes information on growth and the share of countries and major water bodies in inland fisheries production in the Nile Basin. Here we give an overview of fisheries and aquaculture, but further work is necessary to integrate these into the overall WP of the basin.

Lake Victoria, shared among Kenya, Tanzania and Uganda, produces up to a million tonnes of fish a year. The fishery generated about US$600 million a year in 2006 (LVFO, 2006). Lake

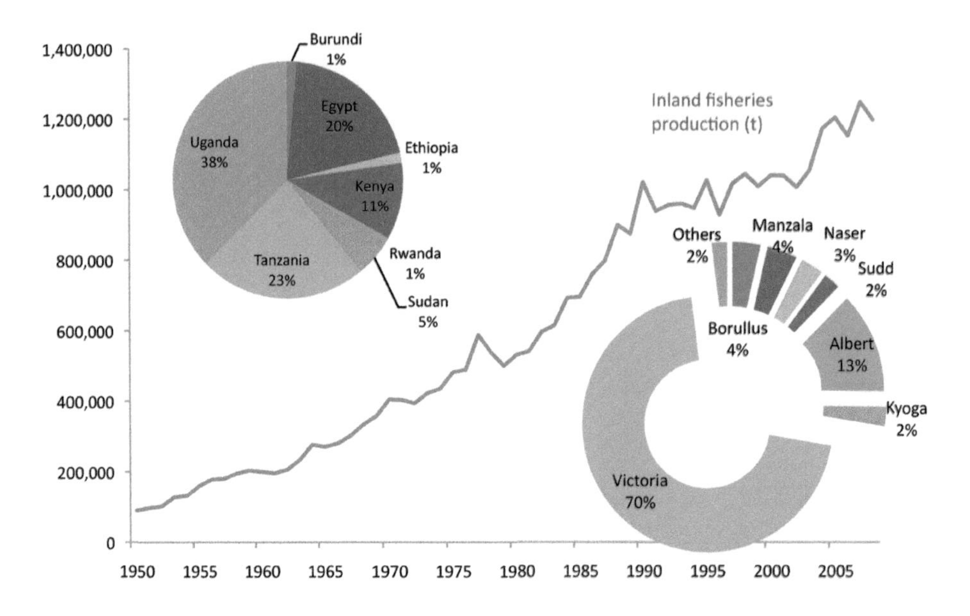

Figure 8.14 Total inland fisheries production in the Nile (excluding Democratic Republic of Congo, in which most of the fishers production takes place outside of the Nile Basin)

Sources: modified from FAO data; Witte *et al.*, 2009

conditions and unsustainable fishing practices have affected the harvest of fresh fish, which has decreased by 40 per cent. New nets and hooks have helped, but still many remove small fish and the stocks are depleted.

The lake basin is used as a source of food, energy, drinking and irrigation water, shelter, transport, and as a repository for human, agricultural and industrial waste. With the populations of the riparian communities growing at rates among the highest in the world, the multiple activities in the lake basin have increasingly come into conflict. The lake ecosystem has undergone substantial and, to some observers, alarming changes, which have accelerated over the last three decades. Recent pollution studies show that eutrophication has increased from human activities mentioned above (Scheren *et al.*, 2000). Policies for sustainable development in the region, including restoration and preservation of the lake's ecosystem, should therefore be directed towards improved land-use practices and control over land clearing and forest burning.

Diminishing water levels and pollution have acute consequences for several economic sectors that depend on the basin lakes. It greatly affects the fishery by changing water levels. Water-level variations affect shallow waters and coastal areas which are of particular importance for numerous fish species, at least in certain stages of their lives. Pollution poses a problem for fishery productivity in the Nile Basin. Some areas of the rivers feeding the lake and the shoreline are particularly polluted by municipal and industrial discharges. Cooperation between all concerned authorities is necessary to search for coherent solutions to ensure the sustainability of the fisheries.

Aquaculture

Aquaculture is the farming of fish, molluscs, crustaceans and aquatic plants in freshwater, brackish water or the marine environment. In 2008, aquaculture production in the Nile Basin countries reached 756,000 tonnes, which represents just over US$1.3 billion. Egypt is the main producer of farmed fish; since the mid-1990s it has rapidly expanded its aquaculture, extending its production from 72,000 tonnes in 1995 to 694,000 tonnes in 2008. Aquacultural expansion has contributed to increasing the total fisheries production in Egypt. The relative importance of Egyptian aquaculture to total fisheries production has increased from 16 to 56 per cent of total fisheries production between the years 1997 and 2007. Aquacultural activities in Egypt are more concentrated in subregions of the Nile Delta, where the water resources are available. Most of the aquacultural production is derived from farmers' use of earthen ponds in production systems.

Uganda is a distant second of the total basin aquacultural production. Kenya, Rwanda and Sudan are developing fisheries with the help of foreign aid to boost production which, together with other basin countries, represents 1 per cent of the farmed fish in the basin. Uganda's aquacultural export market, regional use and employment have risen dramatically over the past 10 years. The Government of Uganda is promoting aquaculture to boost livelihoods and food security of farmers with plans to either capture floodwaters or use groundwater to expand aquacultural production in the northern and eastern areas of the country (see www.thefishsite.com).

Egypt has given support for the development of aquaculture to promote farmers' livelihoods and provide nutritional benefit to poor farm families. The programmes instituted have been provided at minimal cost and often free of charge. Uganda has also started many fish programmes with foreign aid and government support. Egypt's advanced technical knowledge in aquaculture could be used to help train and support development of aquaculture in other basin countries.

Conclusions

The Nile Basin is a large transboundary basin that is home to a population of nearly 160 million, with the majority of them reliant on local agricultural products for their food and on agricultural activities for earning their livelihood. Due to the size, the basin is host for different geographical areas, agro-ecological conditions, environmental characteristics, and farmers' socio-economic assets. As a result, farming systems in the Nile are highly variable in terms of size, distribution and characteristics. The results of the farming system classification exercise show that the most prevailing system in the Nile Basin is the pastoral system, followed by mixed crop-livestock and agro-pastoral systems, covering 45, 36 and 19 per cent of the land area, respectively. Agricultural production in the Nile Basin faces different biophysical constraints. The biophysical constraints of crop productivity include aluminium toxicity, high leaching potential and low nutrient reserves, mainly in mixed rain-fed systems and salinity and poor drainage in some irrigated areas.

However, water scarcity in terms of both physical water scarcity and economic water scarcity remains the major limiting factor for agricultural development in the basin. In the face of this challenge agriculture water sector calls for an improved management in order to increase and maximize WP. With the exception of Egypt, the Nile Basin's agriculture is predominantly rain-fed. Productivity is highly influenced by spatial variations of rainfall in the rain-fed system while in the irrigated areas farm and scheme management is the main determining factor in the productivity variation.

Measures like expansion of irrigated agriculture, implementing water conservation techniques (e.g. rainwater harvesting) for the rain-fed systems, improved scheme management in the irrigated areas, and increased water accessibility through construction of new control and storage infrastructures in areas where inaccessibility to water is the issue rather than unavailability of water, could largely contribute towards increasing productivity in the Nile Basin. However, these interventions have to be considered within a basin context, and further work is required to assess the impact of implementing these interventions on the hydrological cycle and water flows in the basin.

References

Ahmad, M. D., Islam, A., Masih, I., Muthuwatta, L., Karimi, P. and Turral, H. (2009) Mapping basin level water productivity using remote sensing and secondary data in the Karkheh River Basin, Iran, *Water International*, 34, 1, 119–133.

Al-Feel, M. A. and Al-Bashir, A. A. R. (2012) Economic efficiency of wheat production in Gezira scheme, Sudan, *Journal of the Saudi Society of Agricultural Sciences*, 11, 1–5.

Allan, J. (2009) Nile Basin asymmetries: A closed fresh water resource, soil water potential, the political economy and Nile transboundary hydropolitics, in *The Nile*, H. J. Dumont (ed.), Monographiae Biologicae 89, 749–770, Springer, Dordrecht, The Netherlands.

de Fraiture, C., Molden, D. and Wichelns, D. (2010) Investing in water for food, ecosystems, and livelihoods: An overview of the comprehensive assessment of water management in agriculture, *Agricultural Water Management*, 97, 495–501.

de Wit, C. T. (1958) Transpiration and crop yield, in *Verslag van Landbouwk Onderzoek*, 64, 6, Institute of Biological and Chemical Research on Field Crops and Herbage, Wageningen, The Netherlands.

Dixon, J., Gulliver, A. and Gibbon, D. (2001) *Farming Systems and Poverty: Improving Farmers' Livelihoods in a Changing World*, FAO, Rome; World Bank, Washington, DC.

Gamal, K. A. E. M. (2009) Impact of policy and institutional changes on livelihood of farmers in Gezira scheme of Sudan, MSc thesis, University of Gezira, Sudan.

Guvele, C. A. (2001) Gains from crop diversification in the Sudan Gezira scheme, *Agricultural Systems*, 70, 319–333.

Hanks, R. J. (1974) Model for predicting plant yield as influenced by water use, *Agronomics Journal*, 66, 660–665.

Howell, T. A. (1990) Relationships between crop production, and transpiration, evaporation and irrigation, in *Irrigation of Agricultural Crops*, Agronomy Monograph 30, American Society of Agronomy, Madison, WI.

Laki, S. L. (1993) Policy analysis of the irrigated sector of the Sudan, PhD dissertation, Department of Agricultural Economics, Michigan State University, East Lansing, MI.

Loeve, R., Dong, B., Molden, D., Li, Y. H., Chen, C. D. and Wang, J. Z. (2004) Issues of scale in water productivity in the Zhanghe irrigation system: implications for irrigation in the basin context, *Paddy and Water Environment*, 2, 227–236.

LVFO (Lake Victoria Fisheries Organization) (2006) Fisheries development and management with reference to Lake Victoria, in ICEIDA/United Nations University Workshop on Fisheries and Aquaculture in Southern Africa, Development and Management, 21–24 August, Windhoek, Namibia, www.iceida.is/media/pdf/Maembe_Fisheries_Development_and_Management_with_Reference_to_Lake_Victoria.pdf.

Mahir, A. E. A. E. and Abdelaziz, H.H. (2011) Analysis of agricultural production instability in the Gezira scheme, *Journal of the Saudi Society of Agricultural Sciences*, 10, 53–58.

Molden, D., Sakthivadivel, R., Perry, C. J., de Fraiture, C. and Kloezen, W. H. (1998) *Indicators for Comparing Performance of Irrigated Agricultural Systems*, Research Report 20, IWMI, Colombo, Sri Lanka.

Molden, D., Oweis, T. Y., Pasquale, S., Kijne, J. W., Hanjra, M. A., Bindraban, P. S., Bouman, B. A. M., Cook, S., Erenstein, O., Farahani, H., Hachum, A., Hoogeveen, J., Mahoo, H., Nangia, V., Peden, D., Sikka, A., Silva, P., Turral, H., Upadhyaya, A. and Zwart, S. (2007) Pathways for increasing agricultural water productivity, in *Water for Food, Water for Life: A Comprehensive Assessment of Water Management in Agriculture*, Molden, D. (ed), pp279–310. Earthscan, London, UK.

Molden, D., Oweis, T., Steduto, P., Bindraban, P., Hanjra, M. and Kijne, J. (2010) Improving agricultural water productivity: between optimism and caution, *Agricultural Water Management*, 97, 528–535.

Notenbaert, A. (2009) The role of spatial analysis in livestock research for development, *GIScience and Remote Sensing*, 46, 1, 1–11.

OECD (Organisation for Economic Co-operation and Development) (2006) *Promoting Pro-poor Growth – Agriculture,* DAC Guidelines and Reference Series, OECD, Paris, France.

Otte, M. and Chilonda, P. (2002) *Cattle and Small Ruminant Production Systems in Sub-Saharan Africa: A Systematic Review,* Food and Agriculture Organization of the United Nations, Rome, Italy.

Pender, J. (2006) Development pathways for hillsides and highlands: some lessons from Central America and East Africa, *Food Policy*, 29, 339–367.

Plusquellec, H. (1990) T*he Gezira Irrigation Scheme in Sudan: Objectives, Design, and Performance*, World Bank technical paper no. 120, The World Bank, Washington, DC.

Qureshi, A. S., Oweis, T., Karimi, P. and Porehemmat, J. (2010) Water productivity of irrigated wheat and maize in the Karkheh River Basin of Iran, *Irrigation and Drainage*, 59, 264–276.

Scheren, P. A. G. M., Zanting, H. A. and Lemmens, A. M. C. (2000) Estimation of water pollution sources in Lake Victoria, East Africa: application and elaboration of the rapid assessment methodology, *Journal of Environmental Management*, 58, 235–248.

Seleshi, A., Rebelo, L. M. and Molden, D. (2010) The Nile Basin: tapping the unmet agricultural potential of Nile waters, *Water International*, 35, 5, 623–654.

Seré, C. and Steinfeld, H. (1996) *World Livestock Production Systems: Current Status, Issues and Trends*, Animal Production and Health Paper no. 127, FAO, Rome, Italy.

Stanley, D. (1996) Nile delta: extreme case of sediment entrapment on a delta plain and consequent coastal land loss, *Marine Geology*, 129, 189–195.

WaterWatch (2009) Agricultural water use and water productivity in the large scale irrigation (LSI) schemes of the Nile Basin, unpublished project report, provided by W. G. M. Bastiaanssen, 23 October 2009.

Witte, F., de Graaf, M., Mkumbo, O. C., El-Moghraby, A. I. and Sibbing, F. A. (2009) Fisheries in the Nile System, in *The Nile*, H. J. Dumont (ed.), Monographiae Biologicae 89, 723–748, Springer, Dordrecht, The Netherlands.

Woldeamlak, B. (2003) *Land Degradation and Farmers' Acceptance and Adoption of Conservation Technologies in the Digil Watershed, Northwestern 99 Highlands Ethiopia*, Social Science Research Report Series – no 29, Organisation for Social Science Research in Eastern and Southern Africa (OSSERA), Addis Ababa, Ethiopia.

Yasir, M., Thiruvarudchelvan, T., Mamad, N., Mul, M. and van Der Zaag, P. (2011) Performance assessment of large irrigation systems using satellite data: the case of the Gezira scheme, Sudan, in *ICID, 21st Congress on Irrigation and Drainage: Water Productivity towards Food Security*, Tehran, Iran, 19–23 October 2011, ICID, New Delhi, India, *ICID Transactions*, no. 30-A, 105–122.

You, L., Wood S. and Wood-Sichra U. (2009) Generating plausible crop distribution maps for Sub-Saharan Africa using a spatially disaggregated data fusion and optimization approach, *Agricultural Systems*, 99, 126–140.

9

Livestock and water in the Nile River Basin

Don Peden, Tilahun Amede, Amare Haileslassie, Hamid Faki, Denis Mpairwe, Paulo van Breugel and Mario Herrero

Key messages

- Domestic animals contribute significantly to agricultural GDP throughout the Nile Basin and are major users of its water resources. However, investments in agricultural water development have largely ignored the livestock sector, resulting in negative or sub-optimal investment returns because the benefits of livestock were not considered and low-cost livestock-related interventions, such as provision of veterinary care, were not part of water project budgets and planning. Integrating livestock and crop development in the context of agricultural water development will often increase water productivity and avoid animal-induced land and water degradation.
- Under current management practices, livestock production and productivity cannot meet projected demands for animal products and services in the Nile Basin. Given the relative scarcity of water and the large amounts already used for agriculture, increased livestock water productivity (LWP) is needed over large areas of the basin. Significant opportunities exist to increase LWP through four basic strategies. These are: (i) utilizing feed sources that have inherently low water costs for their production; (ii) adoption of the state-of-the-art animal science technology and policy options that increase animal and herd production efficiencies; (iii) adoption of water conservation options; and (iv) optimally balancing the spatial distributions of animal feeds, drinking water supplies and livestock stocking rates across the basin and its landscapes. Suites of intervention options based on these strategies are likely to be more effective than a single-technology policy or management practice. Appropriate interventions must take account of spatially variable biophysical and socio-economic conditions.
- For millennia, pastoral livestock production has depended on mobility, enabling herders to cope with spatially and temporally variable rainfall and pasture. Recent expansion of rain-fed and irrigated croplands, along with political border and trade barriers, has restricted mobility. Strategies are needed to ensure that existing and newly developed cropping practices allow for migration corridors along with water and feed availability. Where pastoralists have been displaced by irrigation or encroachment of agriculture into dry-season grazing and watering areas, feeds based on crop residues and by-products can offset loss of grazing land.

- In the Nile Basin, livestock currently utilize about 4 per cent of the total rainfall, and most of this takes place in rain-fed areas where water used is part of a depletion pathway that does not include the basin's blue water resources. In these rain-fed areas, better vegetation and soil management can promote conversion of excessive evaporation to transpiration while restoring vegetative cover and increasing feed availability. Evidence suggests that livestock production can be increased significantly without placing additional demands on river water.

Introduction

Pastoralists in the Nile River Basin had kept cattle as long as 10,000 years ago (Hanotte *et al.*, 2002). These early bovines (*Bos taurus*) evolved through domestication from wild aurochs (*Bos primigenius*) either in northeastern Africa or in the Near East. Zebu (*Bos indicus*) reached Egypt during the second millennium BC with further introductions of Zebu from the East African coastal region in subsequent centuries. Over thousands of years, livestock-keeping has formed core sets of livelihood strategies and cultural values of the Nile's peoples and nations. Livestock have played a major role in shaping landscapes and land use systems as well as current demands for, and patterns of, use of agricultural water in the basin. Taking into account this history remains paramount for peaceful and sustainable human development in the Nile. Given the rapidly increasing human population in the basin, this requires optimal use of agricultural water resources (Chapter 3).

The contribution of agriculture to total GDP in most Nile countries has declined over the past few decades because of increased income generated in the service and industrial sectors of country economies. Nevertheless, agriculture, including livestock and fisheries, remains an important component of regional food security. Currently, livestock contribute 15–45 per cent of agricultural GDP in the Nile riparian nations (Peden *et al.*, 2009a), although estimated GDP for the actual basin land areas within countries is not known. Vast land areas within the basin are sparsely inhabited and unsuitable for crop production, but livestock-keeping remains the most suitable agricultural livelihood strategy. Non-livestock contributions to agricultural GDP concentrate in higher rainfall areas and near urban market centres. But even there, livestock remain important, particularly in rain-fed, mixed crop-livestock farming. Across the Nile Basin, livestock populations are rapidly growing in response to increasing African demand for meat and milk products. For example, Herrero *et al.* (2010) predicted that total livestock numbers are expected to increase by 59 per cent between 2000 and 2030 with the greatest percentage increase occurring in the swine populations (Table 9.1).

Table 9.1 Estimated and projected population numbers and percentage changes of livestock populations for the period 2000–2030 in Nile riparian countries

Year	*Livestock numbers (thousands)*					
	Cattle	*Chicken*	*Goats*	*Pigs*	*Sheep*	*Total*
2000	66,560	96,540	51,970	1820	53,420	272,310
2030	111,320	17,1510	73,290	6230	68,580	432,960
Projected increase (%)	67.2	77.7	41.0	242.3	28.4	59.0

Source: Data extracted from Herrero *et al.*, 2010

Despite the importance of the livestock sector to poor rural people, animal production has failed to achieve sustainable returns for poor livestock raisers, owing to several key constraints. Chief among them are water scarcity, and the failure of policy-makers to recognize the importance of livestock and to support livestock production through appropriate policies and interventions (IFAD, 2009). Notwithstanding the dependence of livestock and people on water resources, evidence shows that, for the most part, livestock have largely been ignored in water planning, investment, development and management (Peden *et al.*, 2006). Not only does livestock-keeping make important contributions to farm income, but investing in herds of cattle, sheep and goats is also a preferred form of wealth-savings for diverse Nile populations. One consequence of successful investing in agricultural water for poverty reduction is the tendency for farmers to use newly generated income to purchase and accumulate domestic animals. Safeguarding farmers' assets including livestock or alternatives to them is therefore required.

This chapter summarizes research undertaken by the CGIAR Challenge Program on Water and Food (CPWF) on Nile Basin livestock water productivity (Peden *et al.*, 2009a). The starting point is an overview of livestock distributions and production across the entire river basin. The chapter continues with a description of livestock water productivity (LWP), a concept that is useful for identifying opportunities for more effective use of water by animals. Based on CPWF research in Ethiopia, Sudan and Uganda, the chapter then highlights some key water–livestock interactions characteristic of major production systems. It concludes with a discussion of options for making better use of agricultural water through better livestock and water management. The purpose of this chapter is to share insights on livestock–water interactions, with a view to making better integrated use of basin water resources, improving livestock production and LWP, rehabilitating degraded croplands, pastures and water resources, and contributing to improved livelihoods, poverty reduction and benefit-sharing.

A work of this nature could not cover all aspects of livestock-keeping, and thus focuses on cattle, sheep and goats. We recognize that poultry, swine, equines, camels, buffalo and beekeeping are also important, and further consideration of them will be necessary in future research and development. The basin contains many exotic and imported breeds that vary in their effectiveness to use water efficiently and sustainably, but this topic was beyond the scope of this CPWF research. This chapter also does not address, in deserved detail, the increasing trend towards industrialization of livestock production occurring near rapidly growing urban centres and the engagement in international trade.

Livestock distributions, populations, and demand for animal products and services

Livestock-keeping is the most widespread agricultural livelihood strategy in the Nile Basin. Domestic animals are kept within diverse agro-ecologies and production systems. This diversity generates varying animal impacts on water demand and the sustainability and productivity of water resources adjacent to pasturelands and riparian areas. Livestock production systems are defined in terms of aridity and the length of the growing season (Seré and Steinfeld, 1995; van Breugel *et al.*, 2010). Rain-fed production systems cover about 94 per cent of the basin, of which about 61 per cent is classified as livestock-dominated or grazing land and about 33 per cent as mixed crop–livestock production (Table 9.2; Figure 9.1). Livestock are kept virtually wherever crops are grown, but vast areas of rangeland are not suitable for crop production, leaving animal production as the only viable form of agriculture, even if at very low levels of intensity. Irrigated areas are small amounting to less than 2 per cent of the land area, but even there, livestock are typically important assets for irrigation farmers (Faki *et al.*, 2008; Peden *et al.*, 2007). Urban areas, protected forests and parks are also present but make up a minute percentage of the Nile's land area and are not discussed in this chapter.

Table 9.2 Livestock production systems in the Nile River Basin showing their defining aridity classes and lengths of the growing season

Production system		*Unique code*	*Area (km²)*	*Basin land area (%)*	*Aridity*	*Length of the growing season (days yr⁻¹)*
Rain-fed	Grazing	LGHYP	935,132	31.2	Hyper-arid	0–1
	Grazing	LGA	758,593	25.3	Arid–semi-arid	1–180
	Grazing	LGH	123,618	4.1	Humid	>180
	Grazing	LGT	13,749	0.5	Temperate	>180
	Sub-total		1,831,092	61.1		
Rain-fed	Mixed	MRA	608,547	20.3	Arid–semi-arid	1–180
	Mixed	MRT	228,005	7.6	Temperate	>180
	Mixed	MRH	155,575	5.2	Humid	>180
	Mixed	MRHYP	6381	0.2	Hyper arid	0–1
	Sub-total		998,508	33.3		
Irrigated	Mixed	MIHYP	35,322	1.2	Hyper arid	0–1
	Mixed	MIA	2842	0.1	Arid-semi-arid	1–180
	Sub-total		38,164	1.3		
Wetlands, forest and parks		Other	110,512	3.7	Various	Variable
Urban with >450 persons·km⁻²			20,170	0.7	Various	Not relevant
Total land area of the Nile Basin			2,998,446	100.0		

Notes: Unique codes are used in other tables and figures in this chapter. Urban areas are not shown on maps in this chapter. 'Mixed' refers to mixed crop-livestock production. Codes beginning with LG, MR and MI refer to livestock dominated grazing areas, rain-fed mixed crop–livestock systems and irrigated mixed crop–livestock farming, respectively. Codes ending in HYP, A, H and T refer to hyper-arid, arid/semi-arid, humid and temperate climatic regions, respectively. 'Other' refers to lands designated for non-agricultural uses including forests, wetlands, parks, and wildlife reserves

Source: Peden *et al.*, 2009a

The Nile's livestock production systems are dispersed unevenly across the basin, with arid systems concentrated in the northern two-thirds of the basin (Figure 9.1), an area occupied largely by Sudan and Egypt. Mixed crop–livestock production systems are common in the southern countries around the great lakes and in the Ethiopian Highlands. Irrigated systems are found mostly in the Nile Delta and along the banks of the Nile River in Sudan. At the map scale used in Figure 9.1, small-scale household and community-scale irrigation based on water harvesting and stream diversion is not included in irrigation and falls within rain-fed agriculture for the purpose of this chapter.

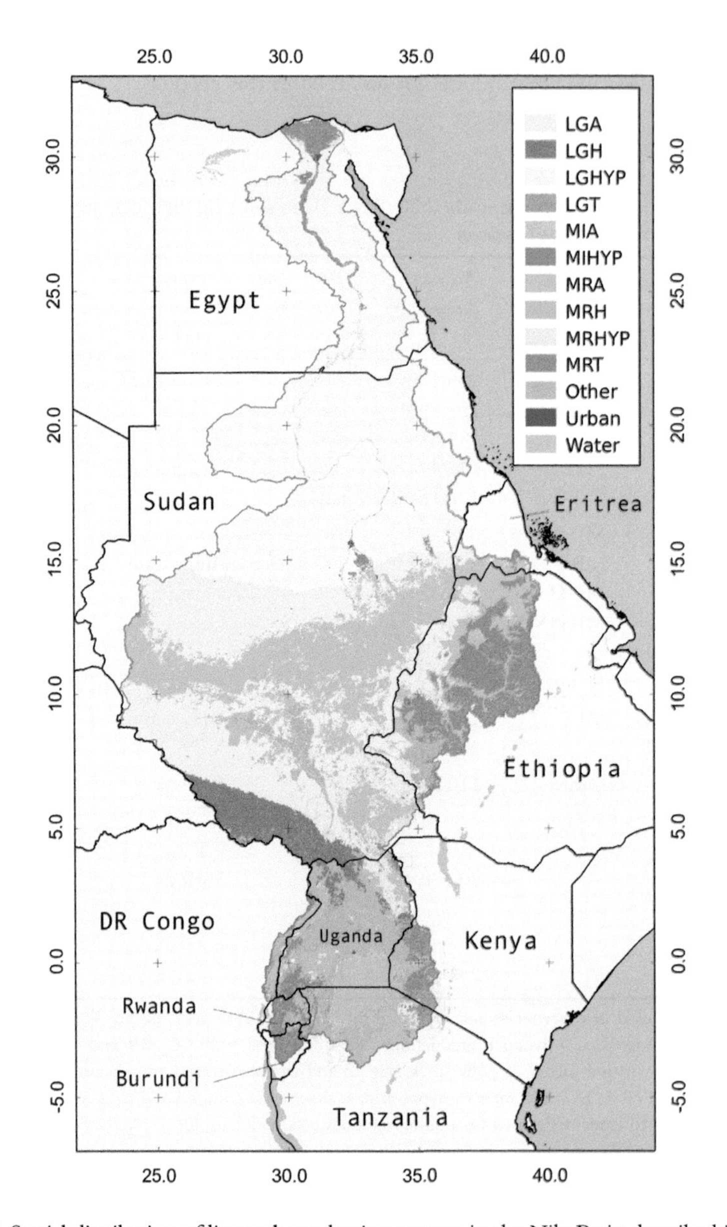

Figure 9.1 Spatial distribution of livestock production systems in the Nile Basin described in Table 9.2
Sources: Peden *et al.*, 2009a, b; van Breugel *et al.*, 2010

In 2000, the Nile Basin was home to about 45 million sheep, 42 million goats, 67 million cattle and 173 million people (Table 9.3). Also present are millions of swine, poultry, camels and buffalo, which, although locally important, are not considered in this chapter. These estimates are totals for animals residing within basin parts of riparian nations and are thus lower than those reported in Table 9.1 for the entire land area of the Nile riparian states. Because animals of different species have different weights, Table 9.3 also shows tropical livestock units (TLU), which give a weighted total for livestock biomass. Overall, about 56 million TLU live within the Nile Basin. Assuming that an average person weighs 50 kg, five persons would be equivalent to one TLU, and the Nile's human population would be equivalent to about 35 million TLU or about 63 per cent of the domestic animal biomass (cattle, sheep and goats). These values suggest that basin-wide animal demand for feed by weight is at least equal to human food requirements. As will be shown, this has implications for agricultural water use in the Nile Basin.

Four production systems (MRA, LGA, MRT, MRH) contain 86 per cent of the animal TLU within about 58 per cent of the Nile's land area. The vast hyper-arid livestock system of Sudan and Egypt has very low animal densities, but because the area is large, the total livestock biomass is large, numbering about 1.2 million TLU. The highest livestock densities (TLU·km^{-2}) are found in the irrigated and urban areas of the basin, while the lowest animal densities are found in the livestock-dominated grazing lands. In general, high livestock and human densities are positively correlated.

Table 9.3 Estimated populations and densities of sheep, goats, cattle and people within the Nile Basin production systems defined in Table 9.2 and ranked in decreasing order by TLU density

LPS	*Number (millions^{-1})*						*Mean density (number km^{-2})*					
	Sheep	*Goats*	*Cattle*	*TLU*	*Persons*	*Persons (TLU)*	*Sheep*	*Goats*	*Cattle*	*Animals (TLU)*	*Persons*	*Persons (TLU)*
MIHYP	1.8	1.3	2.3	1.9	32.7	6.5	51	34	64	53	926	185
MIA	0.1	0.1	0.2	0.2	0.2	<0.1	31	32	63	50	86	17
MRT	5.0	4.1	13.0	10.0	35.0	7.0	22	18	57	44	15	3
URBAN	0.7	0.8	0.8	0.7	43.5	8.7	34	41	38	34	2156	431
MRA	16.1	14.2	22.3	18.6	18.3	3.7	26	23	37	31	30	6
MRH	1.1	3.3	6.1	4.7	20.8	4.2	7	21	39	30	134	27
LGT	0.2	0.3	0.3	0.3	0.2	<0.1	15	20	23	20	15	3
LGA	15.2	12.6	17.1	14.8	9.4	1.9	20	17	22	19	1	<1
OTHER	0.8	1.0	1.8	1.4	6.4	1.3	7	9	16	13	58	12
MRHYP	0.1	0.1	0.1	0.1	0.4	0.1	17	21	11	11	57	11
LGH	1.7	1.7	1.2	1.2	0.8	0.2	14	14	10	9	7	1
LGHYP	2.7	1.9	2.1	1.9	5.5	1.1	3	2	2	2	6	1
Total	45.4	41.5	67.2	55.7	173.2	34.6	15	13	22	18	58	12

Notes: 'Urban' refers to urban and peri-urban areas. Core urban populations have higher and lower human and livestock densities, respectively. Codes beginning with LG, MR and MI refer to livestock-dominated grazing areas, rain-fed mixed crop–livestock systems and irrigated mixed crop–livestock farming, respectively. Codes ending in HYP, A, H and T refer to hyper-arid, arid/semi-arid, humid and temperate climatic regions, respectively. 'Other' refers to lands designated for non-agricultural uses including forests, wetlands, parks, and wildlife reserves

Source: Peden *et al.*, 2009a

Current livestock and human population numbers and densities also vary greatly among Nile riparian nations (Table 9.4; Figure 9.2). Basin-wide, an estimated 67, 45, 41 and 173 million cattle, sheep, goats and people, respectively, were living in the river basin in 2000. Egypt and Ethiopia were the two most populous countries, while Rwanda and Burundi had the highest densities of people (302 and 284 persons km^{-2}, respectively). In terms of livestock, Sudan alone contained more than half of the Nile Basin's cattle, sheep and goats. Ethiopia ranked second in terms of livestock numbers. However, Kenya had the highest animal density. Although not described herein, swine, camels, equines, poultry, fish and bees contribute to human livelihoods and place increasing demands on land and water resources.

Table 9.4 Estimated populations and densities of sheep, goats, cattle and people within the basin portion of Nile riparian countries hand-ranked according to human density

Country	*Land area (km^2)*	*Number (millions^{-1})*				*Density (number km^{-2})*			
		Cattle	*Sheep*	*Goats*	*Persons*	*Cattle*	*Sheep*	*Goats*	*Persons*
Rwanda	20,681	0.7	0.2	0.83	6.2	36	12	40	302
Burundi	12,716	0.2	0.1	0.46	3.6	15	9	36	284
Kenya	47,216	4.2	1.4	1.58	12.1	89	30	34	257
Egypt	285,606	2.8	3.1	1.97	64.9	10	11	7	227
Uganda	204,231	5.0	1.3	2.97	23.6	24	6	15	114
DR Congo	17,384	0.1	<0.1	0.10	2.0	3	2	6	113
Tanzania	85,575	5.5	0.8	2.89	7.4	64	9	34	86
Ethiopia	361,541	14.0	5.4	3.72	25.9	39	15	10	70
Eritrea	25,032	0.9	0.7	0.83	01.1	34	29	33	46
Sudan	1,932,939	33.9	32.2	26.07	27.2	17	17	13	14
Total	2,992,921	67.1	45.2	41.4	173.2	22	15	14	58

Sources: Peden *et al.*, 2009a, b; van Breugel *et al.*, 2010

The rapidly growing human population in the Nile riparian countries drives increasing demand for meat and milk; a force catalysed and amplified by urbanization and increased discretionary income of urban dwellers. In response, animal population projections suggest that livestock numbers will rise from 272 million in 2000 to about 434 million in 2030, a 59 per cent increase in the next 20 years (Table 9.2). In addition, demand for poultry and fish is also increasing. Simultaneously, grazing lands are being cultivated, implying a trend toward intensification of animal production within mixed crop-livestock systems. Without increased efficiency and effectiveness of water use, water demand for livestock will also similarly rise. One key implication is the need for Nile countries to integrate livestock demands on water resources within the larger set of pressures being placed on basin water resources.

Water use and availability for Nile livestock

Without adequate quality and quantity of drinking water, livestock die. Given about 56 million cattle, sheep and goats TLU (Table 9.2) and drinking water requirements of about 50 l day^{-1}TLU^{-1}, their annual intake would amount to about 1 billion m^3 yr^{-1}, or about 0.05 per cent of the total basin rainfall. The actual voluntary and required drinking water intake varies from about 9 to 50 l day^{-1} TLU^{-1} under Sahelian conditions, depending on the type of animal, climatic conditions, feed water content, and animal management practices (Peden *et al.*, 2007). Feed production requires much more water than meeting drinking water demand. Livestock TLU consume about 5 kg day^{-1} TLU^{-1} of feed on a maintenance diet that theoretically utilizes

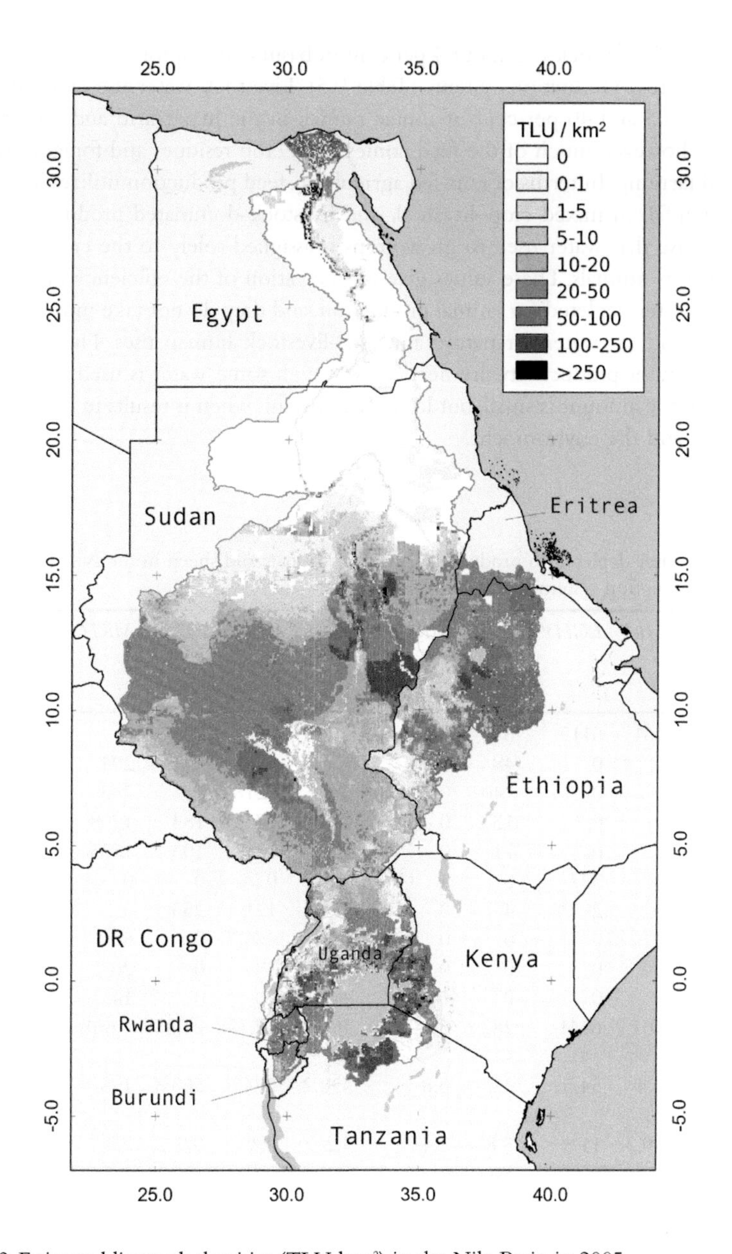

Figure 9.2 Estimated livestock densities (TLU km^{-2}) in the Nile Basin in 2005

Sources: Peden *et al.*, 2009a, b; van Breugel *et al.*, 2010

about 450 m^3 of depleted evapotranspiration for its production (Peden *et al.*, 2007), an amount about 90 times greater than the daily water intake. Additional feed for production, work, lactation and reproduction involves the use of a greater amount of water. The actual water cost of feed production is spatially variable depending on the type of animal, feed and vegetation management, and climatic conditions (Peden *et al.*, 2009a). Under harsh arid conditions in Sudan, water for feed production may reach 400 times the amount of drinking water used (Faki *et al.*, 2008). The Nile's livestock use about 4 per cent of basin's rainfall for feed production, but this varies greatly across production systems (Table 9.5). Livestock water use amounts to the equivalent of about 65 and 40 per cent of annual rainfall in the hyper-arid and arid irrigated areas, respectively; however, much of the feed comes from crop residues and forages produced through irrigated farming. In terms of rain-fed agriculture, feed production utilizes less than 12 per cent of the rainfall in mixed crop–livestock and livestock-dominated production systems based on the premise that water used to grow crops is assigned solely to the crop and not to residues consumed by animals. These values give no indication of the efficiency or productivity of agricultural water depleted for animal production, and they do not take into account the magnitude of demand for water for nature and non-livestock human uses. Therefore, assessments of livestock water productivity are needed. Although some water is used for processing of animal products, the amount is small, but locally important, when it results in contamination harmful to people and the environment.

Table 9.5 Estimated water depleted to produce feed for cattle, goats and sheep in the Nile portion of riparian production systems and countries (million m^3 yr^{-1})

Nile riparian country	*MIHYP*	*MIA*	*LGHYP*	*LGT*	*MRHYP*	*MRA*	*MRT*	*LGA*	*MRH*	*LGH*	*Whole basin*
Sudan	277	161	6112	6	55	14,481	21	20,459	8	1	41,581
Ethiopia	0	0	0	48	0	2203	8464	857	204	26	11,802
Kenya	0	0	0	140	0	163	2218	4	786	1	3312
Uganda	0	0	0	13	0	490	576	183	1708	136	3106
Tanzania	0	0	0	71	0	777	121	103	1835	9	2916
Egypt	1359	0	327	0	4	0	0	0	0	0	1690
Eritrea	0	0	2	4	0	579	121	253	0	0	959
Rwanda	0	0	0	0	0	127	466	0	80	0	673
Burundi	0	0	0	0	0	1	191	0	26	0	218
DR Congo	0	0	0	0	0	4	20	0	14	3	41
Total	1636	161	6441	282	59	18,825	12,198	21,859	4,661	3	66,125
Rainfall (billion m^3)	2.5	0.4	54.5	3.2	0.8	450.2	294.6	565.8	198.1	158.7	1728.8
Percentage of rainfall	65.4	40.3	11.8	8.8	7.4	4.2	4.1	3.9	2.4	<1.0	3.8

Notes: Codes beginning with LG, MR and MI refer to livestock-dominated grazing areas, rain-fed mixed crop–livestock systems and irrigated mixed crop–livestock farming, respectively. Codes ending in HYP, A, H and T refer to hyper-arid, arid/semi-arid, humid and temperate climatic regions, respectively. 'Other' refers to lands designated for non-agricultural uses including forests, wetlands, parks, and wildlife reserves

Source: Peden *et al.*, 2000a

Providing water for livestock depends on the availability of water within the context of competing uses, especially for meeting other human needs as well as those of nature. The United Nations World Health Organization indicates that an acceptable minimum renewable freshwater threshold (in terms of both blue and green water) to satisfy human food production and domestic needs is 2000 m^3 $person^{-1}$ yr^{-1} (Khosh-Chashm, 2000). Average annual rainfall per capita across the Nile is about 11,000 m^3, but it varies greatly among the countries and livestock production systems (Figure 9.3). Egypt is the only riparian country that falls below the 2000 m^3 person^{-1} yr^{-1} threshold, demonstrating its reliance on river inflow to meet water demand. Sudan and Ethiopia have the highest levels of renewable freshwater per capita. Population pressure can be expected to push several countries toward the threshold, regardless of any changes in rainfall caused by climate change. Rwanda, Burundi and Kenya may be most vulnerable.

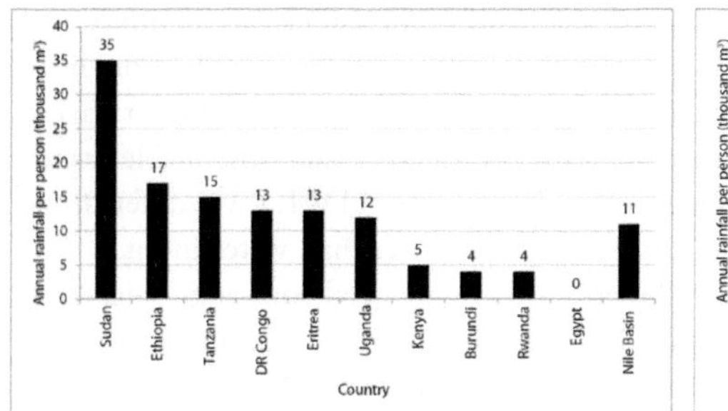

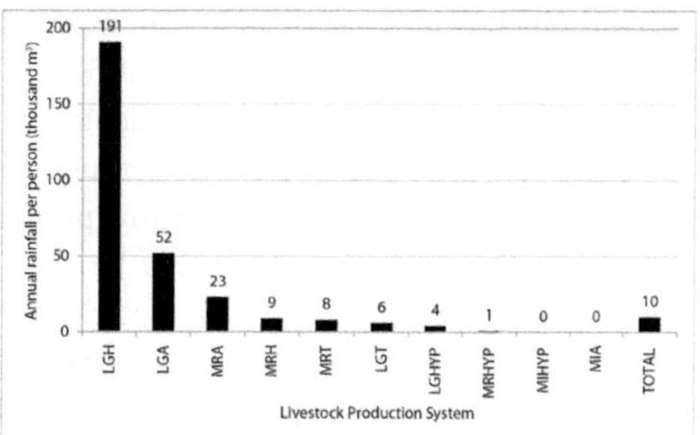

Figure 9.3 Annual rainfall per capita within the basin part of the Nile's countries and livestock production systems

Source: Derived from van Breugel *et al.*, 2010

The highest per capita rainfall occurs in the humid grazing lands (LGH) and the arid and semi-arid grazing areas (LGA), while the lowest amounts are in the more densely populated, mixed crop–livestock systems, especially in humid and temperate regions. The key point is that water scarcity in rain-fed areas reflects both the abundance of rainfall and the human population density. Although access to, and the cost of, developing rainwater resources may be constraints, the greatest potential for livestock development may exist in livestock-dominated arid, semi-arid and humid landscapes. The future of livestock development will depend on rainwater management that promotes high agricultural water productivity.

Livestock water productivity

Livestock water productivity (LWP) is the ratio of net beneficial livestock-related products and services to the amount of water depleted in producing these benefits (Peden *et al.*, 2007, 2009a, b). LWP is a systems concept based on water accounting principles, integrates livestock-water interactions with our collective understanding of agricultural water use, and is applicable to all agricultural systems ranging from farm to basin scales. The distinction between production and productivity is important but often confused. Here, we are concerned with productivity, the benefits gained per unit of water depleted, whereas production is the total amount of benefits produced. Both are important, but high levels of LWP and animal production are not necessarily correlated. LWP differs from water or rain use efficiency because it looks at water depletion rather than water input. As a concept, LWP is preferable to water or rain use efficiency because it does not matter how much water is used as long as that water can be used again for a similar or a higher-value purpose. The water productivity concept helps to focus on preventing or minimizing water depletion.

Livestock provide multiple benefits, including production of meat, milk, eggs, hides, wool and manure, and provision of farm power. Although difficult to quantify, the cultural values gained from animals are important. Accumulating livestock is also a preferred means for people to accumulate wealth. CPWF research in the Nile used monetary value as the indicator of goods and services derived from livestock.

Water enters an agricultural system as rain or surface inflow. It is lost or depleted through evaporation, transpiration and downstream discharge. Depletion refers to water that cannot be easily reused after prior use. Degradation and contamination deplete water in the sense that the water may be too costly to purify for reuse. Transpiration is the primary form of depletion without which plant growth and farm production are not possible. Livestock production is not possible without access to feed derived from plant materials. Thus, like crop production, livestock production results in water depletion through transpiration. In the Nile riparian countries, strategies are needed to ensure that effective, productive and sustainable water management underpins crop and animal production through increased LWP. LWP differs from water or rain use efficiency because it looks at water depletion rather than water input.

Four basic strategies help to increase LWP directly: improving feed sourcing, enhancing animal productivity, conserving water, and optimal spatial distribution of animals, drinking water and feed resources over landscape and basin mosaics (Figure 9.4; Peden *et al.*, 2007). Providing sufficient drinking water of adequate quality also improves LWP. However, drinking water does not factor directly into the LWP calculations because water consumed remains temporarily inside the animal and its production system.

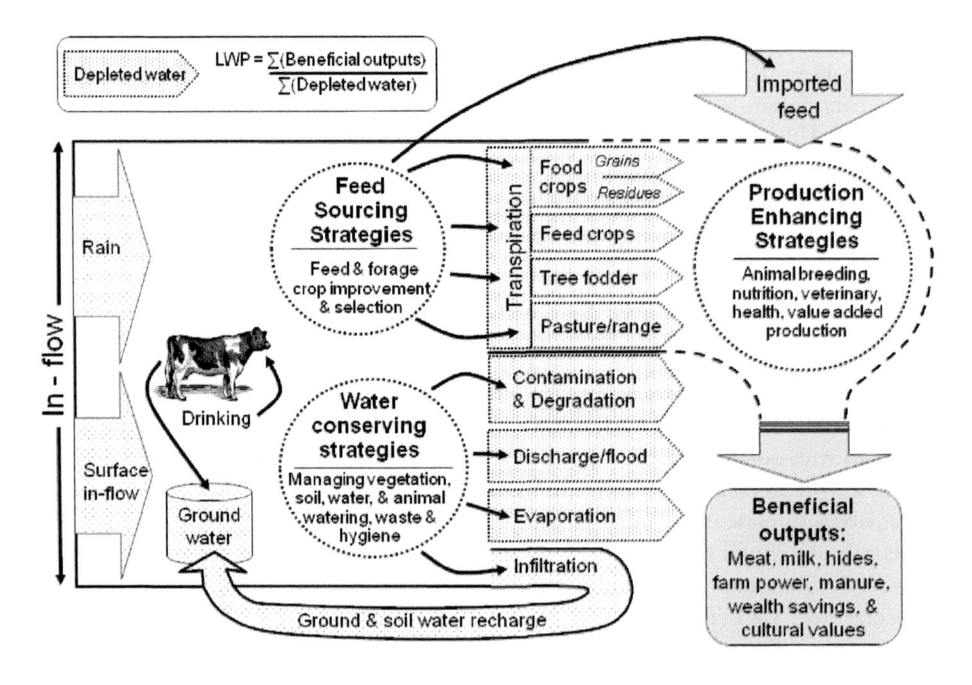

Figure 9.4 Livestock water productivity assessment framework based on water accounting principles enables identification of key strategies for more sustainable and productive use of water

Feed sourcing

The first strategy for enhancing LWP is feed sourcing and management. The photosynthetic production of animal feed is the primary water cost associated with livestock-keeping. Thus, increasing LWP requires selecting quality pasture, feed crops, crop residues and by-products that have high crop water productivity (CWP). Any measures that help to increase CWP will also lead to higher LWP (Chapter 8). However, estimates of water used for feed production are highly variable, context-specific and limited in number. Science-based knowledge of water use for feed remains contradictory.

Maximum practical feed water productivity in rain-fed dry matter production is about 8 kg m^{-3}, but in practice it is often less than 0.5 kg m^{-3} (Table 9.6; Peden *et al.*, 2007). Variability is due to many factors such as inconsistent methodologies, varying concepts of water accounting and the reality of particular production systems. For instance, examples from the literature often estimate CWP on the basis of fresh rather than dry weights, a practice that will overestimate water productivity and make comparisons meaningless. Typically, the water cost of producing below-ground plant materials is ignored, failing to recognize water's role in maintaining soil fertility.

Table 9.6 Example of estimates of dry matter water productivity of selected animal feeds

Feed source	*Feed water productivity ($kg\ m^{-3}$)*	*Reference*
Various crops and pastures, Ethiopia	4	Astatke and Saleem, 1998
Pennisetum purpureum (1200 mm yr^{-1} ET)	4.3	Ferris and Sinclair, 1980
Irrigated alfalfa, Sudan	1.2–1.7	Saeed and El-Nadi, 1997
Grasslands, United States	0.5–0.6	Sala *et al.*, 1988
Various food crops and residues, Ethiopia	0.3–0.5	Haileslassie *et al.*, 2009a
Mixtures of maize, lab-lab and oats vetch	<0.5	Gebreselassie *et al.*, 2009

Source: Summary of examples cited in Peden *et al.*, 2007

The prime feed option for increasing LWP is the use of crop residues and by-products. When crops are grown for human food, taking advantage of their residues and by-products imposes little or no additional water cost beyond what the crop itself requires. In contrast, using irrigation water to produce forage results in a comparatively high water cost and, thus, relatively low LWP.

Importation of feed effectively transfers the water cost of feed production to distant areas, reducing local demand for agricultural water. If insufficient water is available for feed production, animals must move to new feed sources or rely on imported feed associated with virtual water (Chapagain and Hoekstra, 2003). For example, dairy production in Khartoum takes advantage of crop residues produced in large-scale irrigation on the Blue Nile. The low nutritional value of crop residues limits their effectiveness as animal feeds. However, this can be overcome with modest supplements of high-quality feed such as grain and forage legumes and using technologies that increase digestibility of roughages.

Enhancing animal productivity

Water transpired to produce maintenance feed is a fixed input for animal production. Livestock require additional water to produce the feed required to gain weight, produce milk, work and reproduce. Enhancing LWP requires a higher ratio of energy use for production than that used for maintenance. Traditional off-the-shelf Animal Science interventions of nutrition, genetics, veterinary health, marketing and animal husbandry help to increase LWP (Peden *et al.*, 2007). Typical interventions include:

- Providing continuous quality drinking water (Muli, 2000; Staal *et al.*, 2001).
- Selecting and breeding livestock for improved feed conversion efficiency (Basarab, 2003).
- Providing veterinary health services to reduce morbidity and mortality (Peden *et al.*, 2007; Descheemaeker *et al.*, 2010a, 2011) and meet safety standards for marketing animals and animal products (Perry *et al.*, 2002).
- Adding value to animal products, such as farmers' production of butter from liquid milk.

Conserving water resources

Agricultural production and the sustainability of natural and agro-ecosystems are dependent on transpiration (T). Here we define water conservation as the process of reducing water loss through non-production depletion pathways. Water depleted through evaporation and discharge does not contribute to plant production, although it may do so downstream. One effective way to increase agricultural water productivity, including LWP, is to manage water, land, vegetation and crops in ways that convert evaporation and excessive run-off to T. Proven means to increase transpiration, CWP and LWP include maintaining high vegetative ground cover and soil-water-holding capacity, water harvesting that enables supplemental irrigation of feeds including crops that produce residues and by-products, terracing and related measures that reduce excessive run-off and increasing infiltration, and vegetated buffer zones around surface water bodies and wells.

Sheehy *et al.* (1996), in a comprehensive overview of the impact of grazing livestock on water and associated land resources, conclude that livestock must be managed in ways that maintain vegetative ground cover because vegetation loss results in increased soil erosion, down slope sedimentation, reduced infiltration, and less production of pasture. While they find that low to moderate grazing pressure has little negative impact on hydrology, they also find that there is an optimal or threshold site-specific level of grazing intensity above which water and land degradation become problematic and animal production declines. Within this limit, LWP can be maximized by balancing enhanced leaf-to-land area ratios that shift water depletion from evaporation to transpiration (Keller and Seckler, 2005), with profitable levels of animal production and off-take.

The type, mix and density of grazing animals affect the species composition of the vegetation (Sheehy *et al.*, 1996). High grazing pressure causes loss of palatable and nutritional species, but very low grazing pressure encourages encroachment of woody vegetation. Maintaining higher water productivity depends on having both palatable vegetation and the presence of domestic animals that utilize the pasture.

Strategic allocation of livestock, feed and drinking water across landscapes

At landscape and large river-basin scales, suboptimal spatial allocation of livestock, pasture and drinking water leads to unnecessarily low LWP and severe land and water degradation. Faki *et al.* (2008) and Peden *et al.* (2007, 2009a, b) show, especially for cattle, that LWP is low near drinking water sources because of animal weight loss, morbidity and mortality associated with shortages of feed and pastures and increased risk of disease transmission. LWP is also low far from water because trekking long distances between feeding and watering sites reduces animal production (van Breugel *et al.*, 2010; Descheemaeker *et al.*, 2010b). Over vast areas, moving animals, feed and water from areas of surplus to places of scarcity can help maintain an optimal spatial balance and maximize LWP.

Livestock demand management

The four LWP enhancing strategies depicted in Figure 9.4 take a *supply-side* approach to animal production. If livestock keepers respond to increased LWP by simply keeping larger herds, requiring more feed, water and other inputs, no additional water will become available to meet other demands that can benefit people or nature. Ultimately, some limits to livestock production are needed. Two *demand-side* approaches to production require further research. The first is restricting livestock use of land to levels above which environmental sustainability and positive synergies with other livelihood strategies are lost. This might involve including environmental costs of livestock production in the prices of animal products and services. The second is adopting policy that ensures equitable access to animal products and services and limits consumption of meat and milk to levels that enhance human nutrition while discouraging increasing incidence of diabetes and obesity. Promotion of alternatives to animal products and services also helps to limit demand. Livestock provide energy for farm power and fuel. Alternative energy sources can be procured that will reduce demand for animals and, by implication, water resources. Throughout the Nile Basin, livestock keepers maintain large herds for drought insurance, social status and wealth savings (Faki *et al.*, 2008). Finding alternatives for these culturally important livestock services could reduce pressure on feed and water resources.

Case studies from the Nile Basin

The water required to produce food for the Nile's population of 173 million (Table 9.5) is about 1300 m^3·per capita, or 225 billion m^3 for the entire basin, assuming all food is produced within the basin.

Six livestock production systems cover 60 per cent of the Nile Basin's land area (1799 million km^2) and support about 50 and 90 per cent of its human and livestock (cattle, sheep and goats) biomass, respectively. They receive about 1680 billion m^3, of rain or about 85 per cent of the basin's total. Of this, about 1027 billion m^3 are depleted as evapotranspiration (ET), water that does not enter the Nile's blue water system. Making the best use of this ET in rain-fed areas affords great opportunity to reduce agricultural demand on the Nile's water resources. Production of feed for cattle, sheep and goats utilizes about 77 billion m^3, or 3.9 per cent of the total basin rainfall for feed production through ET. This implies that about 1190 billion m^3 are directly lost to the atmosphere without contributing directly to production of these three animal species. Some of this water supports crops, poultry, equines, swine, and crops in mixed crop-livestock systems. Maintenance of ecosystems also requires up to 90 per cent of green water flow (Rockström, 2003). However, the high degree of land degradation with its attendant low level of vegetative ground cover implies that much of the 1190 billion m^3 is lost as non-productive evaporation. *Vapour shifts* (the conversion of evaporation to T) can realize 50 per cent increases in CWP from about 0.56 to 0.83 kg m^{-3} in rain-fed tropical food crop production (Rockström, 2003) by maximizing infiltration and soil water-holding capacity and by increasing vegetative cover. This increase in CWP would also lead to a similar increase in crop residues for animal feed without additional water use. Similar increases are possible in LWP in rangelands (Peden *et al.*, 2009a). In highly degraded landscapes, WP gains may be higher as suggested by Mugerwa (2009). Thus rehabilitating degraded grazing lands and increasing provision of livestock and ecosystems goods and services are possible. Combining feed, animal and water management could lead to a doubling of animal production without placing extra demand on the Nile's blue water resources.

CPWF research assessed LWP at four sites in Ethiopia, Sudan and Uganda that represent four of the basin's major production systems (Peden *et al.*, 2009a). Due to agro-ecological diversity, LWP varied greatly among sites. The highest LWP was observed in the densely populated mixed crop-livestock systems of the Ethiopian highlands while the lowest was found in Uganda's Cattle Corridor (Figure 9.5) These analyses suggest that LWP increases as a result of agricultural intensification. The following sections highlight selected conditions and potential interventions that may help improve LWP and more generally make more effective and sustainable use of water in the Nile Basin.

Ethiopia

Temperate rain-fed mixed crop–livestock systems (MRT) dominate the highlands of Ethiopia, Kenya, Uganda, Rwanda and Burundi. MRT accounts for 7.6, 18, 24 and 17 per cent of the Nile Basin area, TLU (cattle, sheep and goats), rural human population and rainfall, respectively. Annual rainfall exceeding 800 mm, dense human and animal populations, intensified rain-fed cropping, high levels of poverty and food insecurity, and vulnerability to severe soil erosion and loss of water through excessive run-off and downstream discharge prevail. These areas, along with remnants of forests and montane pastures, serve as water towers for the entire Nile Basin and provide Sudan and Egypt with significant amounts of water.

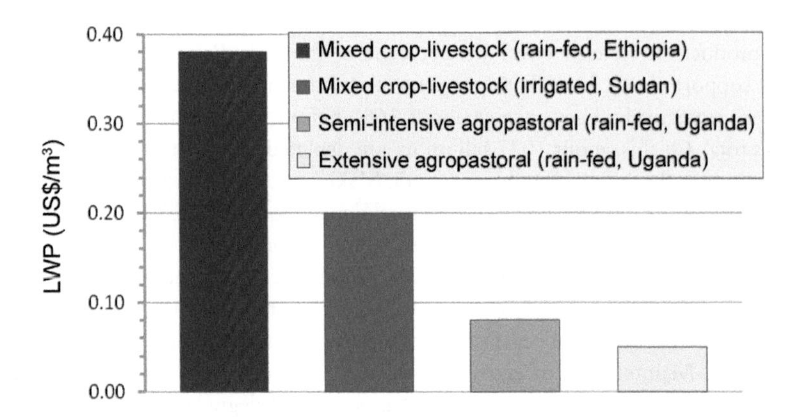

Figure 9.5 LWP estimates for four production systems in Ethiopia, Sudan and Uganda
Source: Peden *et al.*, 2009a

Livestock-keeping is an integral part of Ethiopian rain-fed grain farming. Cattle, sheep, goats, equines and poultry contribute to rural livelihoods. Livestock productivity, in terms of production per animal, is low. For example, milk yields range from 0.6 to 1.8 $l \cdot cow^{-1} \cdot day^{-1}$ and the average live weight of mature cattle reaches only about 210 $kg \cdot head^{-1}$ (Peden *et al.*, 2009a).

CPWF research focused on farming systems in the Gumera watershed which drains into the eastern shore of Lake Tana (Alemayehu *et al.*, 2008) and spans elevations ranging from 1900 to 3700 m above sea level. This watershed contains four farming systems: rice, teff–millet, barley–wheat and potato–barley, which occupy lower to higher elevations in that order. Cattle are dominant in lower areas while sheep are more prevalent at higher elevations. Equines and goats are found throughout. Although both human and animal densities are high (Tables 9.3 and 9.4), households are poor typically owning less than 4 TLU each. Farm production is low and land degradation severe. Rainwater is plentiful, but not well utilized.

Haileslassie *et al.* (2009a, b) assessed LWP in the Gumera watershed based on multiple livestock benefits including meat, milk, traction and manure (Table 9.7). These estimates are consistent with other estimates and suggest that in monetary terms, LWP compares favourably with crop water productivity. However, observed physical water productivity for crops (CWP) was low, averaging about 0.4 kg m^{-3} implying substantive scope for improvement that would translate into a corresponding increase in LWP. Observed monetary CWP and LWP in the Gumera watershed were low ranging from US\$0.2 to US\$0.5$\cdot m^{-3}$ for crops and US\$0.1 to US\$0.6 m^{-3} for livestock.

Relatively wealthy farmers appeared to exhibit higher CWP and LWP than poor farmers. Although researchers have operated on the premise that increasing agricultural water productivity contributes to poverty reduction, evidence also suggests that farmers with greater wealth are better able to make investments in farming that lead to higher profits. Numerous opportunities exist to increase LWP in the Ethiopian Highlands.

Table 9.7 Livestock and water productivity by farming household health class in three farming systems of the Gumera watershed, Blue Nile Highlands and Ethiopia

	Production system	*Wealth group*			*Weighted mean*
		Rich	*Medium*	*Poor*	
CWP (kg m^{-3})	Potato–barley	0.5a	0.3b	0.4b	0.5
	Barley–wheat	0.5a	0.3b	0.4b	0.4
	Teff–millet	0.4a	0.3a	0.3b	0.3
	Rice	0.5	0.5	0.5	0.5
CWP (US\$ m^{-3})	Potato–barley	0.5a	0.2b	0.3b	0.3
	Barley–wheat	0.5a	0.3b	0.3b	0.3
	Teff–millet	0.2ab	0.3a	0.2b	0.2
	Rice	0.4a	0.3b	0.2c	0.3
LWP (US\$ m^{-3})	Potato–barley	0.5a	0.5a	0.4a	0.5
	Barley–wheat	0.5a	0.5a	0.6a	0.5
	Teff–millet	0.6a	0.3b	0.2c	0.3
	Rice	0.4a	0.3a	0.1b	0.3

Note: Letters a, b and c indicate sets of estimated water productivity values within which differences were not significant ($p = 0.01$)

Source: Haileslassie *et al.*, 2009b

One opportunity to increase water productivity focuses on mitigating the impact of traditional cultivation and livestock keeping on run-off and erosion. These two production constraints vary with scale, cropping patterns, land-use and tenure arrangements of the pasturelands. The most severe run-off and erosion are commonly linked to cultivation of annual crops because bare soil is highly vulnerable to erosive forces of rainfall (Hellden, 1987; Hurni, 1990). Communally owned pasture with unrestricted grazing was the next most vulnerable (Table 9.8) with rainy season run-off and soil loss estimated at 10,000 $m^3\ ha^{-1}$ and 26.3 $t\ ha^{-1}$, respectively. Effective by-laws controlling stocking rates on community pastures reduced run-off and erosion by more than 90 per cent. Privately owned grazing land fared even better. In all cases, the severity of resource degradation was correlated with the steepness of hillsides. These observed trends confirm other studies (such as Taddesse *et al.*, 2002) and support the view that water depleted through downslope discharge does not contribute in a positive way to increasing water productivity. Descheemaeker *et al.* (2010a, 2011) also concluded that providing drinking water to livestock at individual households, increasing feed availability and quality and promoting land rehabilitation would greatly increase LWP.

Table 9.8 Run-off volume and sediment load of the main rainy season from pastures having different ownership patterns and slopes

Pattern of pastureland ownership	*Slope (%)*	*Run-off volume ($m^3\ ha^{-1}$)*	*Sediment load ($t\ ha^{-1}$)*
Communally owned and open unrestricted grazing	<10 15–25	10,125.0 12,825.0	26.3 45.27
Communally owned pasture supported with local by-laws	<10 15–25	3307.5 4927.5	7.84 14.24
Privately owned enclosed pasture	<10 15–25	1147.5 1687.5	1.65 3.39
Cropland (Hellden, 1987)	<10 10–15		29.4 69.6
Standard error of the mean		607.5	1.47

Source: Alemayehu *et al.*, 2008

Sudan

Most of the Nile Basin's livestock reside in Sudan (Table 9.4), where they sustain millions of poor farmers and herders, contribute about 20 per cent of national GDP, and form a significant part of Sudan's non-oil exports. This section describing livestock in Sudan takes a broad-brush review of secondary information and includes selected specific surveys, in contrast to more detailed work undertaken in smaller areas in Ethiopia and Uganda. The majority of the country's domestic animals (including sheep, goats, cattle, camels and equines) are found in the Central Belt of Sudan (Figure 9.6), an area composed of arid and semi-arid livestock-dominated and mixed crop–livestock systems (LGA and MRA), irrigated (MIH and MIA) and urban livestock production in the Nile Basin (Figure 9.1). Rainfall ranges from below 100 mm yr^{-1} in its far north to about 800 mm yr^{-1} in its far south. Limited surface water is locally available from the Nile, its tributaries and other seasonal rivers. The Central Belt encompasses 13 states, covers 75 per cent of the area of the Sudan, accommodates 80 per cent of its people and 73 per cent of its total livestock, and sustains most of the crop production. The belt's link to the Nile Basin is strong in terms of livestock production in schemes irrigated from the Nile, livestock mobility between rain-fed and irrigated areas and livestock trade with other Nile Basin countries (Faki *et al.*, 2008). For example, the only practical way livestock can access vast grazing lands during the more favourable rainy season is by having access to the relatively nearby Nile's blue water system in dry periods. Transhumance and nomadic modes of production, thriving on natural pastures, is the ruling practice, but cropland expansion increasingly impedes pastoral mobility. Modern sedentary dairy farms exist in the vicinity of towns and big settlement areas.

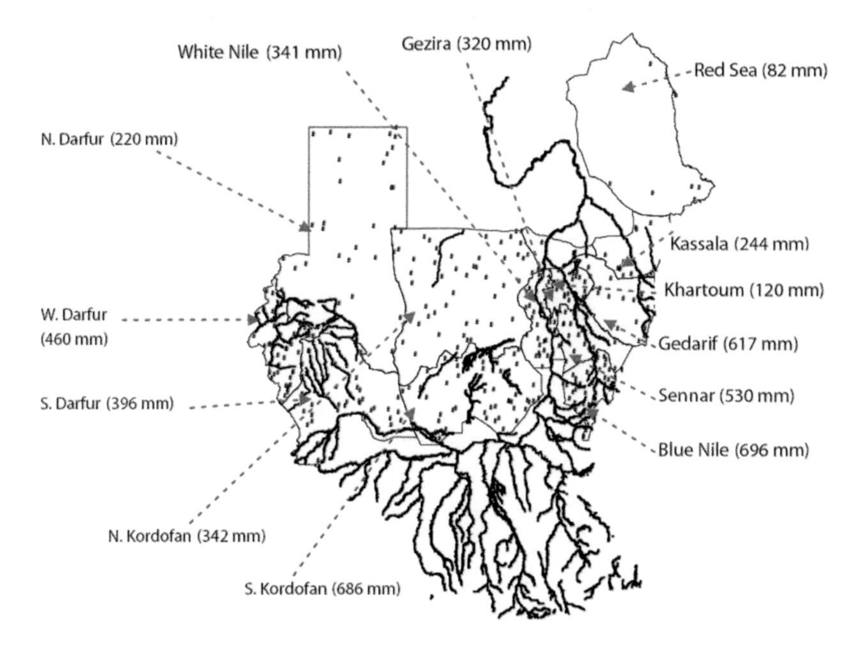

Figure 9.6 Sudan's Central Belt with spatial distribution of livestock (TLU), rivers and streams, and average rainfall from 1978 to 2007 in states' capitals

Milk and meat productivity and production are low and variable due to lack of feed and drinking water, low fertility, and high morbidity and mortality rates (Faki and van Holst Pellekaan, 2007; Mekki, 2005; Wilson, 1981; Mufarrah, 1991). In general, animal production underperforms relative to the potential of both the breeds of animals kept and the capacity of the environment where they are raised, implying considerable potential for improvement. Low off-take confirms the legacy of animal hoarding by pastoral communities, a tendency motivated by perceived need for prestige, insurance against drought, and a wealth savings strategy. Poor market access for transhumant herders also discourages investment in animal production. Other important constraints to animal production include increasing barriers to pastoral migration, lack of secure land tenure and water rights, water regulatory and management institutions that largely ignore the needs of the livestock sector, encroachment of irrigated and mechanized rain-fed agriculture into rangelands, and a breakdown in the traditional means for conflict resolution.

Feed and water shortages are the major biophysical constraints to livestock production. For example, in 2009, feed balances were negative in nine of the 13 states and in surplus in only North Darfur and Red Sea states (Figure 9.7). Access to drinking water is vital for livestock production. Without drinking water, livestock die. All states in the Central Belt except for Khartoum and Red Sea also suffer from shortages of drinking water for livestock for at least part of the year (Table 9.9). In brief, shortages in drinking water and long treks in high temperatures to find watering sites increase heat stress, consume excessive metabolizable energy and expose animals to increased feed shortages and disease risks, when large numbers concentrate around the few available water sources, including the Nile's lakes and rivers, wells and hafirs.

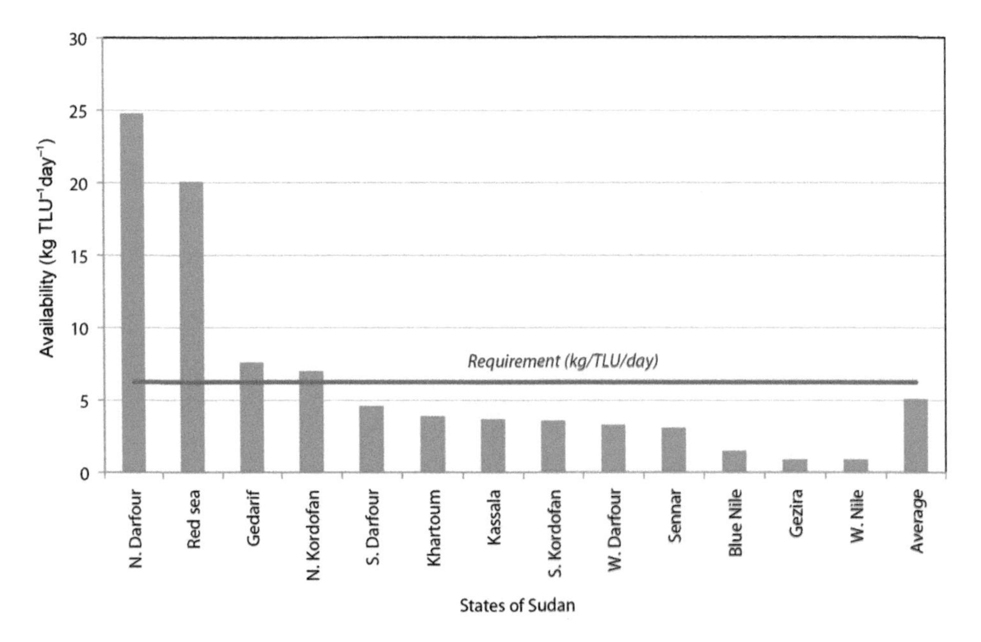

Figure 9.7 Feed balances in terms of dry matter feed by state across Sudan's Central Belt in terms of requirements versus availability

Notes: Feed balances are calculated according to daily feed dry matter (DM) requirements. Dotted line, 6.25 kg TLU^{-1} day^{-1}, assuming 2.5 per cent DM of animal weight per day

Sources: DM assumption based on Ahmed El-Wakil, personal communication. Data on pasture availability are from the Range Department of the Ministry of Agriculture, provided by Mr Mohamed Shulkawi

LWP in Sudan is much lower than its potential. In monetary terms, LWP derived from live animals and milk sales provides a useful overall productivity indicator, although it does not include other benefits that domestic animals provide. For example, in 2009, LWP was substantially higher than irrigated production in rain-fed areas (Table 9.10). In rain-fed areas, LWP was higher in good seasons compared with drought years. In irrigated areas, LWP may increase slightly during good years compared with drought periods. With estimates of US\$0.17 and US\$0.23 mm^{-1} of rain in Kordofan and Gezira, respectively, crop water productivity was lower

Table 9.9 Average daily rural livestock drinking water availability, demand and balance (m^3 day^{-1}) in different states within Sudan's Central Belt (2007)

	Available water	*Average drinking demand*	*Peak drinking demand*	*Balance at average demand*	*Balance at peak demand*
Red Sea	126,410	20,075	31,677	106,335	94,733
Khartoum	83,210	24,979	28,083	58,231	55,127
Gedarif	55,096	66,417	85,896	–11,321	–30,800
Kassala	43,972	61,441	86,709	–17,469	–42,737
Sennar	32,839	71,622	92,136	–38,783	–59,297
North Darfur	52,448	87,478	115,947	–35,030	–63,499
White Nile	48,184	118,823	156,805	–70,639	–108,621
Gezira	61,507	140,928	170,469	–79,421	–108,963
Blue Nile	19,133	151,871	203,441	–132,738	–184,309
South Darfur	51,088	187,184	235,637	–136,096	–184,549
West Darfur	29,495	172,336	229,290	–142,842	–199,795
Greater Kordofan	244,488	335,245	464,446	–90,757	–219,959
Total	847,870	1,438,399	1,900,536	–590,530	–1,052,669

Note: Average demands are 25, 30, 4 and 4 l day^{-1} for cattle, camels, sheep and goats, respectively; at peak summer months, the corresponding values are 35, 65, 4.5 and 4.5 l day^{-1}. Human rural requirements are 20 l day^{-1} $person^{-1}$, according to the Ministry of Irrigation

Source: Available water computed from data of the Ministry of Irrigation; livestock in 2007 estimated from data of MoARF, 2006; requirements are calculated according to Payne, 1990

than LWP. The prime entry point for increasing LWP in Gezira is through improved water use efficiency in irrigation. Increased expansion of watering points and better management of adjacent grazing lands constitute a key starting point for increasing LWP in Kordofan. High LWP relative to CWP in the good and normal seasons partially reflects the higher prices for animal sourced food products relative to crops. In poor seasons, higher mortality and morbidity and lower prices reduce LWP. In all cases, lack of feed and water along with poor animal and rangeland management contributes to lower than optimal LWP whereby most water used supports animal maintenance rather than reproduction and growth or is lost through non-productive depletion such as excessive evaporation and surface water run-off.

Table 9.10 Monetary rainwater use efficiency (RUE) for livestock in selected rain-fed and irrigated areas (US$ $mm^{-1}$of rain equivalent TLU^{-1})

Area (state)	*RUE (good season)*	*RUE (normal season)*	*RUE (poor season)*	*Mean crop RUE*
Kordofan (rain-fed)	0.75	0.42	0.26	0.17
Gezira (irrigated)	0.20	0.23	0.28	0.23

Notes: Exchange rate of 2.2 Sudanese pounds per United States dollar. Methodological difference implies these estimates are not directly comparable to other LWP estimates in this chapter based on US$ m^{-3}. Figure 9.4 provides an estimate of LWP from Gezira in 2009 using units of US$ m^{-3}. RUE serves as a useful proxy for LWP in very dry areas where ET and rainfall are almost equal. These estimates are based on a synthesis of secondary information and are not comparable with LWP estimates based on volume of water depleted; error estimates are not available

Source: Peden *et al.*, 2009a

One key challenge faced by both the water and livestock sectors in Sudan's economy is increasing the water productivity of livestock. Multiple intervention options and opportunities exist and must be tailored to meet specific local needs. Some focus on water management. Others focus on livestock development. Yet others act indirectly on water and livestock. All address four LWP strategies (feed sourcing, enhancing animal productivity, conserving water, and spatially optimizing the distribution of animals, feed and drinking water), plus the adoption of livestock and water demand management practices. Selected examples of each follow.

Feed sourcing

Prior to the 1900s, pastoralism prevailed in the Central Belt of Sudan. In the early 1900s, irrigation development took place along the Nile River systems, particularly in the Gezira state. More recently, large-scale mechanized grain production evolved in Gederif state. Both of these developments displaced herding practices. However, both also provide opportunities for livestock development through the use of crop residues (Figure 9.8).

Figure 9.8 Large quantities of crop residues produced in Sudan's large-scale irrigation schemes and rain-fed, mechanized grain farms support animal production in feedlots near Khartoum

One advantage of using crop residues for feed lies in the fact that this feed source requires little or no additional water for production compared with that used to produce the crop. Dual-purpose crops are more water-productive than either food crops or feed crops by themselves (Peden *et al.*, 2007). As previously noted, the Nile's large-scale irrigated agriculture supports the basin's highest livestock densities, largely due to the abundance of crop residues and to a lesser extent due to irrigated forage. The Gezira's large human population generates a high demand for livestock and livestock products. Thus, demand for animal feed exceeds supply. In Gederif, large quantities of crop residues are also available, but, in contrast to Gezira, lack of drinking water restricts livestock-keeping so that feed supply exceeds local demand. Options to increase water productivity through the use of crop residues include provision of nutritional supplements to enable greater residue digestibility by ruminants. In Gezira, strengthening irrigation water management policy and practice that accommodate livestock and crop production is needed in large-scale irrigation. In Gederif, there is a need to either provide drinking water for livestock or transport the feed to locations where animal demand for feed is high.

Enhancing animal productivity

Maximum LWP is only possible when individual animals and herds are productive, healthy and kept in stress-free environments. Premature death and disease result in reduced or zero benefits from animals and animal products. Throughout the Central Belt of Sudan, high levels of mortality and morbidity keep LWP low. Long treks during dry seasons for drinking water subject animals to heat stress and exertion that divert energy from weight gains, lactation and reproduction. A primary LWP-enhancing option lies in the provision of husbandry practices that prevent disease transmission, veterinary care that improve animal health, and living conditions that reduce stress and unnecessary expenditure of energy. Examples of interventions include measures that protect herders' migration routes (Peden *et al.*, 2009a), provision of safe drinking water by use of troughs that spatially separate animals from drinking water sources (Figures 9.9 and 9.10) and veterinary care for waterborne diseases such as fascioliasis (Goreish and Musa, 2008).

Conserving water

Inappropriate watering practices (Figure 9.10a) lead to high risk of disease transmission. Separating animals from the hafir (reservoir) and pumping water to drinking troughs help maintain high-quality drinking water (Figures 9.10b, c).

Optimizing the spatial distribution of animals feed and drinking water over landscapes

Water depleted through evapotranspiration during production of feed sources on rangelands is lost for productive purposes when underutilized by livestock (Faki *et al.*, 2008; Peden *et al.*, 2009a). Consequently, LWP is low in grazing lands located far from drinking points because livestock, particularly cattle, cannot utilize otherwise available feed resources. We recognize that this water not used by livestock may contribute in other ways to ecosystem services. Conversely, LWP is also low in grazing lands with poor feed availability near watering points. Without adequate feed, they lose weight and become more susceptible to waterborne diseases characteristic of large numbers of animals concentrated around watering points in dry seasons. The primary intervention opportunity lies in distributing livestock, grazing land and drinking water resources over large areas in a manner that maximizes animal production but does not lead to degradation of feed and water resources. Collaborative multi-stakeholder action to place effective limits on herd sizes can help to ensure maximum productivity and sustainability accompanied by investments in optimally distributed watering points and satellite monitoring of seasonally variable rangeland conditions.

Figure 9.9 Sudan's pastoralists trek a long distance to find drinking water. Thirsty animals queue for extensive periods waiting for a chance to quench their thirst. High concentrations of animals quickly deplete feed resources near watering points while water deprivation, energy loss through trekking, limited feed and heat stress all lower animal production and LWP

Demand management

The Central Belt of Sudan is a prime example of the tendency of some livestock keepers to hoard animals as means for securing wealth and enhancing prestige, culturally important processes that consume large quantities of water. Strictly in terms of achieving goals for water development for food production and environmental sustainability, we argue that water used to enable hoarding is suboptimal. In terms of the LWP assessment framework (Figure 9.4), we hypothesize that, in future, LWP should be evaluated based on the value rather than volume of water depleted. In effect, introduction of water pricing could serve as an important option of increasing agricultural water productivity.

Need for better integration of livestock and water development

Growing domestic and export demand for livestock products, now encouraged by high-level policy, will place substantial new demands on agricultural water resources. This, however, provides opportunities to farmers, but may also increase competition for agricultural water, provoke conflict and aggravate poverty (Peden *et al.*, 2007). Increased livestock water productivity through application of the foregoing four strategies is required. To achieve a positive future outcome, there is great need for better integration of livestock, crop and water development and management. In Sudan's Central Belt, water supply is, on the whole, more limiting than fodder, particularly because fodder production and utilization are also highly dependent on access to water. In turn, evidence suggests current investment returns from water development in Africa are suboptimal (World Bank, 2008). Yet, proactive inclusion of livestock in irrigation development can significantly and sustainably increase farm income. Integration must take into account diverse disciplines such as hydrology, agronomy, soil science, animal nutrition, veterinary medicine, water engineering, market development, diverse socio-economic sciences and financial planning. Integration must simultaneously address local, watershed, landscape and basin scales and accommodate the need for biophysical, spatial and livelihood diversity as it seeks to establish prosperity and environmental sustainability.

a

b

c

Figure 9.10 In Sudan, water harvesting systems based on reservoirs, known as hafirs, and adjacent catchments are important sources of drinking water for livestock. (a) In some cases, uncontrolled free access to water creates hot spots for transmission of waterborne diseases and causes rapid sedimentation of water bodies. (b, c) Restricting animal access to open water and pumping water into drinking troughs help prevent animal disease and extend the useful lifespan of the infrastructure

Uganda

CPWF research on livestock and water in Uganda focused on the country's Cattle Corridor that comprises rain-fed, mixed crop-livestock systems in a relatively humid area (Figure 9.1). The Cattle Corridor stretches from the north-east, through central to south-east Uganda, covering about 84,000 km^2, or 40 per cent of the country's land area, and mostly falling within the Nile Basin. This area is highly degraded and the stocking rate is 80 per cent less than the land's potential carrying capacity. Overgrazing and charcoal production led to loss of much pasture and soil. Without adequate vegetation cover, rain washes soils downslope filling water bodies such as ponds and lakes. When small amounts of rain fall on bare clay soil, the clay expands sealing the soil surface preventing infiltration. Lost rainwater provides little value locally, but contributes to downstream flooding. Termites are prevalent and quickly consume virtually all useful plant materials including forage on which livestock depend. Damage from termites is most serious during the dry season, creating extensive patches of bare ground which often force the cattle owners to migrate in search of new pasture.

Some livestock keepers have constructed ponds known as *valley tanks* to provide domestic water and livestock drinking. Sediments from degraded upslope pastures fill valley tanks, reducing their water-holding capacity and limiting their usefulness in dry seasons. Without drinking water, herders are forced to migrate with their animals to the Nile's Lake Kyoga for watering where they are at high risk to waterborne diseases and, at high densities, they quickly deplete feed supplies. Resulting feed deficiency aggravates disease impact while overgrazing threatens riparian habitats and water quality.

The Makerere University team, in collaboration with the Nakasongola District administration and livestock-keeping communities, undertook an integrated systems approach to (i) reseeding degraded pastures, (ii) managing the valley tank and pasture complex, and (iii) enhancing quality and quantity of water in the valley tanks for livestock and domestic use.

Reseeding degraded pastures

Reseeding degraded pastures in 2006 was the first attempted intervention designed to restore pasture productivity and to prevent sedimentation of downstream valley tanks, but it completely failed due to the aggressive and destructive power of termites that devoured all new plant growth. After consultation with CPWF partners, the researchers learned that a similar problem was solved in Ethiopia by applying manure to the land before reseeding activities commenced. The second experimental reseeding began in 2007 using a formal replicated field experiment. The innovation included treatments with two weeks of night corralling of cattle in fenced areas located on former, highly degraded pastureland that lost all vegetative ground cover. This new study included six treatments with three replicates each:

- fencing plus manuring (FM);
- fencing exclosures only (FO);
- fencing plus reseeding (FR);
- fencing plus manure left on soil surface plus reseeding (FMRs);
- fencing plus manure incorporated into the soil plus reseeding (FMRi); and
- the control with no manuring, fencing or reseeding (C).

Estimates of pasture production were made during the following 15 months.

Application of manure, combined with fencing and reseeding enabled pasture production to increase from zero to about 4.5 and 3.1 t ha^{-1} in the wet and dry seasons, respectively (Table 9.11; Figure 9.11). The yearly total biomass production reached 7 t ha^{-1}. Fencing and reseeding had no lasting positive impact without prior application of manure through night corralling of cattle. Vegetative ground cover was higher in the wet season than in the dry season. Vegetative ground cover and production in the controls were zero in both dry and wet seasons. Numerous efforts have been made in the past to reseed degraded pasture in the Cattle Corridor. The tipping point that led to successful rehabilitation of the rangeland was the application of manure. Ironically, while overgrazing played a key role in degrading the system, manure from livestock was the key that enabled system recovery.

Figure 9.11 Night corralling of cattle prior to reseeding degraded rangeland (left) enabled the establishment of almost complete ground cover and annual pasture production of about 7 t ha^{-1} within one year in Nakasongola, Uganda.

Table 9.11 Impact of reseeding, fencing and manuring on rehabilitation of degraded pastures in Nakasongola, Cattle Corridor, Uganda

Season	*Treatment*			*Vegetative ground cover (%)*	*Dry matter production*
	Fencing	*Manure*	*Reseeding*	*(t ha^{-1} season^{-1})*	
Wet	No	No	No	0	0
	Yes	Yes	Yes	98	4.5
	Yes	Yes	Yes	77	2.7
	Yes	Yes	No	88	3.7
	Yes	No	Yes	50	1.9
	Yes	No	No	29	1.6
Dry	No	No	No	0	0
	Yes	Yes	Yes	71	3.1
	Yes	Yes	Yes	51	2.5
	Yes	Yes	No	71	3.3
	Yes	No	Yes	28	1.7
	Yes	No	No	14	1.2

Notes: Significant differences are as follows. Vegetative cover: seasons ($p<0.001$), treatments ($p>0.05$) and season treatment interaction ($p<0.001$). Dry matter production: seasons ($p<0.001$) and treatments ($p<0.05$)

Source: Mugerwa, 2009

The mechanisms by which manuring led to rangeland recovery in Uganda are complex. Responses may differ elsewhere. Termite damage is most apparent in overgrazed rangelands during dry seasons where loss of vegetation cover greatly reduces infiltration of rainwater and constrains plant growth even though termite activity enhances infiltration (Wood, 1991). This Ugandan experience suggests that termites prefer to consume non-living organic materials, but feed on live seedlings when land has become highly degraded. One hypothesis is that, during the dry season, natural die-back of pasture species' roots provides sustenance to termites averting their need to consume living plant materials. Once rangelands have been restored, the newly established pasture supports termite activity, processes that actually generate ecosystem services.

Agricultural water productivity of the grazing land was essentially zero because rainfall was depleted through evaporation or downstream discharge rather than through transpiration, a key driver of primary production. Within one year, vegetative ground cover increased from zero to almost 100 per cent, implying a huge opportunity to capture and utilize rainwater more effectively in the upper catchments.

Catchment and valley tanks management

Valley tanks are a major source of water for both livestock and people in the Cattle Corridor. Seasonal siltation and fluctuations in the quality and quantity of water greatly affect livestock production and LWP. Prevailing grazing and watering practices led to water depletion through enhanced contamination, run-off, discharge and evaporation, and decreased LWP and ecosystem health. Reseeding pastures aided by manuring affords a great opportunity to increase feed production and reduce sedimentation of the valley tanks and soil movement farther downstream. Providing a year-round drinking water supply mitigates the need for counterproductive and hazardous treks to alternative drinking sites along the Nile River.

The Makerere research team assessed the impact of improving vegetative cover and pasture production on water volume and quality down slope in valley tanks (Zziwa, 2009). Valley tanks with upslope vegetation retained water throughout the year-long study while those down slope from degraded pasture dried out before the end of the dry season (Figure 9.12). Water harvested from unvegetated catchments and open gullies had higher turbidity, faecal coliforms and sediment loads compared with vegetated ones. The unvegetated valley tanks received 248 m^3 of silt reducing the storage capacity by 18 per cent during the study whereas the vegetated reservoir received only 7 m^3 of sediment. Correcting for the volume of the reservoirs, the ratio of the storage capacity at the start of the study to the volume of sediment accumulated was 5.6 and 266. This implies that vegetated, well-managed catchments might sustain reservoirs for many decades, but sediments from degraded upslopes could completely fill valley tanks within 5 or 10 years. Zziwa (2009) also indicated that vegetated catchments help maintain the quality of water in valley tanks in terms of NH_4^+, NO_2^-, NO_3^-, turbidity and faecal coliform counts. Zziwa (2009) also concluded that the presence of abundant duck weed (*Lemna* spp.) is associated with higher water quality and suggests that livestock keepers could harvest *Lemna* spp. and use it as a high-quality feed supplement (Leng *et al.*, 1995).

Opportunities to increase water productivity of livestock

The Ugandan case study shows that increasing LWP is possible through conservation of water resources that enables regeneration of feed supplies and sustains drinking water supplies. Moreover, effective pasture-reservoir systems mitigate the need for herders to trek to the Nile

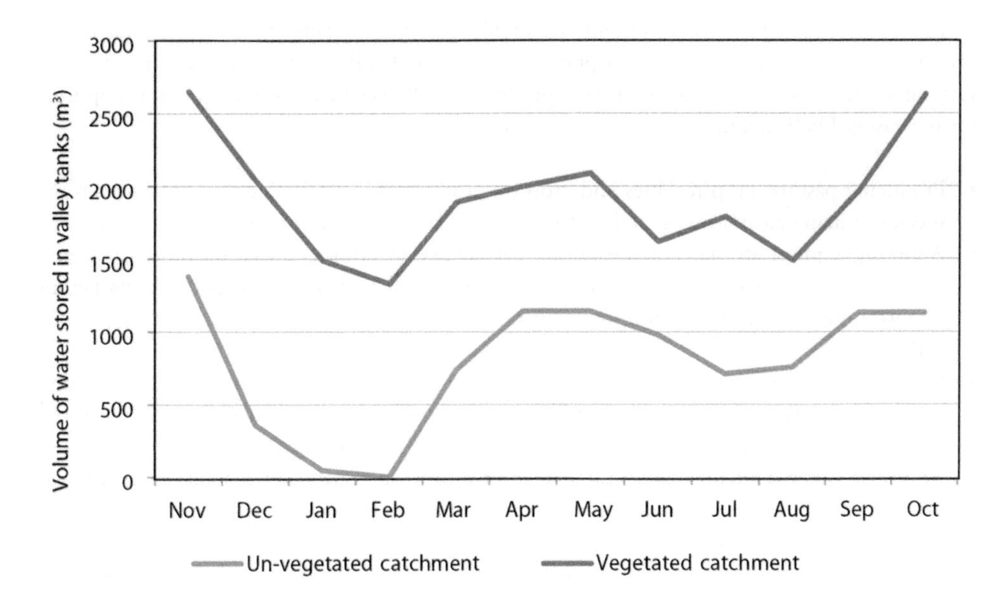

Figure 9.12 Comparison of impact of vegetated and un-vegetated catchments on water storage in valley tanks (November 2006 to October 2007)

River during the dry season thereby safeguarding animal health and reducing pressure on the river's riparian resources. Overgrazing aggravated by charcoal production has eliminated ground cover flipping the agro-ecosystem into a state of low productivity from which recovery was difficult. Identifying the ecological lever enabling rehabilitation of the system was a major breakthrough enabling the conversion of evaporation and excessive run-off into transpiration and vegetative production. Large-scale adoption of the lessons learned in the Cattle Corridor could make a major contribution to reversal of desertification and improvement of livelihoods in the Nile Basin.

Conclusions

Livestock water productivity (LWP) is a systems concept. Increasing LWP requires understanding of the structure and function of agro-ecological systems. In most of the Nile's production systems, livestock are raised on already degraded land and water resources. Often, human livelihood systems are vulnerable or broken due to poverty, inequity and lack of access to livelihood assets. Degraded systems have frequently passed tipping points and are trapped in states of low productivity. Thus, increasing LWP is a matter of rehabilitation rather than one of sustainable management of the status quo. For example, the Cattle Corridor case study demonstrated the value of identifying key constraints to system improvement. In this case, providing manure for termites unlocked the potential for agro-ecosystem restoration.

Although controlling termites is not a quick fix for the broad challenges of land and water degradation, this innovation from Uganda serves as an example suggesting that opportunities exist to increase water productivity in ways that promote agriculture, improve livelihoods and contribute to combating desertification.

Because livestock production systems are highly variable in terms of biophysical and socio-economic conditions, intervention options to increase LWP must be tailored to spatially varying local, regional, national and basin-specific scales. Nevertheless, key intervention options for increasing LWP include:

- Producing pasture, crop residues and crop by-products using palatable and nutritious plant species utilizing agronomic practices that foster high crop water productivity (CWP).
- Adopting appropriate state-of-the-art animal science technologies that promote high feed conversion efficiencies, low mortality and morbidity, efficient herd management, marketing opportunities for livestock, and provision of essential farm inputs such as veterinary drugs and credit.
- Managing rain-fed croplands and pasturelands to maximize production, subject to maintaining high levels of transpiration, infiltration, biodiversity and soil health along with low levels of excessive run-off, evaporation and soil water-holding capacity.
- Adopting water demand management planning tools such as water pricing to encourage rational use of water for livestock production and provision of alternatives to livestock hoarding as a means to secure wealth.
- Ensuring coherent policies, institutional and financial arrangements at local, regional and national scales are conducive to increasing LWP and more importantly to ensuring equitable and sustainable food security and livelihoods.

CPWF research suggests that it is not sufficient to focus on single interventions aimed at increasing LWP. In most cases, multiple interventions addressing two or more LWP enhancing strategies are needed. Selecting appropriate feed sources, enhancing animal production and conserving water resources are required simultaneously. In most cases, interventions will require expertise from diverse academic and practical disciplines. For example, water conservation will involve governance, gender analyses, economics, soil science, crop science, animal sciences, engineering and hydrology.

LWP is a characteristic of livestock production systems ranging from local basin scales. Improving LWP at one scale may decrease LWP at another. By reducing water depletion due to downstream discharge or down-slope run-off, upstream areas may increase LWP. However, such action may reduce LWP downstream. At the level of the whole basin, spatial allocation of benefits derived from water may make it possible to increase overall benefits for diverse stakeholders. For example, the historical development of large-scale irrigation systems in Sudan marginalized herders while providing new opportunities for crop production. Nearly a century later, the irrigation systems show capacity to greatly strengthen the livestock sector through animal production within the schemes, provision of crop residues and by-products to nearby herders, and supply of quality feeds for dairy production near Khartoum. Realizing this potential will require greater integration of the water, crop and livestock sectors at national, state and local levels.

Finally, emerging research on livestock in the Nile Basin suggests that activities undertaken to increase LWP are highly compatible with, and perhaps identical to, the two priority development goals of reversing desertification and providing more water for agricultural production. The relatively arid rain-fed livestock-based and mixed crop–livestock systems receive about 10^{12} m^3 of rainfall. Much of this water never reaches the blue water systems of the Nile because it is depleted by evaporation. Shifting this evaporative loss to transpiration is potentially a major pathway for not only increasing LWP but also driving primary production to enable greater rain-fed crop production and rehabilitation of degraded lands.

References

Alemayehu, M., Peden, D., Taddesse, G., Haileselassie A. and Ayalneh, W. (2008) Livestock water productivity in relation to natural resource management in mixed crop-livestock production systems of the Blue Nile River Basin, Ethiopia, in *Fighting Poverty Through Sustainable Water Use: Volume II: Proceedings of the CGIAR Challenge Program on Water and Food 2nd International Forum on Water and Food, Addis Ababa, Ethiopia, 10–14 November*, CGIAR Challenge Program on Water and Food, Colombo, Sri Lanka.

Astatke, A. and Saleem, M. (1998) Effect of different cropping options on plant-available water of surface-drained vertisols in the Ethiopian Highlands, *Agricultural Water Management*, 36, 2, 111–120.

Basarab, J. (2003) Feed Efficiency in Cattle, press release, Alberta Beef Industry Development Fund, Calgary, Canada, https://mail.une.edu.au/lists/archives/beef-crc-technet/2003-November/000016.html, accessed 27 November 2011.

Chapagain, A. K. and Hoekstra, A.Y. (2003) *Virtual Water Flows between Nations in Relation to Trade in Livestock and Livestock Products*, Value of Water Report Series no 13, UNESCO, Delft, The Netherlands.

Descheemaeker, K., Amede, T. and Haileslassie, A. (2010a) Improving water productivity in mixed crop-livestock farming systems of sub-Saharan Africa, *Agricultural Water Management*, 97, 579–586.

Descheemaeker, K., Mapedza, E., Amede, T. and Ayalneh, W. (2010b) Effects of integrated watershed management on livestock water productivity in water scarce areas in Ethiopia, *Physics and Chemistry of the Earth*, 35, 723–729.

Descheemaeker, K., Amede, T., Haileslassie, A. and Bossio, D. (2011) Analysis of gaps and possible interventions for improving water productivity in crop-livestock systems of Ethiopia, *Experimental Agriculture*, 47, S1, 21–38.

Faki, H. and van Holst Pellekaan, J. (2007) *Background Paper on Trade in Agricultural Products: Revitalizing Non-Oil Exports*, Diagnostic Trade Integration Study (DTIS) for the Integrated Framework Program, The World Bank, Khartoum, Sudan.

Faki, H., El-Dukheri, I., Mekki, M. and Peden, D. (2008) Opportunities for increasing livestock water productivity in Sudan, in *Fighting Poverty Through Sustainable Water Use: Volume II*: Proceedings of the CGIAR Challenge Program on Water and Food 2nd International Forum on Water and Food, Addis Ababa, Ethiopia, 10–14 November, CGIAR Challenge Program on Water and Food, Colombo, Sri Lanka.

FAO (Food and Agricultural Organization of the United Nations) (2004) *Tropical Livestock Units (TLU)*, Virtual Livestock Centre, Livestock and Environment Toolbox, Rome, Italy.

Ferris, R. and Sinclair, B. (1980) Factors affecting the growth of *Pennisetum purpureum* in the wet tropics. II: Uninterupted growth, *Australian Journal of Agricultural Research*, 31, 5, 915–925.

Gebreselassie, S., Peden, D., Haileslassie, A. and Mpairwe. D. (2009) Factors affecting livestock water productivity: animal scale analysis using previous cattle feeding trials in Ethiopia, *The Rangeland Journal*, 31, 2, 251–258.

Goreish, I. A. and Musa, M. T. (2008) Prevalence of snail-borne diseases in irrigated areas of the Sudan, 2008, in *Fighting Poverty Through Sustainable Water Use: Volume I*: Proceedings of the CGIAR Challenge Program on Water and Food 2nd International Forum on Water and Food, Addis Ababa, Ethiopia, 10–14 November, CGIAR Challenge Program on Water and Food, Colombo, Sri Lanka.

Haileslassie, A., Peden, D., Gebreselassie, S., Amede, T. and Descheemaeker, D. (2009a) Livestock water productivity in mixed crop-livestock farming systems of the Blue Nile basin: assessing variability and prospects for improvement, *Agricultural Systems*, 102, 1–3, 33–40.

Haileslassie A., Peden, D., Gebreselassie, S., Amede, T., Wagnew, A. and Taddesse, G. (2009b) Livestock water productivity in the Blue Nile Basin: Assessment of farm scale heterogeneity, *The Rangeland Journal*, 31, 2, 213–222.

Hanotte, O., Bradley, D. G., Ochieng, J. W., Verjee, Y., Hill, E. W. and Rege, J. E. O. (2002) African pastoralism: genetic imprints of origins and migrations, *Science*, 296, 336–339.

Hellden, U. (1987) An assessment of woody biomass community forests, land use and soil erosion in Ethiopia, *Regional Geography*, no. 14, Lund University Press, Lund, Sweden.

Herrero, M. and Philip K. Thornton, P. (2010) Mixed crop livestock systems in the developing world: present and future, *Advances in Animal Biosciences,* 1, 481–482.

Hurni, H. (1990) Degradation and conservation of soil resources in the Ethiopian Highlands, in *African Mountains and Highlands: Problems and Perspectives*, Hans Hurni and Bruno Messerli (eds), 51–63, African Mountains Association, Marceline, MO.

IFAD (International Fund for Agricultural Development) (2009) *Water and Livestock for Rural Livelihoods*, International Fund for Agricultural Development, Rome, Italy, www.ifad.org/english/water/innowat/topic/Topic_2web.pdf, accessed19 January 2011.

Keller, A. and Seckler, D. (2005) Limits to the productivity of water in crop production, in *California Water Plan Update 2005*, 4, Arnold Schwarzenegger, Mike Chrisman and Lester A. Snow (eds), pp177–197, Department of Water Resources, Sacramento, CA.

Khosh-Chashm, K. (2000) Water-conscious development and the prevention of water misuse and wastage in the Eastern Mediterranean Region, *Eastern Mediterranean Health Journal*, 6, 4, 734–745.

Leng, R. A., Stambolie, J. H. and Bell, R. (1995) Duckweed: a potential high-protein feed resource for domestic animals and fish, *Livestock Research for Rural Development*, 7, 1, www.lrrd.org/lrrd7/lrrd7.htm, accessed11 September 2010.

Mekki, M. (2005) Determinants of market supply of sheep in the Sudan: case study of Kordofan area, PhD thesis, University of Khartoum, Sudan.

MoARF (Ministry of Animal Resources and Fisheries) (2006) *Statistical Bulletin for Animal Resources*, no 15–16, General Administration for Planning and Economics of Animal Resources, Ministry of Animal Resources and Fisheries, Sudan.

Mufarrah, M. E. (1991) Sudan desert sheep: their origin, ecology and production potential, *World Animal Review*, 66, 23–31.

Mugerwa, Swidiq (2009) Effect of reseeding and cattle manure on pasture and livestock water productivity in rangelands of Nakasongola District, Uganda, MSc thesis, Makerere University, Kampala, Uganda.

Muli, A. (2000) Factors affecting amount of water offered to dairy cattle in Kiambu District and their effects on productivity, BSc thesis, University of Nairobi, Kenya.

Payne, W. (1990) *An Introduction to Animal Husbandry in the Tropics*, 4th edition, Longman Scientific and Technical York, Harlow, UK.

Peden, D., Freeman, A., Astatke, A. and Notenbaert, A. (2006) *Investment Options for Integrated Water–Livestock–Crop Production in Sub-Saharan Africa*, Working Paper 1, International Livestock Research Institute, Nairobi, Kenya.

Peden, D., Tadesse, G. and Misra, A. K. (2007) Water and livestock for human development, in *Water for Food, Water for Life: A Comprehensive Assessment of Water Management in Agriculture*, D. Molden (ed.), pp485–514, Earthscan, London.

Peden, D., Alemayehu, M., Amede, T., Awulachew, S. B., Faki, F., Haileslassie, A., Herrero, M., Mapezda, E., Mpairwe, D., Musa, M. T., Taddesse, G., van Breugel, P. (2009a) *Nile Basin Livestock Water Productivity*, CPWF Project Report Series, PN37, Challenge Program on Water and Food (CPWF), Colombo, Sri Lanka.

Peden, D., Taddesse, G. and Haileslassie, A. (2009b) Livestock water productivity: implications for sub-Saharan Africa, *The Rangeland Journal*, 31, 2, 187–193.

Perry, B., Randolph, T., McDermott, J., Sones, K. and Thornton, P. (2002) *Investing in Animal Health to Alleviate Poverty*, International Livestock Research Institute, Nairobi, Kenya.

Rockström, J. (2003) Water for food and nature in drought-prone tropics: vapour shift in rain-fed agriculture, *Philosophical Transactions of the Royal Society of London B*, 358, 1997–2009.

Saeed, I. and El-Nadi, A. (1997) Irrigation effects on the growth, yield and water use efficiency of alfalfa, *Irrigation Science*, 17, 2, 63–68.

Sala, O., Parton, W., Joyce, A. and Lauenroth, W. (1988) Primary production of the central grasslands of the United States, *Ecology*, 69, 1, 40–45.

Seré, C. and Steinfeld, H. (1995) *World Livestock Production Systems: Current Status, Issues and Trends*, FAO, Rome, Italy, www.fao.org/WAIRDOCS/LEAD/X6101E/X6101E00.HTM, accessed 28 November 2011.

Sheehy, D., Hamilton, W., Kreuter, U., Simpson, J., Stuth, J. and Conner, J. (1996) *Environmental Impact Assessment of Livestock Production in Grassland and Mixed Rainfed Systems in Temperate Zones and Grassland and Mixed-Rainfed Systems in Humid and Subhumid Tropic and Subtropic Zones, Volume II*, FAO, Rome, Italy.

Staal, S., Owango, M., Muriuki, G. Lukuyu, B., Musembi, F., Bwana, O., Muriuki, K., Gichungu, G., Omore, A., Kenyanjui, B., Njubi, D., Baltenweck, I. and Thorpe, W. (2001) *Dairy Systems Characterization of the Greater Nairobi Milk-Shed*, SDP Research Report, Ministry of Agriculture and Rural Development, Kenya Agricultural Research Institute, and International Livestock Research Institute, Nairobi, Kenya.

Taddese, G., Saleem, M. A., Abyie, A. and Wagnew, A. (2002) Impact of grazing on plant species richness, plant biomass, plant attribute, and soil physical and hydrological properties of vertisols in East African Highlands, *Environmental Management*, 29, 2, 279–289.

Van Breugel, P., Herrero, M., van de Steeg, J. and Peden, D. (2010) Livestock water use and productivity in the Nile basin, *Ecosystems,* 13, 2, 205–221.

Wilson, T. (1981) Management and productivity of sheep and goats in traditional systems of tropics, *International Symposium on Animal Production in the Tropics*, University of Gezira, Wad Madani, Sudan, 21–24 February1984.

Wood, T. (1991) Termites in Ethiopia: the environmental impact of their damage and resultant control measures, *Ambio*, 20, 3–4, 136–138.

World Bank (2008) *Investment in Agricultural Water for Poverty Reduction and Economic Growth in Sub-Saharan Africa*, World Bank, Washington, DC.

Zziwa, E. (2009) Effect of upper catchment management and water cover plants on quantity and quality of water in reservoirs and their implications on livestock water productivity (LWP), MSc thesis, Makerere University, Kampala, Uganda.

10

Overview of groundwater in the Nile River Basin

Charlotte MacAlister, Paul Pavelic, Callist Tindimugaya, Tenalem Ayenew, Mohamed Elhassan Ibrahim and Mohamed Abdel Meguid

Key messages

- Groundwater is gaining increasing recognition as a vital and essential source of safe drinking water throughout the Nile Basin, and the demands in all human-related sectors are growing. The technical and regulatory frameworks to enable sustainable allocation and use of the resource, accounting for environmental service requirements, are largely not in place.
- The hydrogeological systems, and the communities they support, are highly heterogeneous across the basin, ranging from shallow local aquifers (which are actively replenished by rainfall recharge, meeting village-level domestic and agricultural needs) through to deep regional systems (which contain non-replenishable reserves being exploited on a large scale). A uniform approach to management under such circumstances is inappropriate.
- The database and monitoring systems to support groundwater management are weak or non-existent. With few exceptions, groundwater represents an unrecognized, shared resource among the Nile countries.
- Most Nile countries have strategic plans to regulate and manage groundwater resources but, so far, these largely remain on paper, and have not been implemented.

Introduction

Groundwater has always been essential for human survival throughout Africa, and this is the case in the Nile River Basin (NRB; UNEP, 2010). Traditionally, groundwater was accessed first at naturally occurring springs and seepage areas by humans and animals; later, as human ingenuity increased, it was accessed via hand-dug wells, advancing to hand-pumps and then to boreholes and mechanized pumps. Throughout the NRB we can see all of these forms of groundwater access in use today. As the population's ability to develop and use technologies to access groundwater has grown, the scale of abstraction and human demand on groundwater resources have also increased. Masiyandima and Giordano (2007) provide a good overview of the exploitation of groundwater in Africa. Groundwater use the NRB includes domestic water supply in rural and urban settings for drinking and household use and small commercial activities; industrial use and development for tourism; agricultural use for irrigation and livestock production, from subsistence through to commercial scales; and large-scale industrial activities, such as mineral exploitation.

The overall type and distribution of the primary aquifers in the region have been quite well known since the mid-twentieth century (Foster, 1984). However, quantitative information on characteristics such as recharge rates, well yields and chemical quality is less consistent and depends largely on specific surveys largely generated by prospecting within a particular area. The same can be said for knowledge on the groundwater resource at the national level: groundwater is extensively utilized within the Egyptian part of the NRB, and therefore there is an abundance of data at the local and national level (although this may not be compiled to best manage the resource as a whole). This can be compared with upstream Uganda, where until recently most groundwater use was traditional hand-dug shallow wells for domestic use and supplementary irrigation, and there is limited quantitative data to enable management of the resource. However, groundwater in Uganda, and Africa as a whole, is essential for domestic water supply.

This chapter provides an overview of the groundwater within the NRB, and the uses, monitoring, policy and regulations relating to groundwater in four of the Nile countries (Egypt, Ethiopia, Sudan and Uganda), based on the current situation and available literature. The sections on regional hydrogeology, groundwater recharge rates, distribution and processes provide a summary of the known physical characteristics of the NRB aquifers. We discuss current and potential utilization of groundwater in the NRB in the section on groundwater utilization and development; the section on monitoring and assessment of groundwater resources briefly examines the current state of groundwater monitoring, while the section on policy, regulation and institutional arrangements for groundwater resource management provides an overview of some of the policy and regulatory arrangements and constraints. As statistics on utilization and development plans are normally based on national boundaries, and given the wide range in the type and form of data available, we address current groundwater use, potential, monitoring and regulation on a country basis. The final section offers some concluding remarks on groundwater in the basin.

Regional hydrogeology

The regional hydrogeological framework for the NRB and surrounding regions is well-defined as a result of several decades of effort resulting in the development of hydrogeological maps at the continent scale. Within the NRB (and the continent as a whole), there are four generalized types of hydrogeological environments: crystalline/metamorphic basement rocks, volcanic rocks, unconsolidated sediments and consolidated sedimentary rocks (Figure 10.1; Table 10.1; Foster, 1984; MacDonald and Calow, 2008).

Basement rocks comprise crystalline igneous and metamorphic rocks of the Precambrian age and are present across the area, but mainly in the upstream parts of the basin. With the exception of metamorphic rocks the parent material is essentially impermeable, and productive aquifers occur where weathered overburden and extensive fracturing are present. Consolidated sedimentary rocks are highly variable and can comprise low permeability mudstone and shale, as well as more permeable sandstones, limestones and dolomites. They tend to be present in the lower parts of the basin, forming some of the most extensive and productive aquifers. In the more arid regions, large sandstone aquifers have extensive storage, but much of the groundwater can be non-renewable, having originated in wetter, past climates. Unconsolidated sedimentary aquifers are present in many river valleys. Volcanic rocks occupy the NRB uplands (mainly the Ethiopian Highlands), where they form highly variable, and usually highly important, productive aquifers.

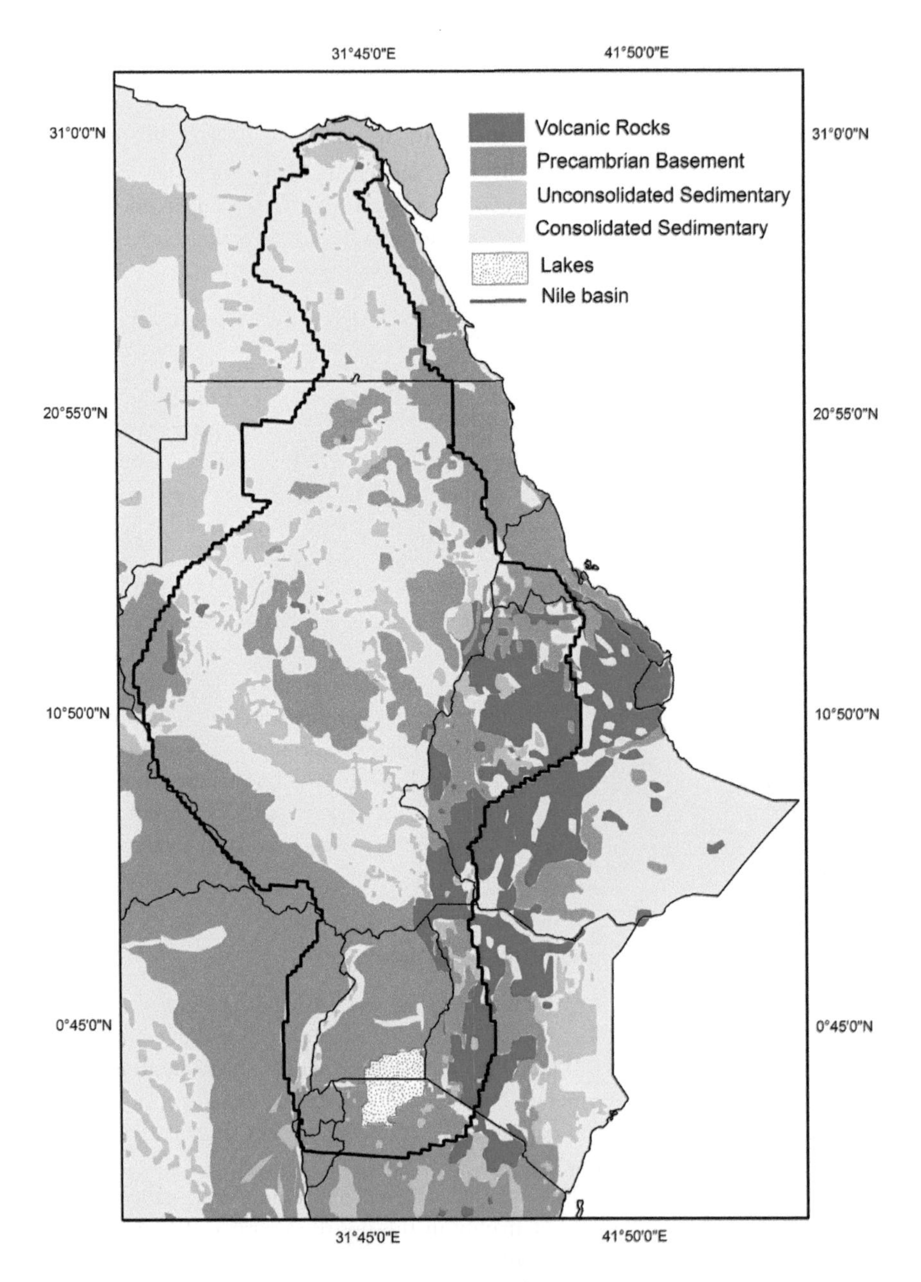

Figure 10.1 Generalized hydrogeological domains of the Nile River Basin

Source: Adapted from MacDonald and Calow, 2008

The upstream NRB reaches of Uganda are characteristic of a crystalline bedrock setting. Here aquifers occur in the regolith (weathered rock) and in the fractured rock (unweathered), typically at greater depth. If the regolith thickness is large, weathered rock aquifers may have a good well yield; however, generally, the more productive aquifers are found in the contact zone

Table 10.1 General characteristics of the aquifers within the Nile River Basin

Country	*Basin/region name*	*Hydrogeological environment*	*Depth (m)*	*Depth to SWL (m)*	*Yield* ($l\ s^{-1}$)	*T*[2] ($m^2\ d^{-1}$)	*S*[3]
UGA	Country-wide	Basement rock + alluvial			0.2–13.9	16–34	0.011–0.21
ETH	Abay Basin	Volcanic and basement	60–252	AR[4]–138	0–2.5	31–2157	
ETH	Abay and Baro-Akobo basins (northwest and west)	Dominantly volcanic	56–100	0.8–73	0.8–30	1–2630	
ETH	Tekeze (north and northwest)	Volcanic, sedimentary and basement		51–180	32–168	0.0–2.5	32–240
SUD	El Gash	Unconsolidated (alluvial)			0.6–1.2	1000	10^{-1}–10^{-2}
SUD	Bara	Detrital Quaternary & Tertiary deposits			0.1–5.8	35–210	10^{-2}
SUD	Baggara	Detrital Quaternary & Tertiary deposits			0.2	130–880	10^{-3}–10^{-5}
SUD	Seleim	Consolidated (NSAS)			2.3–5.8	1500	10^{-2}
SUD	Khartoum	Consolidated (NSAS)			0.5–1.6	250	10^{-3}–10^{-4}
EGY	Nile Delta	Unconsolidated		0–5		500–1400	
EGY	Nile Valley	Unconsolidated		0–5		<50,000	
EGY	Kharga	Consolidated (NSAS)		0–30		1000–2800	
EGY	Natron/Qattara	Consolidated (Mohgra)		100			
EGY	Wadi Araba	Consolidated (Carbonate)		AR[4]			

Notes: SWL = standing water level; UGA = Uganda; ETH = Ethiopia; SUD = Sudan (North and South); EGY = Egypt; T[2] = transmissivity; S[3] = storativity; AR[4] = artesian conditions; NSAS = Nubian Sandstone Aquifer System.

Sources: Authors' data; Tindimugaya, 2010; El Tahlawi and Farrag, 2008

between the regolith and bedrock due to higher aquifer transmissivity associated with the courser grain sizes and less secondary clay minerals. The highest yielding aquifers are the fractured bedrock if the degree of fracturing is high and hydraulically connected to the overlying regolith which, although low in permeability, provides some degree of storage and replenishment. The weathered aquifer is unconfined whereas the fractured-bedrock aquifer is leaky and the two aquifers form a two-layered aquifer system (Tindimugaya, 2008). Alluvial and fluvial aquifers are found adjacent to major surface water courses. The aquifers are found at relatively shallow depths, with average depths for shallow wells of 15 m and boreholes of 60 m.

The hydrogeological setting in the Ethiopian part of the basin is extremely complex, with rock types ranging in age from Precambrian to Quaternary, with volcanic rocks most common in the highlands and the basement, and complex metamorphic and intrusive rocks in peripheral lowlands and a few highland areas (Chernet, 1993; Ayenew *et al.*, 2008). Sedimentary rocks cover incised river valleys and most recent sediments cover much of the lowlands of all the major river sub-basins. The areas of Precambrian basement terrain are particularly complex due to various tectonic events. Groundwater flow systems, known from studies conducted in sub-basins such as Tekeze and Abay, suggest an intricate interaction of recharge and discharge, operating at local, intermediate and regional scales (Kebede *et al.*, 2005). Springs are abundant at different topographic elevations, suggesting that the shallow groundwater operates under local flow systems controlled by static ground elevation. However, the thickness and lateral extent of the aquifers indicate that deeper, regional flow systems operate mainly in the volcanic and sedimentary rocks. Most of the Precambrian rocks have shallow aquifers. In these aquifers depth to groundwater level is not more than a few tens of meters. From a database of 1250 wells from across the country, Ayenew *et al.* (2008) showed that the yields of most shallow and intermediate aquifers do not exceed 5 l s^{-1}, whereas the highly permeable volcanoclastic deposits and fractured basalts of Addis Ababa and Debre Berehan areas, for example, can yield between 20 and 40 l s^{-1}, respectively. Recent drilling in deep volcanic aquifers has located highly productive aquifers, yielding over 100 l s^{-1}. Depth to the static water level in the unconfined aquifers in alluvial plains and narrow zones close to river beds do not normally exceed 10 m except in highland plains, where it is around 30 m. Seasonal water table fluctuations rarely exceed 2 m.

Groundwater in the Sudanese part of the NRB lies within a multi-structural system of rifts, which range in age from the Paleozoic through to the most recent Quaternary and have resulted from the accumulation and filling with consolidated and unconsolidated sediments. Rift structures in Sudan also act as reservoirs for hydrocarbon reserves at greater depths. The major hydrogeological formations in Sudan include the Nubian Sandstone Aquifer System (NSAS), the Umm Ruwaba, Gezira sedimentary aquifer, the unconsolidated alluvium *khors* (seasonal streams) and *wadis*, and the Basement Complex aquifers. The NSAS may attain a thickness of 500 m, and is found under water table (unconfined) conditions or semi-confined artesian conditions. In some areas (e.g. northern Darfur), the NSAS is overlain by volcanic rocks. The Umm Ruwaba sediments are characterized by thick deposits of clay and clayey sands under semi-confined to confined conditions. The Basement Complex, extending over half of Sudan, is a very important source of groundwater. Unless subjected to extensive weathering, jointing and fracturing the parent rock is largely impervious. In the White and Blue Nile sub-basin sands and gravels in the Gezira and El Atshan Formations constitute important aquifers. Quaternary and recent unconfined aquifers tend to comprise a few metres of sand, silt and clay as well as gravel.

In Egypt, the major aquifers are generally formed of either unconsolidated or consolidated granular (sand and gravel) material or in fissured and karstified limestone. The hydrogeological provinces present within the NRB include the Nile Valley and Delta aquifers, Nubian sandstone aquifer, Moghra aquifer, tertiary aquifer, carbonate rock aquifers and fissured basement aquifers. The hydrogeological characteristics and extent of each hydrogeological unit are generally well known. The Nile Valley aquifer, confined to the floodplain of the Nile River system, consists of fluvial and reworked sand, silt and clay under unconfined or semi-confined conditions (Omer and Issawi, 1998). The saturated thickness varies from a few metres through to 300 m. This high storage capacity, combined with high transmissivity (5000–20,000 m^2 day^{-1}) and active replenishment from the river and irrigation canals makes the aquifer a highly valued resource. The Nile Delta consists of various regional and sub-regional aquifers with thicknesses of up to 1000 m. Much like the Nile Valley aquifer, the delta aquifers are composed of sand and gravel with intercalated clay lenses and are highly productive with transmissivities of 25,000 m^2 day^{-1} or more (El Tahlawi and Farrag, 2008). The NSAS is an immense reservoir of non-renewable (fossil) fresh groundwater that ranks among the largest on a global scale, and consists of continental sandstones and interactions of shales and clays of shallow marine of deltaic origin (Manfred and Paul, 1989). The 200–600 m thick sandstone sequence is highly porous with an average bulk porosity of 20 per cent, in addition to fracture-induced secondary porosity. Aquifer transmissivities vary from 1000 to 4000 m^2 day^{-1}. The Moghra aquifer is composed of sand and sandy shale (500–900 m thick) and covers a wide tract of the Western Desert between the Delta and Qattara Depression. The water in this aquifer is a mixture of fossil and renewable recharge. Discharge takes place through evaporation in the Qattara and Wadi El Natron depressions and through the lateral seepage into carbonate rocks in the western part of the Qattara Depression. The fissured and karsified carbonate aquifers generally include three horizons (lower, middle and upper) separated by impervious shales. Recharge to the aquifer is provided by upward leakage from the underlying NSAS and some rainfall input. The flow systems in the fissured limestone are not well understood, but it is known that surface outcrops create numerous natural springs. Hard rock (metavolcanic) outcrops are found in the Eastern Desert and beyond the NRB in South Sinai.

Groundwater quality and suitability for use

Data on groundwater quality in the NRB vary widely from country to country, but are generally restricted to the major constituents with a few exceptions. Time series are largely absent, except in Egypt where a groundwater quality monitoring network is well established (Dawoud, 2004), and more generally when associated with monitoring of public water supply wells (Jousma and Roelofsen, 2004). Spatial coverage is limited. Based on the available data, groundwater quality is known to be highly variable and influenced by the hydrogeological environment (granular, hard rock), type of water sources (tube wells, dug wells, springs) and level of anthropogenic influence.

In the Ugandan part of the basin, groundwater quality in most areas meets the guideline requirements for drinking water with the exception of iron and manganese in highly corrosive low pH groundwater, and nitrates in densely populated areas associated with poor sanitation (BGS, 2001). In hydrochemical terms, the groundwater is fresh, and contains a mixture of calcium–magnesium sulphate and calcium–magnesium bicarbonate types of water, which result from differences in the water–rock interactions. Generally, calcium bicarbonate groundwaters are younger and found under phreatic conditions, whereas calcium sulphate waters are older (Tweed *et al.*, 2005).

Generally groundwater quality is naturally good throughout the Blue Nile Basin (BNB), with freshwater suitable for multiple uses (Table 10.2). There are some localized exceptions, including salinity due to mineralization arising from more reactive rock types or from pollution due to urbanization, particularly underlying areas of highly permeable unconsolidated sediments in waters drawn from hand-dug wells and unprotected springs (Demlie and Wohnlich, 2006). The groundwater is dominantly fresh with total dissolved solid levels less than 200 mg l^{-1}, with pockets of elevated salinity evident in deep boreholes due to the presence of gypsum in sedimentary rocks of the Tekeze sub-basin and in the Tana sub-basin (Asfaw, 2003; Ayenew, 2005). Hydrochemical facies include bicarbonate, sulphate and chloride types, with calcium and magnesium being the dominant cations bringing associated hardness to the water.

The amount of the solute content depends on the residence time of groundwater and the mineral composition of the aquifer resulting in, for example, elevated mineral/salinity content in deep sedimentary aquifers due to extended residence times. Naturally high levels of hydrogen sulphide and ammonia can be present in deep anaerobic environments or shallow organic carbon-rich (swampy) areas and cause problems of taste and odour.

Fluoride is a major water-related health concern and is present at levels above drinking water standards in a number of localities, particularly in the western highlands, including waters emanating from hot springs (Kloos and Tekle-Haimanot, 1999; Ayenew, 2008) and within the Ethiopian rift volcanic terrain, adjacent to the NRB. Ethiopia recognizes the issue of high localized levels of fluoride in groundwater (e.g. Jimma) and is hosting the National Fluorosis Mitigation Project (NFMP). According to the Ministry of Health and United Nations Children's Fund (UNICEF), 62 per cent of the country's population are iodine-deficient. Nitrate contamination of groundwater, derived mainly from anthropogenic sources including sewerage systems and agriculture (animal breeding and fertilizers), is already a problem in rural and urban centres. This is worst in urban areas close to shallow aquifers. High nitrate concentrations thought to originate from septic tank effluents have been detected in several urban areas including Bahirdar, Dessie and Mekele (Ayenew, 2005). Several small towns and villages utilizing shallow groundwater via hand-dug wells have reported problems of nitrate pollution from septic pits (Alemayehu *et al.*, 2005).

The most common source of poor water quality in groundwater (and surface water) in Ethiopia is microbiological contamination, primarily by coliform bacteria. Poor management of latrine pits and septic systems in both rural and urban areas continues to lead to faecal contamination of groundwater, for example, digging septic pits too close to drinking water wells. Many urban populations still rely on hand-dug wells and unprotected springs as a drinking water source and these are frequently contaminated. The Ministry of Water Resources is developing national water quality guidelines. However, enforcement of any guidelines will need to be backed up by extensive training and education campaigns at all levels in rural and urban areas.

Within North and South Sudan, Nubian aquifers are considered to contain the best quality groundwater and are generally suitable for all purposes. The salinity of the groundwater varies from 80 to 1800 mg l^{-1}. More saline water is associated with down-gradient areas having enhanced residence times; shallow water table areas due to enrichment from evaporation and evapotranspiration, mineralization from claystones, mudstones, basalts, dissolution from salt-bearing formations and mixing with overlying Tertiary and Quaternary aquifers. Nubian groundwater is mainly sodium bicarbonate type, with calcium or magnesium bicarbonate waters common near the recharge zones. The salinity of the Umm Ruwaba sedimentary formation, the second most important groundwater source after the NSAS, is generally good but may rise to over 5000 mg l^{-1} along the margins. Groundwater quality is a major determinant in location and type of groundwater development that can take place. In one of the few reported studies on groundwater quality in Sudan, groundwater production wells in Khartoum State, east of the Nile and the Blue Nile rivers, reveal the NSAS groundwater is largely fit for human and irrigation purposes except at a few localities, due to elevated major ion levels (Ahmed *et al.*, 2000). Groundwater quality tends to be measured only in association with development activities to test suitability and ensure human health.

Within the Nile Valley region in Egypt, groundwater is of good quality (<1500 mg l^{-1} total dissolved solids, TDS, and mainly used for irrigation and domestic purposes; El Tahlawi and Farrag, 2008). In the valley margins remote from surface water systems to the east and west, the groundwater salinity tends to be more elevated (Hefny *et al.*, 1992). The groundwater in the

Nile Delta, which is primarily fed by the Nile River, is of higher quality in the southern part (<1000 mgl^{-1}), as compared with the north close to the Mediterranean Sea coast where there is a marked increase in salinity due to seawater intrusion. Water quality variations are complex and affected by various physical and geochemical processes. Wadi El-Natrun, situated within the Western Desert adjacent to the delta has moderate salinity (1000–2000 mg l^{-1} TDS) in the south, rising and deteriorating to the southwest (2000–5000 mg l^{-1}); whereas the Nubian waters of the Dakhla Oasis are fresh (Table 10.2). Domestic water is obtained from deep wells (800–1200 m) and is generally of very good quality except for naturally elevated levels of iron and manganese. Contamination of groundwater with nitrates in the Valley and Delta by industrial wastes around Cairo and other industrial cities and from sewer drain seepage poses a threat to public health, especially in areas where shallow hand pumps are used.

Table 10.2 Groundwater quality at three locations in the Nile Basin

Parameter[1]	*Blue Nile (Abay) sub-basin, Ethiopia*[2]	*Blue Nile sub-basin, Sudan*[3]	*Western Desert, Egypt*[4]
pH	6.99	8.0	6.39
TDS	366	340	351
Sodium	10	46	–
Calcium	49	27	25
Magnesium	10	20	13
Potassium	5	6	–
Bicarbonate	160	200	58
Chloride	20.5	24	90.8
Carbonate	9.5	–	–
Sulphate	9	18	53
Fluoride	47	0.5	–
Nitrate	47	–	1.0
Silica	40	–	7.5
Phosphate	–	–	0.04

Notes: [1] Units are mg.l^{-1}, except for pH
[2] median value quoted, n = 13
[3] Al-Atshan aquifer from Hussein, 2004
[4] NSAS, Dakhla Oasis, n = 10 from Soltan, 1999

Groundwater recharge rates, distribution and processes

Sustainable development of groundwater resources is strongly dependent on a quantitative knowledge of the rates at which groundwater systems are being replenished. A reasonably clear picture of the distribution of recharge rates across the NRB has recently begun to emerge. Using satellite data from the Gravity Recovery and Climate Experiment (GRACE), supported by recharge estimates derived from a distributed recharge model, Bonsor *et al.* (2010) found values ranging from less than 50 mm yr^{-1} in the semi-arid lower (as well as upper) catchments, and a mean of 250 mm yr^{-1} in the subtropical upper catchments (Figure 10.2). Along the thin riparian valley strips recharge from surface water and irrigation seepage may be up to 400 mm yr^{-1}. The total annual recharge within the basin has been estimated at about 130 mm, or 400 km^3 in volumetric terms. High temporal and spatial rainfall variability within the basin, when combined with the contrasting surface geology, accounts for this large range and generally low rates of groundwater replenishment. Values derived from the handful of local field studies, used as independent checks, are within this range (0–200 mm yr^{-1}). At the African scale, based on a 50×50 km grid resolution, Döll and Fiedler (2008) determined recharge to range from 0 to 200 mm yr^{-1} across the NRB with similar magnitudes and patterns to those later reported by Bonsor *et al.* (2010).

There have been a number of regional and local-scale recharge studies employing a variety of methods to arrive at groundwater recharge fluxes. An annual groundwater recharge in the order of 200 mm yr^{-1}, for the 840 km^2 Aroca catchment of the Victoria Nile, central Uganda, was determined by Taylor and Howard (1996) using a soil moisture balance model and isotope data. In several of the upper subcatchments of the Blue Nile, Ethiopia, recharge was estimated at less than 50 mm yr^{-1} in arid plains and up to 400 mm yr^{-1} in the highland areas of north-western Ethiopia, using a conventional water balance approach and river discharge analysis, chloride mass balance, soil–water balance methods and river or channel flow losses (Ayenew *et al.*, 2007). Bonsor *et al.* (2010) report groundwater recharge in the Singida region of northern Tanzania to be 10–50 mm yr^{-1}. Lake Victoria river-basin average estimate of just 6 mm yr^{-1} is reported by Kashaigili (2010). Within southern Sudan, Abdalla (2010) determined recharge from direct infiltration of rainfall through the soil to be less than 10 mm yr^{-1} at distances 20–30 km away from the Nile River. Farah *et al.* (1999) examined the stable isotope composition of the groundwater at the confluence of the Blue and White Nile sub-basins and determined the contribution of modern rainfall to groundwater recharge to be minimal, with much of the recharge derived from the cooler Holocene period. In the eastern desert region of Egypt, Gheith and Sultan (2002) deduced that around 21–31 per cent of the rainfall in high rainfall years (average recurrence interval of 3–4 years) is concentrated in wadis that replenish the alluvial aquifers.

Rainfall intensity, more than amount, is often a key determinant of groundwater recharge. In the upper catchments of central Uganda, Taylor and Howard (1996) showed that recharge of groundwater is largely determined by heavy rainfall events, with recharge effectively controlled more by the number of heavy rainfall events (>10 mm day^{-1}) during the monsoon, than the total volume of rainfall. This was further supported by the more recent work of Owor *et al.* (2009).

Aquifers in the proximity of the Nile River and its tributaries receive preferentially high recharge from the base of those watercourses, and from seepage return flows in areas under irrigation. Bonsor *et al.*, (2010) estimate the values to be in the order of around 400 mm yr^{-1} (Figure 10.2).

In the Ethiopian Highland subcatchments, studies consistently revealed that groundwater recharge varies considerably in space and time in relation to differences in the distribution and amount of rainfall, the permeability of rocks, geomorphology and the availability of surface water bodies close to major unconfined and semi-confined aquifers that feed the groundwater. Across the landscape, large differences are observed in recharge between the lowlands, escarpments and highlands (Chernet, 1993; Ayenew, 1998; Kebede *et al.*, 2005).

Within the Nile Valley areas of Egypt, the Quaternary aquifer is recharged mainly from the dominant surface water, especially from the irrigation canals that play an essential role in the configuration of the water table. The aquifer is recharged by infiltration from the irrigation distribution system and excess applications of irrigation water, with some of this returned to the Nile River.

Palaeo-groundwater is a vast resource in the more arid lower reaches of the basin. The NSAS

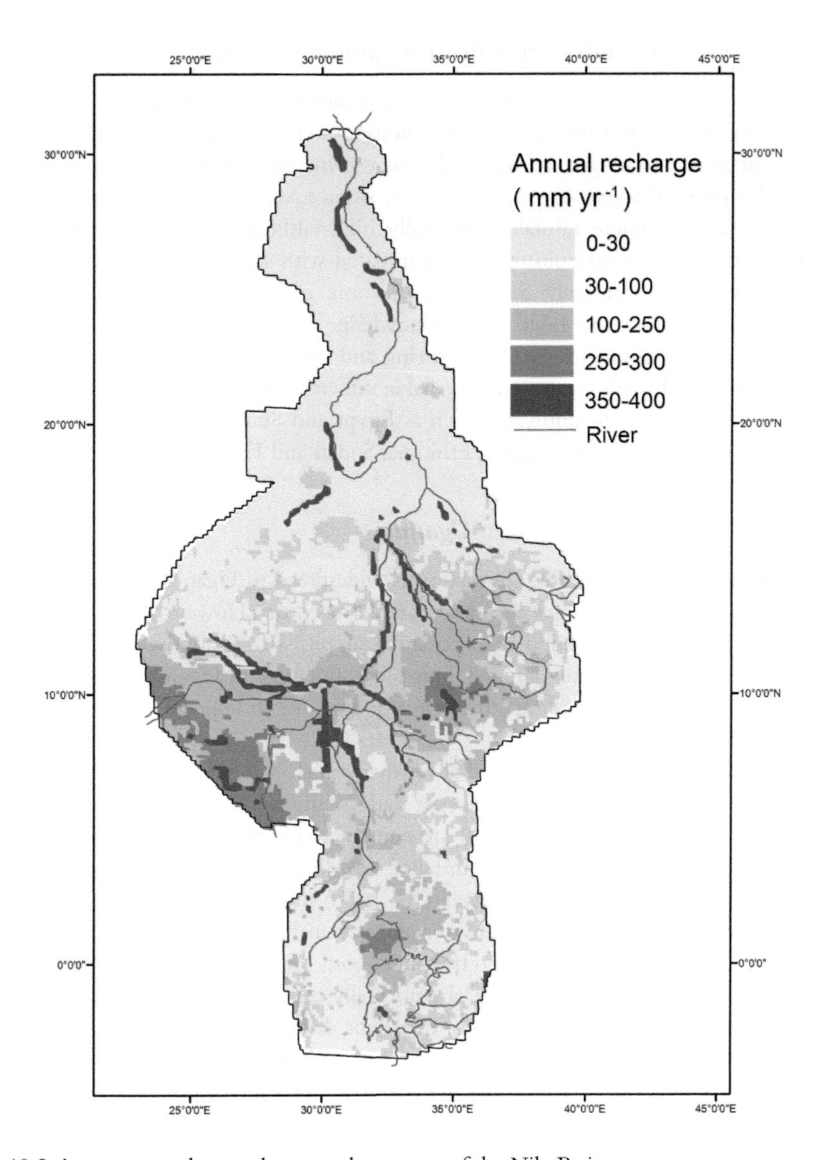

Figure 10.2 Average annual groundwater recharge map of the Nile Basin

Source: Adapted from Bonsor *et al.*, 2010

is considered an important groundwater source, but this is fossil groundwater and non-renewable due to both limited modern-day recharge and the long travel time. It has been suggested that in Pleistocene times, when more humid climatic conditions prevailed, that the NSAS was recharged by meteoric waters (Isaar *et al.*, 1972). The NSAS is found at depths and so is expensive to develop and the pumping and delivery infrastructures are also expensive to maintain. Under circumstances where the groundwater resource is poorly replenished or non-renewable, as is common across more arid environments, concepts of sustainable development must be revisited, with the intensive use of groundwater a contentious issue (Abderrahman, 2003).

Groundwater utilization and development

Throughout the NRB as a whole, the level of use or exploitation of groundwater varies widely. Groundwater is essential for drinking and for domestic water supply in most of the basin, while the use of groundwater to irrigate agricultural areas is primarily driven by the amount of rainfall and by the ease of access to, and supply of, surface waters. In the Upper Blue Nile catchment of Ethiopia, where rainfall is generally high (although seasonal droughts occur), groundwater extraction for agriculture is low compared with some areas of Egypt and Sudan where the resource is extensively developed. In some cases, the abstraction rate exceeds recharge (e.g. Gash, Sudan; see Table 10.1). Knowledge and data on groundwater use vary widely within the Nile countries (see 'Monitoring and assessment of groundwater resources' later in this chapter), although more readily available information naturally exists in those countries which rely heavily on groundwater, such as Egypt and Sudan. In the following sections we focus on four Nile countries: Uganda, Ethiopia, Sudan and Egypt.

Uganda

Historically, small-scale groundwater abstraction is widespread in Uganda, and more intensive development has been ongoing since the early twentieth century. However, abstraction remains relatively small scale when compared with potential supply, with groundwater utilized largely to satisfy rural and urban domestic demand. This is because most of Uganda has a ready supply of rainfall and surface water including large water bodies and widespread wetland areas, many of which are groundwater-fed.

Throughout Uganda, aquifers are found at relatively shallow depths (average 15 m) and the 'deep' boreholes are small-diameter wells deeper than 30 m (average 60 m). Shallow wells (less than 30 m, with an average depth of 15 m) are constructed in the unconsolidated formation. Boreholes and shallow wells are normally installed with hand pumps with a capacity of 1 m^3 hr^{-1} and their yields commonly range between 0.5 and 5 m^3 hr^{-1}. Since the early 1990s, there has been an increase in intensive groundwater abstraction for urban water supplies, utilizing high-quality groundwater with little or no treatment costs when compared with surface water. Boreholes with yields greater than 3 m^3 hr^{-1} are normally considered as suitable for piped water supply and installed with motorized pumps. In recent drilling of high-yielding boreholes in former river channels, yields of more than 20 m^3 hr^{-1} have been achieved. There are an estimated 20,000 deep boreholes, 3000 shallow wells and 12,000 protected springs in the country, utilized mainly for rural domestic water supply. Approximately 40,000 additional boreholes and 20,000 shallow wells are needed to provide 100 per cent rural water supply coverage (Tindimugaya, 2010).

While agriculture dominates the Ugandan economy, this is mostly smallholder rain-fed subsistence farming and irrigation is not widespread largely due to high investment costs, unsure returns, and lack of capacity. Most irrigation utilizes surface water with some limited upland horticulture crop irrigation using small-scale pumping systems. Traditionally, mineral-rich groundwater-fed wetlands with shallow water tables were utilized for rice production (with water tables managed in some cases for other crops) but this is now limited by the Ministry of Water, Lands and Environment (MWLE) in recognition of the ecosystem services and biodiversity value of wetlands.

Agricultural use of groundwater in Uganda is predominantly for the watering of livestock. However, figures which quantify this supply are very limited. According to MWLE figures (MWLE, 2006), approximately 20 deep boreholes, yielding an average of 8 m^3 hr^{-1}, have been constructed and installed with pumping windmills for livestock watering in northeastern Uganda. There is very little evidence of the utilization of groundwater to irrigate fodder crops or crop with residue used as animal feed, which would otherwise constitute the largest part of livestock water demand.

Ethiopia

Throughout the Ethiopian BNB groundwater is the most common source of domestic water, supplying at least 70 per cent of the population. In the rugged mountainous region of the Ethiopian Highlands settlement patterns of rural communities are determined largely by the distribution of springs. The depth to water table in most highland plains is less than 30 m. In the alluvial plains and narrow zones close to river beds, depth to the static water level in the unconfined aquifers does not normally exceed 10 m. These aquifers are the most commonly utilized water source for rural communities. In both cases, seasonal water table fluctuation is not thought to exceed 2 m (Ayenew *et al.*, 2008). Groundwater for domestic use is commonly utilized via springs, shallow hand-dug wells, sometimes fitted with manual pumps, and in some cases, boreholes. In urban centres deep boreholes normally provide water for both drinking and industrial purposes. Over 70 per cent of the large towns in the basin depend on intermediate to deep boreholes fitted with submersible pumps and in some cases, large fault-controlled high discharge springs.

Generally, groundwater quality is naturally good and suitable for multiple uses throughout the BNB. There are some naturally occurring areas of high total dissolved solids and locally high salinity, sulphides, metals and arsenic, but the main concerns for water quality are fluoride, iodine and man-made, point source pollution including nitrates and coliform bacteria.

Despite the importance of groundwater to the majority of the population, it was given very limited attention in planning and legislation in the past, although this is now changing. The demand for domestic water supply from groundwater has been increasingly achieved over the last two decades but will continue to increase and plans to expand access are ongoing (Table 10.3).

Table 10.3 Estimate of rural population supplied with domestic water from groundwater in Ethiopia: 2008 figures and planned improvements to be implemented by 2012

	Region					*National*
	Tigray	*Gambella*	*Benishangul Gumuz*	*Amhara*	*Oromia*	
Population in 2007 (million)[1]	4.3	0.31	0.67	17.2	27.2	74
Population supplied from groundwater (million)[2]	2.05	0.1	0.26	8.6	12.6	34.4
Percentage of population supplied[2]	58	39	44	56	51	54
Percentage supply planned for 2012[2]	109	94	89	118	95	100
Number of groundwater schemes planned 2009–12[2]	7928	610	1428	37,468	26,093	110,460

Sources: [1]2007 National Census, CSA/UNFPA, 2008; [2]MWR/GW-MATE, 2011

Currently, direct groundwater utilization for irrigated agriculture is marginal. This mostly takes the form of shallow wells close to rivers, and in some cases, downstream of micro-dams and sand dams, constructed to effectively recharge groundwater. Generally, a well yield of 2 l s^{-1} is considered adequate to irrigate one hectare. This is mostly supplemental irrigation of cash crops. While surface water irrigation is more common, the groundwater baseflow contribution to river flow should not be ignored. This component is significant and, without it, abstraction for irrigation, especially supplemental irrigation in dry periods, would be impossible. There is some extraction of groundwater for commercial horticulture and fruit production but as these wells are privately managed and largely unregulated, it is difficult to estimate their contribution. Groundwater pumping for commercial agriculture is likely to increase in the future, especially close to urban centres, such as the areas around Addis Ababa, where demand for fresh vegetables all year-round can be satisfied using groundwater and growing more sensitive crops inside (in greenhouses/polytunnels).

Groundwater is essential for livestock production in Ethiopia, primarily for drinking as opposed to feed or forage production. Farmers and pastoralists access groundwater all year-round to water livestock. When this is combined with domestic access, contamination of human drinking water with coliform bacteria is common. Outside of the Nile Basin, the 'singing wells' of the Borena people, pastoralists in southern Ethiopia, are one well-known example of shallow hand-dug wells (<10 m) where water is lifted by hand on a series of ladders, into livestock watering troughs.

A significant number of industrial sectors in Ethiopia rely on groundwater. Ethiopia has several bottled water producers, including naturally carbonated mineral water drawn from the Antalolimestones in the Takeze Basin of Tigray. In general the beverage industry, food processing, textile and garment, and cement producers are heavily reliant upon groundwater. The expanding mineral exploration sector requires significant access to water (both mining and opencast workings), and with prospecting ongoing, there is significant potential for further growth in groundwater demand. The construction 'boom' in many urban areas of Ethiopia also poses a major strain on current 'domestic' supplies: whether domestic supply is drawn from ground or surface water, the construction industry uses (and wastes) vast amounts of water, causing a significant stress on the existing domestic network. The growth in urban populations and migration to towns and cities, in many cases driving the construction boom, must be met by expansion in domestic water supply from groundwater (see 'Policy' section).

Sudan (North and South)

The Nile Basin drainage encompasses most of the major groundwater basins of North and South Sudan with total groundwater storage estimated at around 16,000 billion m^3. A range of values can be found for annual abstraction rates (Ibrahim, 2010), from 1 billion m^3 to more than double this when agricultural and domestic uses are combined. Annual recharge is estimated at around 2.3 billion m^3. Throughout Sudan, groundwater is accessed for drinking and domestic water supplies. While 70 per cent of groundwater abstraction in Sudan is reported to be for irrigation, groundwater constitutes around 50 per cent of urban and 80 per cent of rural domestic water supply (Ibrahim, 2010). At the time of writing, North and South Sudan have just undergone a process of separation after years of civil war. While government structures remain in place in North Sudan, and new Ministries are evolving in South Sudan, it is very difficult to access official 'government' figures externally, and most figures which can be accessed are in an unpublished form. Much of the recent published information relates to aid and donor missions, with a particular focus on water supply (e.g. Michael and Gray, 2005; Pact,

2008). Where data exist there are a number of discrepancies in reported figures and conflicting sources of information. For example, rural domestic water supply from groundwater was estimated at 63 million m^3 yr^{-1} in 2002 (including livestock; Omer, 2002) to 300 million m^3 yr^{-1} in 2010 (Ibrahim, 2010; Table 10.4), constituting a rise of 500 per cent at a time when the country was subject to severe conflict, which would naturally limit development.

The Nubian aquifers, underlying the Sudanese Nile, are generally considered to contain good-quality groundwater for all uses: the NSAS, Umm Ruwaba sediments, basement complex, the Gezira sands and gravels and alluvial formations are all important sources of both drinking and irrigation water, with the exception of a few saline pockets. The main constraint to supply of adequate safe drinking water is lack of management and provision of infrastructure, with under-investment and symptomatic poverty still obstructing water supply throughout Sudan. There are a number of historical conflicts over water and water supply points, which continue to be an issue, especially in rural areas (e.g. Darfur and Abyei regions).

Reliance on groundwater for domestic water supply, illustrated in Table 10.4, is likely to be an underestimate as many traditional or informal methods can go unrecorded. Domestic wells range in design and level of technology from simple holes in or close to banks of seasonal streams, lined with grass or tree branches, open large-diameter hand-dug wells lined with brick or concrete slabs, slim low-depth boreholes fitted with hand-pumps or electrical submersible pumps, to deep boreholes fitted with pumps. Even the more formal methods of abstraction can go unrecorded.

Table 10.4 Groundwater utilization for domestic supply throughout North and South Sudan

Region	*Urban (%)*	*Rural (%)*
Khartoum	50	90
Northern	50	60
Eastern	70	90
Central	50	60
Kordofan	60	70
Darfur	70	80
South Sudan	10	100
Average	51	79
Total supply (million m^3 yr^{-1})	800	300

Source: Ibrahim, 2009

Groundwater is essential for agriculture in Sudan. According to the Ministry of Irrigation and Water Resources, around 875 million m^3 are utilized annually for irrigation. The fertile soils found along the Nile floodplain and seasonal streams are irrigated using both the Nile surface water and groundwater. Groundwater is used to irrigate fruit and vegetables via a range of abstraction methods. Where individual plots do not exceed 4.2 ha, hand-dug wells or '*matars*' (hand-dug wells with pipes driven in) fitted with centrifugal pumps are common. Most plots are associated with fertile and easily cultivated floodplains with shallow, annually replenished water tables and low construction costs. Boreholes are also used for irrigation in some areas, either exclusively or to supplement surface water and rainwater.

There are numerous methods of groundwater abstraction in Sudan, determined mostly by the depth to water level, intended use, access to main electric supply and resources available. In rural areas open hand-dug wells are common, with water drawn by a rope-and-bucket system by hand, animal-power or windlass depending on depth. Technologies in drilled wells for domestic or agricultural use range from reciprocating hand-pumps, centrifugal pumps driven by electrical motors or diesel engines and electrical submersible diesel-engine-driven vertical turbine pumps.

Livestock contribute a significant additional agricultural water demand, particularly in less-fertile areas away from river valleys. The livestock population is estimated at around 140 million head (compared with a population of 45 million) concentrated mainly in southern, central and western Sudan, and annual groundwater abstracted for livestock watering is estimated at 400 million m^3. Groundwater also contributes to livestock water demand through production of fodder crops and crop residue for feed.

Table 10.5 illustrates the distribution of abstraction rates and well type in different areas irrigated with groundwater in both North and South Sudan.

Table 10.5 Areas irrigated with groundwater in North and South Sudan

State	*Locality*	*Area (ha)*	*Abstraction (million m^3 yr^{-1})*	*Well type*
Northern	El Seleim	20,000	345	Matar
Northern	Lat'ti Basin	5000	115	Matar
Nile	Lower Atbara Basin	1430	40	Matar
Kassala	Gash Basin	6200	145	Matar + borehole
Gezira	North Gezira	1500	40	Matar + borehole
Khartoum	Khartoum area	5000	120	Matar + borehole
North Kordofan	Bara area	1430	8	Hand-dug
North Kordofan	Khor Abu Habil	1400	5	Hand-dug
South Kordofan	Abu Gebeiha	4000	7	Hand-dug
South Kordofan	Abu Kershola	1000	3	Hand-dug
Greater Darfur	W. Azoum	3000	10	Matar + borehole
Greater Darfur	Jebel Marra area	2900	15	Hand-dug
North Darfur	Kabkabiya	1430	5	Matar
North Darfur	Wadi Kutum	1400	5	Matar
West Darfur	Wadi Geneina	950	5	Matar
South Darfur	Wadi Nyala	1400	8	Matar + borehole
Total		58,040	876	

Source: Ibrahim, 2009

North and South Sudan combined are estimated to have approximately 82 million ha of land suitable for arable production (one-third of the total combined area), of which around 21 per cent is currently under cultivation. In addition to the irrigated areas included in Table 10.5, 1.4 million ha of agricultural land across North and South Sudan were classified by the previous government as eligible for supplementary or complete irrigation by groundwater. Given that only 0.06 million ha out of 82 million ha of suitable arable land seem to be irrigated with groundwater currently, it is likely that, at some point in the near future, resources will be found to develop this resource and groundwater abstraction will increase significantly. A number of schemes were planned in the 1990s to produce food for export to Gulf states, and with the end to civil war, it is likely that such ventures will once more become viable.

Egypt

Currently, the total annual water requirement of all socio-economic sectors in Egypt is estimated to be 76 billion m^3 yr^{-1}, of which the agriculture sector alone requires 82 per cent (Attia, 2002). Egypt relies heavily on surface water from the Nile, with an annual quota of 55.5 billion m^3 yr^{-1} allocated according to the 1959 agreement between Egypt and Sudan. The total harvestable national run-off is around 1.3 billion m^3 yr^{-1} and the remainder of the demand must be satisfied by using groundwater. The two most important groundwater aquifers are the NSAS of the Western Desert, and Nile Valley and Delta system. The deep and non-renewable fossil water of the NSAS covers about 65 per cent of Egypt and extends into Libya, Sudan and Chad.

Clearly, Egypt's demand already exceeds its apparent supply and this is likely to be exacerbated in the future with increasing demand for expanding agriculture, population growth, urbanization and higher living standards. As the volume of surface water from the Nile cannot be guaranteed with shifting regional politics and uncertainties, groundwater exploitation will undoubtedly accelerate.

There are known to be more than 31,410 productive deep wells and 1722 observation wells distributed throughout the Nile Delta, Nile Valley, coastal zone, oases and Darb El Arbain, and Eastern Oweinat. These include wells for domestic and agricultural supply. Tables 10.6 and 10.7 illustrate extraction rates, use and potential, and distribution of wells.

Table 10.6 Current and potential groundwater use in the Egyptian Nile River Basin (2004 and 2010 values)

Location	*Production wells*[1]	*Abstraction (million m^3 yr^{-1})*	*Observation wells*[1]	*Potential (million m^3 yr^{-1})*
Northern West Coastal Zone and Siwa[2]	1000	149	1	194
Nile Delta and Nile Valley	27,300	5	1704	500
Zone of Lake Nasser	–	0.05	–	20
Western Desert, Oases and Darb El Arbain[2]	3100	1108	13	2246
Eastern Oweinat[2]	50	390	4	1210
Toshka[2]	–	59	–	101
Total		1711.05		4271

Sources: [1]Hefny and Sahta, 2004; [2]MWRI, 2010

Generally, domestic water is obtained from deep wells (>800 m) of naturally good quality. In urban areas, all houses are connected to a mains supply, while around 40 per cent of rural communities are reported to be connected, but large portions of the rural population still depend on water collected from small waterways. Small-capacity private wells are also common at the household level although many are in a poor condition. In the newly settled areas, most wells are managed by the Ministry of Housing and New Communities or are under the local unities in old towns and villages.

The 82 per cent of Egypt's water demand required for agriculture refers to the irrigation of existing cultivated areas, newly irrigated land reclaimed from the desert, and improved drainage and irrigation conditions. While approximately 70 per cent of this demand is satisfied by surface water diverted in the Nile Valley, the contribution from groundwater is most commonly on the fringes of, or outside of, irrigation project command areas. Around 25 per cent of the total volume allocated to irrigation is thought to contribute to return flow and groundwater recharge via agricultural drainage and deep percolation. Management of groundwater is often fragmented among different stakeholders which may include government agencies, NGOs, farmer organizations, the private sector and investors, depending on the scale of the project. In the newly settled areas around oases and other depressions in the desert, the water supply systems on a subregional and local level include *mesqas* (small/tertiary canals used for water supply and irrigation), wells (government and private), collectors and field drains.

In the newly settled areas of the Western Desert (west of the Nile), agriculture is mainly dependent on groundwater abstracted through deep wells from the NSAS. Shallow aquifers in the mid- and southern desert are contiguous with the deep aquifer providing potential for further groundwater development, and there are plans to expand agricultural land around oases in the western desert with irrigation from both shallow and deep wells. The main obstacles to utilizing this resource are the great depths to the aquifer (up to 1500 m in some areas), and deteriorating water quality at increasing depths. While development of groundwater in the NSAS is naturally limited by pumping costs and economies of scale, there are also transboundary considerations for this shared resource. This is formalized in a multilateral agreement between Egypt, Libya, Sudan and Chad, and an extensive monitoring network exists (see below).

The shallow aquifer of the Nile Valley and Delta is considered nationally as a renewable water source with extraction largely from shallow wells with a relatively low pumping cost. This aquifer is considered as a reservoir in the Nile River system by the Ministry of Water Resources, with a large capacity but with a rechargeable live storage of only 7.5 billion $m^3 yr^{-1}$. The current abstraction from this aquifer is estimated at 7.0 billion m^3 in 2009 (MWRI, 2010). Conjunctive use of surface water and groundwater is practised widely by farmers, especially during periods of peak irrigation demand and at the fringes of the surface water irrigation network, where groundwater can be the only source. In the Nile Delta areas, a distinction is made between 'old' and 'new' lands facing a shortage of irrigation water. In the old land, the main source of irrigation water is the Nile River but towards the end of irrigation canals, groundwater is in many cases the only source. As the shallow aquifer is in hydraulic contact with both the surface water irrigation system and the Nile River system, it can receive both recharge and pollution from surface water sources and is therefore vulnerable. The aquifer is also affected by programmes which reduce conveyance losses in waterways.

In the Eastern Desert (between the east bank of the Nile and the Red Sea) most groundwater development is confined to shallow wells within wadi aquifer systems and to desalination of groundwater. Total groundwater usage was estimated to be 5 million $m^3 yr^{-1}$in 1984 and the current extraction rate is likely to be closer to 8 million $m^3 yr^{-1}$. Potential for further development is largely based on deep wells (200–500 m), accessing the NSAS and large wadis of the Nile Valley and Lake Nasser catchments. There is also some potential for development of brackish groundwater, especially in the Red Sea coastal areas.

In addition to agricultural and domestic use, industry in Egypt is also highly dependent on groundwater. Factories may receive water from mains water supply system or their own wells. A few small factories may depend on surface water from mesqas. The tourism sector generally depends on a mains supply from the government system; private wells also exist along with desalination plants in coastal zones for both groundwater and sea water purification.

Monitoring and assessment of groundwater resources

The major shortcomings associated with groundwater monitoring systems in the NRB are symptomatic of much of Africa as a whole (Foster *et al.*, 2008; Adelana, 2009) and can be summarized as follows:

- Lack of a clear institutional/legal base and fragmented organizational responsibilities.
- Inadequate technical capacity and expertise, and lack of sustainable financing and resources to monitor and manage groundwater.
- Poorly coordinated groundwater development activities with little or no linkage to groundwater monitoring systems, and database management and retrieval systems.

As noted above, knowledge and data on groundwater use vary widely within the Nile countries but, in general, more information tends to be available in those countries which rely heavily on groundwater. All countries of the Nile are trying to improve their management of groundwater and this requires mapping of aquifers, groundwater monitoring, analysis of extraction and recharge rates, and proper data management. National efforts are also broadly supported by the research, NGO and donor community. The Groundwater Management Advisory Team (GW-MATE) of the World Bank Water Partnership Program has provided technical support throughout Africa over the last decade, and continues to do so, with very positive results (Tuinhof *et al.*, 2011).

Despite the heavy reliance on groundwater for domestic supplies in Uganda, there is no national monitoring network in place, and this is needed as a matter of priority.

Throughout Ethiopia, relatively extensive hydrogeological field surveys provide sufficient information to classify the major aquifers and their characteristics (Ayenew *et al.*, 2008). The Ministry of Water and Energy is now compiling an integrated database, the National Groundwater Information System (NGIS), which will replace the earlier ENGDA (Ethiopian National Groundwater Database) system hosted by the Ethiopian Geological Survey (EGS) and Addis Ababa University. Recent well-drilling campaigns for water in the Addis Ababa vicinity have revealed highly productive deep aquifers at more than 300 m and, in a number of cases, recent wells drilled up to 500 m have revealed highly productive artesian aquifers. It is likely that the aquifers close to urban and more developed areas, such as those near Addis Ababa, will be developed for agriculture and industrial uses in the near future. In some areas such as Gonder and Mekele over-extraction has already led to decline of the groundwater level (Ayenew *et al.*, 2008). Careful monitoring and regulation are needed to prevent long-term negative impacts on groundwater resources by the inevitable expansion of groundwater utilization in the basin.

There is no systematic monitoring of groundwater recharge in Sudan. Localized investigations are usually performed on a case by case basis for a limited time period. Plans have been made for a nationwide observation network but they have not been implemented so far mainly due to lack of funding, coordination and recent civil unrest. Apart from urban centres, abstraction data are estimated and do not account for the numerous traditional wells. Therefore, actual abstraction is likely to be higher than the estimated volumes.

Egypt has a highly developed groundwater extraction network, and data on the distribution of wells have been compiled by several agencies over the past two decades, including the Ministry of Water Resources and Irrigation (Ramy *et al.*, 2008; Table 10.6).

Policy, regulation and institutional arrangements for groundwater resource management

The development of policies and the design of regulations and institutional arrangements are the first steps to managing and regulating groundwater. From this perspective of the Nile countries we have considered in this chapter, all have initiated this process to either a greater or lesser extent, broadly in line with their general level of economic development. The governments of Uganda and Ethiopia have a strong focus on domestic supply from groundwater as a most urgent concern, and are pushing ahead with policies expanding this service in both rural and urban areas, while Egypt has a well-established groundwater-fed domestic water supply system with associated regulation at different local levels, and plans for development and expansion of settlements which rely on conjunctive use of groundwater and surface water. In comparison, prior to separation, Sudan had established mandates for water policy generally, but to date in post-separation Sudan, especially South Sudan, there is no clear framework, with groundwater exploitation occurring on an ad-hoc and completely unregulated basis. Development organizations are also involved in the creation of frameworks for groundwater regulation and management to varying degrees across the NRB, and it is difficult to know the extent to which these frameworks are currently implemented (see section on Ethiopia below). As the countries we have considered are at quite different levels of policy development and implementation, more detail is provided below.

Uganda

Most groundwater utilization in Uganda is for domestic demand in both rural and urban areas. The Government of Uganda views the water sector as vital for poverty eradication, and by working with a number of development partners, it has set a target of providing safe water and sanitation for the entire population by 2025. Currently, water supply and sanitation rates for rural populations are 63 and 58 per cent, and for urban dwellers 68 and 60 per cent, respectively. Groundwater development is key to achieving this target and in the 1990s this began with the formation of the Rural Towns and Sanitation Programme, supported by the World Bank. The Ministry of Water and Environment implements this programme which is ongoing and has a high success rate in providing communities with domestic water from groundwater. Under this initiative 60 urban centres were identified for piped water supply (MWLE, 2006). By July 2006, 180 small towns had been included in the scheme for piped water supply. By that time 98 had functional water supply systems, 24 were under construction and 44 were at the design stage (MWLE, 2006). Out of the 98 operational water supply systems, 73 were based on groundwater from a total of 66 deep boreholes and 24 springs. At that time a further 684 small towns were identified to be provided with piped water during the next 15-year period, of which it is estimated that over 550 will be based on groundwater from deep boreholes (MWLE, 2006). The MWLE also regulates groundwater-fed wetlands and prevents destruction of habitat by conversion to agricultural land.

Ethiopia

In the past there was little recognition of the importance of groundwater in Ethiopia. The existing river basin master plans for example, which were developed for all the major basins of Ethiopia, include very limited groundwater data sets or analyses. However, recognition of groundwater is growing, and the Government of Ethiopia has determined to provide domestic water supply to 70 per cent of its population by 2015 as a key millennium development goal, primarily based on development of its groundwater resources (see Table 10.3). Working with GW-MATE, the Ministry of Water and Energy has developed its Strategic Framework for Managed Groundwater Development (MWR/GW-MATE, 2011). The framework aims to build an enabling environment with policy adjustments, regulatory provisions and user engagement, so that effective measures can be taken in managing groundwater quality, and promoting demand-side as well as supply-side management. Within this framework action plans will be developed according to national and local priorities within the 'resource setting' (hydrogeological and socioeconomic) and using a range of management tools. At this stage the first action plan has been developed for the Addis Ababa region (MWR/GW-MATE, 2011). The government also recognizes groundwater as a key instrument for economic growth and livelihood enhancement, with groundwater a major component in the ambitious target to increase the area under irrigation by six-fold by 2015 (MWR/GW-MATE, 2011).

According to the MWR/GW-MATE report, the most pressing policy issues are a stronger integration of groundwater development and land use planning, selection of target areas for intense groundwater development and combining groundwater development (both recharge, retention and reuse) with other water resource programmes, including watershed programmes, drainage, and floodplain development (MWR/GW-MATE, 2011). The framework highlights target areas with proven high reserves, high-potential areas with the most accessible aquifers and areas where climate change predictions indicate the need for supplementary irrigation. The MWR/GW-MATE report acknowledges the need to scale up regulation of groundwater, clarifying responsibilities and mandates of different organizations from federal and regional to river-basin level in the private and public sectors, to include the wide range of stakeholders involved in managed groundwater development. Currently, regulations exist but are rarely enforced, and mandates seem to overlap. The 1960 Civil Code established that groundwater is public property and strictly limits the development of private wells while the 1999 Water Resource Planning Policy provides a set of guidelines for water resources development. These regulations need to be enforced by an organization with a clear mandate, and individual cases should be incorporated within a broader development plan, locally and regionally. There is considerable scope for sustainable development of the resource in Ethiopia if the management measures included in Table 10.7 can be implemented.

Stakeholder participation, capacity-building and the promotion of private-sector capacity in addition to capacity within government departments are key to achieving all of the above institutional and non-institutional targets.

Sudan (North and South)

It is difficult to talk about groundwater policy and regulation without considering recent political events and the formation of two separate states of North and South Sudan in July 2011. Prior to separation there were four levels of government in Sudan (interim constitution 2005), all of which had water-policy making mandates:

1. The Federal (National Unity) Government with one ministry for water affairs and a draft water policy.
2. The Government of South Sudan (GOSS) with an approved water policy and three water affairs ministries.
3. The 26 State Governments each with, at least, two water affairs ministries.
4. The Local '*Magalia*' or council level.

Table 10.7 Proposed institutional responsibilities for the development and management of groundwater resources in Ethiopia

Institution	*Responsibility/Activity/Mandate*
Ministry of Water and Energy	Develop policies, standards and criteria for groundwater management plans, drilling standards and well designs, fluoride/iodine treatment; maintain information base, initiate and support interregional groundwater management plans and oversee water allocation
Ethiopian Geological Survey	Plan and guide groundwater assessments
Federal Environment Protection Authority (EPA)	Review possible impacts of national investments on groundwater quality and quantity; strategic environmental assessments linked to groundwater management plans
Regional governments	Integrate groundwater management into other development programmes
Regional Water Resources Development Bureaus	Adopt policies, standards and criteria; initiate groundwater management plans for selected areas and supervise quality of monitoring; licensing in low-density areas
Regional EPAs	Licensing in high-density areas (need to be upgraded); and review possible impacts of investments on groundwater quality and quantity
River Basin Organizations (need to be created)	Coordinate surface water and groundwater allocation and supply
Water user associations (need to be created)	Local regulation and efficiency measures; support and engagement in groundwater management plans
Well field operators	Monitoring
Universities management	Courses in drilling, drilling supervision and groundwater
Technical and vocational education and training (TVETs), NGOs	Courses in manual drilling and pump development
Private-sector educational services	Specialist courses
Public-sector technical services	Design and supervision
Private-sector technical services	Design and supervision
Corporate private-sector drilling services	Drilling of shallow and deep wells (to be strengthened)
Artisanal private-sector drilling services	Drilling development of very shallow wells (to be strengthened)

Source: MWR/GW-MATE (2011)

Almost all the levels had identical empowerment but no institutional capacity, resulting in general confusion and widespread infringements of existing principles. Therefore, the Water Resources Act was passed in 1995 but remains unimplemented to date. This situation is further complicated by the recent formation of two states. The reinstitution of a legal framework is yet to be implemented and currently the development of groundwater resources in Sudan remains unregulated. Wells are drilled without permits or regulation, often close to septic pits dug for the disposal of household sewage, and there is no accountability for any negative impact on the groundwater resources.

A further and major challenge faced by regulation of groundwater in North and South Sudan is the lack of accurate information on groundwater potential and the absence of quantitative and qualitative monitoring. The institution responsible for this (in the pre-separation Sudan) was an under-resourced, small department of the Ministry of Irrigation and Water Resources, mandated to provide both water resource management and water services. Almost the entire ministry budget is required to deliver irrigation and drinking water services. The significant potential for groundwater development is recognized regionally and there has been interest in investing in groundwater irrigation in the Nile, Northern, Central and Khartoum States, particularly from the Gulf States, but lack of coordination and clear regulation at this point has the potential to cause further conflicts.

Egypt

The challenge of managing scarce water resources, including groundwater, for sustainable development incorporating medium and long-term use for a range of stakeholders is recognized as priority by the Egyptian government. In most water resources management situations, some form of planning already exists, varying according to the resources in question, planning tradition, administrative structure and technical issues. The Ministry of Water Resources and Irrigation (MWRI) has a management plan which aims to address the challenges of water scarcity it considers to be of concern. In relation to groundwater, the plan highlights groundwater development for agricultural expansion into 'new' areas, relocating people from the Nile Valley and Delta to initiate new communities in areas currently desert. This will clearly intensify demands on groundwater.

In the 'Renewable Aquifer Underlying the Nile Valley and Delta' the MWRI plan focuses on the conjunctive use of surface water and groundwater by:

1. Utilizing aquifer storage to supplement surface water during peak periods and artificially recharging the groundwater during the minimum demand periods.
2. Employing sprinkler or drip irrigation from groundwater in the 'new lands' to prevent waterlogging and rising water tables.
3. The use of vertical well drainage systems in Upper Egypt to prevent waterlogging and rising water tables.
4. Utilizing groundwater for artificial fish ponds (high quality and consistent temperature).
5. Pumping groundwater from low-capacity private wells at the end of long *mesqas* to supplement canal water supply.

In the deep aquifers of the Western Desert (and Sinai), which require a large investment to be viable, future strategies outlined in the plan include:

1. Intensive survey to determine the main characteristics of each aquifer including its maximum capacity and safe yield, and monitoring to prevent abstraction beyond sustainable yields.
2. The development of new small communities in the desert areas designed to use all available natural resources through integrated planning.
3. Utilizing renewable energy sources, including solar and wind, to minimize the pumping costs.

4. Application of new irrigation technologies in desert areas minimizing losses, especially deep percolation due to the high porosity.

If implemented, all of these strategies would help to reduce pressure on increasing demand for groundwater. However, at this stage the plan is on paper and subject to investment finance.

The Egyptian government is also considering the use of the brackish groundwater (3000–12,000 mg l^{-1} TDS), such as is found at shallow depths in the Western and Eastern Deserts and at the fringes of the Nile Valley following desalination treatment. Renewable energy sources are proposed to reduce the cost of the treatment process, with the resulting 'fresh' water used for supplemental irrigation of a second-season crop.

Overall, future strategies and policies for groundwater development assessment and utilization identified by the MWRI plan include:

1. Utilization of technologies from the water resources management sector, especially remote sensing and GPS techniques; numerical modelling of groundwater and surface water models; information and decision support systems to integrate the ministry's water resources information; use of geophysical methods (e.g. electromagnetic and electrical resistivity, and use of environmental isotopes).
2. Water quality monitoring and management to prevent transport and contamination by pollutants.
3. Raising awareness with the general population and with policy-makers, of water resource issues and achievements in water management, via the media and by demonstration of positive water saving consequences; achieving public participation and commitment of policy-makers to water policies and programmes; increasing knowledge and capacity on new technologies to conserve water in irrigation and domestic use.
4. Continuous monitoring and evaluation to enable strategic adjustments needed to correct deviations from the original objectives.
5. Coordinating and enabling different water users and water-user groups.
6. Institution building and strengthening, linking the public and private sectors, transferring knowledge and human resources within the water sector; providing management training and technical skills.
7. Strengthening coordination between ministries to avoid overlapping mandates, and enhance exchange of data, knowledge, experience and technical expertise in the different fields of water resources between different authorities.
8. Providing a detailed review of all existing water resources laws and decrees and their relation to water management to ensure that up to date laws and regulations reflect the long-term objectives of water resources management.
9. International cooperation – in the case of groundwater this particularly refers to the NSAS, shared by Chad, Egypt, Libya and Sudan.
10. Ensuring that the planning and policy formulation process is based on the most up to date research and development outcomes and comprehensive planning studies.

Overall and individually, these strategies are highly desirable for the sustainability of the groundwater resources. It can only be hoped that the plans can be implemented.

In considering the information provided above, it is useful to note that the groundwater resources of the NRB, like the river itself, do not conform to national and other political boundaries. Wise management of groundwater requires governments and related agencies to work together to achieve sustainable utilization of this often shared resource.

Conclusions

Large parts of the NRB are prone to high rainfall variability and seasonal and periodic droughts. Further, climate change predictions indicate that this situation may worsen. It is broadly accepted that groundwater can provide some degree of buffering to this threat, supplementing surface water supplies, reducing risk and strengthening resilience, and reducing the vulnerability of the poor to water shortages. In fact, adequate reliable water supplies are essential to economic development at all levels. It is increasingly acknowledged that wise management of groundwater resources can provide this security. In addition, increasing water stored in this reserve during times of high rainfall by actively diverting surface water to groundwater recharge can support sustainable groundwater use.

There are many challenges facing sustainable management of the groundwater resources of the NRB. With the exception of Egypt, most of the ten countries are relatively undeveloped in terms of industry and commercial agriculture and the role of groundwater is primarily for the provision of domestic supplies. This will change in the future, and so the demand for water from groundwater for these uses will increase. Africa's population is growing rapidly and the 1999 population of 767 million is projected to nearly double by 2035 (UNFPA, 2011). Although fertility rates do differ across the continent, these predictions are largely applicable to the NRB, with obvious implications for domestic use, the consumption of water for increased food-agriculture demand, and all other high-water-demand and human-related activities. In parts of the basin where surface waters are already severely stressed, such as Egypt, growing populations pose a severe challenge for groundwater management. In Egypt's case, the authorities are already relying strongly on groundwater to meet the needs of new communities in marginal areas and to relieve pressure on surface water in more densely populated areas close to the Nile River.

In parallel to industrial and commercial development, the building and strengthening of governance and regulatory structures are also in an early stage of development in many Nile countries. Generally, natural resources policies and regulations can be seen to lag behind other sectors (such as law and health). Many countries have developed policies and strategic frameworks, often with outside help (e.g. Ethiopia and GW-MATE), but implementing the policies may take several more years. Naturally, without clear mandates and regulatory structures, monitoring and planning cannot be well implemented. While the importance of sustainable management of groundwater is now broadly acknowledged within the NRB, conjunctive management of groundwater and surface waters is some way off.

References

Abdalla, S. H. (2010) Assessment and analysis of the water resources data and information of Sudan, report by Mohamed Elhassan Ibrahim, Consultant Hydrogeologist to IWMI, January.

Abderrahman, W. A. (2003) Should intensive use of non-renewable groundwater resources always be rejected? in *Intensive Use of Groundwater: Challenges and Opportunities*, R. Llamas and E. Custodio (eds), pp191–206, Swets and Zeitlinger B.V., Lisse, the Netherlands.

Adelana, S. M. A. (2009) *Monitoring Groundwater Resources in Sub-Saharan Africa: Issues and Challenges. Groundwater and Climate in Africa (Proceedings of the Kampala Conference, June 2008)*, International Association of Hydrological Sciences, Rennes, France.

Ahmed, A. L. M., Sulaiman, W. N., Osman, M. M., Saeed, E. M. and Mohamed, Y. A. (2000) Groundwater quality in Khartoum State, Sudan, *Journal of Environmental Hydrology*, 8, 12, p7.

Asfaw, B. (2003) *Regional Hydrogeological Investigation of Northern Ethiopia*, Technical Report, Geological Survey of Ethiopia, Addis Ababa, Ethiopia.

Attia, F. A. R. (2002) National report on groundwater protection in Egypt, in *Status of Groundwater Protection in the Arab Region (Egypt, Bahrain, Yemen, Tunisia)*, Khater, A. R. and Radwan Al Weshah (eds), IHP/No. 13, UNESCO, Cairo, Egypt.

Ayenew, T. (1998) *The Hydrogeological System of the Lake District Basin, Central Main Ethiopian Rift*, Free University of Amsterdam, The Netherlands.

Ayenew, T. (2005) Major ions composition of the groundwater and surface water systems and their geological and geochemical controls in the Ethiopian volcanic terrain, *Sinet, Ethiopian Journal of Science*, 28, 2, 171–188.

Ayenew, T. (2008) Hydrogeological controls and distribution of fluoride in the groundwaters of the central Main Ethiopian rift and adjacent highlands, *Environmental Geology*, 54, 1313–1324.

Ayenew, T., Kebede, S. and Alemyahu, T. (2007) Environmental isotopes and hydrochemical study as applied to surface water and groundwater interaction in the Awash river basin, *Hydrological Processes*, 22, 10, 1548–1563.

Ayenew, T., Demlie, M. and Wohnlich, S. (2008) Hydrogeological framework and occurrence of groundwater in the Ethiopian volcanic terrain, *Journal of African Earth Sciences*, 52, 3, 97–113.

BGS (British Geological Survey) (2001) *Groundwater Quality: Uganda*, information sheet prepared for WaterAid, www.bgs.ac.uk/downloads/start.cfm?id=1292, accessed 22 December 2011.

Bonsor, H. C., Mansour, M. M., MacDonald, A. M., Hughes, A. G., Hipkin, R. G. and Bedada, T. (2010) Interpretation of GRACE data of the Nile Basin using a groundwater recharge model, *Hydrology and Earth System Science Discussions*, 7, 4501–4533.

Chernet, T. (1993) *Hydrogeology of Ethiopia and Water Resources Development*, unpublished report, Ethiopian Institute of Geological Surveys, Addis Ababa, Ethiopia.

CSA/UNFPA (Central Statistical Agency/United Nations Population Fund) (2008) *Summary and Statistical Report of the 2007 Population and Housing Census, Population Size by Age and Sex*, December, Federal Democratic Republic of Ethiopia Population Census Commission, UNFPA, Addis Ababa, Ethiopia.

Dawoud, M. A. (2004) Design of national groundwater quality monitoring network in Egypt, *Environmental Monitoring and Assessment*, 96, 99–118.

Demlie, M. and Wohnlich, S. (2006) Soil and groundwater pollution of an urban catchment by trace metals: Case study of the Addis Ababa region, central Ethiopia, *Journal of Environmental Geology*, 51, 3, 421–431.

Demlie, M., Wohnlich, S. and Ayenew, T. (2008) Major ion hydrochemistry and environmental isotope signatures as a tool in assessing groundwater occurrence and its dynamics in a fractured volcanic aquifer system located within a heavily urbanized catchment, central Ethiopia, *Journal of Hydrology*, 353, 175–188.

Döll, P. and Fiedler, K. (2008) Global-scale modeling of groundwater recharge, *Hydrology and Earth System Sciences*, 12, 863–885.

El Tahlawi, M. R. and Farrag, A. A. (2008) Groundwater of Egypt: An environmental overview, *Environmental Geology*, 55, 3, 639–652.

Farah, E. A., Mustafa, E. M. A. and Kumai, H. (1999) Sources of groundwater recharge at the confluence of the Niles, Sudan, *Environmental Geology*, 39, 6, 667–672.

Foster, S. S. D. (1984) African groundwater development – the challenges for hydrogeological science, in *Challenges in African Hydrology and Water Resources: Proceedings of the Harare Symposium, July* 1984, D. E. Walling, S. S. D. Foster and P. Wurzel (eds), pp3–12, International Association of Hydrological Sciences, Rennes, France.

Foster, S., Tuinhof, A. and Garduño, H. (2008) Groundwater in Sub-Saharan Africa – a strategic overview of developmental issues, in *Applied Groundwater Studies in Africa*, S. M. A. Adelana and A. M. MacDonald (eds), IAH Selected Papers on Hydrogeology 13, CRC Press/Balkema, Leiden, The Netherlands.

Gheith, H. and Sultan, M. (2002) Construction of a hydrologic model for estimating Wadi runoff and groundwater recharge in the Eastern Desert, Egypt, *Journal of Hydrology*, 263, 36–55.

Hefny, K. and Sahta, A. (2004) Underground water in Egypt, Ministry of Water Supplies and Irrigation, Egypt, Annual Meeting, Egypt, Commission on Mineral and Thermal Waters, Handouts, 25, 11–23, Cairo, Egypt.

Hefny, K., Samir Farid, M. and Hussein, M. (1992) Groundwater assessment in Egypt, *International Journal of Water Resources Development*, 8, 2, 126–134.

Hussein, M. T. (2004) Hydrochemical evaluation of groundwater in the Blue Nile Basin, eastern Sudan, using conventional and multivariate techniques, *Hydrogeology Journal*, 12, 144–158.

Ibrahim, M. E. (2009) Status of urban and rural water supply, Public Water Corporation (in Arabic), unpublished report to the Public Water Corporation, available from the Public Water Corporation of the Republic of Sudan.

Ibrahim, M. E. (2010) *Groundwater and Its Use in the Nile Basin of Sudan*, January, report prepared for the International Water Management Institute, Colombo, Sri Lanka.

Isaar, A. S., Bein, A. and Michaeli, A. (1972) On the ancient water of the Upper Nubian sandstone aquifer in central Sinai and southern Israel, *Journal of Hydrology*, 17, 353–374.

Jousma, G. and Roelofsen, F. J. (2004) *World-wide Inventory on Groundwater Monitoring*, Report No. GP 2004-1, International Groundwater Resources Assessment Centre (IGRAC), Utrecht, The Netherlands.

Kashaigili, J. J. (2010) *Assessment of Groundwater Availability and Its Current and Potential Use and Impacts in Tanzania*, report prepared for the International Water Management Institute, Colombo, Sri Lanka.

Kebede, S., Travi, Y., Alemayehu, T. and Ayenew, T. (2005) Groundwater recharge, circulation and geochemical evolution in the source region of the Blue Nile River, Ethiopia, *Applied Geochemistry*, 20, 1658–1676.

Kloos, H. and Tekle-Haimanot, R. (1999) Distribution of fluoride and fluorosis in Ethiopia and prospects or control, *Tropical Medicine and International Health*, 4, 5, 355.

MacDonald, A. M. and Calow, R. C. (2008) Developing groundwater for secure rural water supplies in Africa, in *Water and Sanitation in International Development and Disaster Relief*, University of Edinburgh, Edinburgh, UK.

Manfred, H. and Paul, J. B. (1989) A groundwater model of the Nubian aquifer system, *Hydrological Sciences Journal*, 34, 4, 425–447.

Masiyandima, M. and Giordano, M. (2007) Sub-Saharan Africa: opportunistic exploitation, in *The Agricultural Groundwater Revolution: Opportunities and Threats to Development*, Giordano, M. and Villholth, K. (eds), Comprehensive Assessment of Water Management in Agriculture Series 3, CABI, Wallingford, UK.

Michael, M. and Gray, K. (2005) *Evaluation of ECHO Funded/UNICEF Managed Emergency Health/Nutrition and Water and Environmental Sanitation Action For Conflict Affected Populations in North Darfur, Blue Nile and Unity States of Sudan*, http://unsudanig.org/workplan/mande/reports/docs/evaluations/ECHO%20Blue%20Nile-Unity-North%20Darfur%20Health-WES%20Evaluation%20%20_1_.pdf, accessed January 2010.

MWLE (Ministry of Water, Lands and Environment) (2006) *Status of Implementation of Urban Water and Sanitation Projects and Rural Growth Centres*, unpublished report of the Ministry of Water, Lands and Environment, Directorate of Water Development, Government of Uganda.

MWR/GW-MATE (Ministry of Water Resources/Groundwater Management Advisory Team) (2011) *Ethiopia: Strategic Framework for Managed Groundwater Development*, Ministry of Water Resources, Federal Democratic Republic of Ethiopia and the Groundwater Management Advisory Team, World Bank Water Partnership Program.

MWRI (Ministry of Water Resources and Irrigation) (2010) *Strategic framework for development and management of water resources in Egypt to 2050*, Report prepared by Ministry of Water Resources and Irrigation..

Omer, A. M. (2002) Focus on groundwater in Sudan, *Environmental Geology*, 41, 972–976.

Owor, M., Taylor, R. G., Tindimugaya, C. and Mwesigwa, D. (2009) Rainfall intensity and groundwater recharge: Empirical evidence from the Upper Nile Basin, *Environmental Research Letters*, 4, July–September, p035009.

Pact (2008) *Pact Sudan's Water for Recovery and Peace Program (WRAPP)*, Technical Needs Assessment for the Ministry of Water Resources and Irrigation (MWRI), Government of Southern Sudan (GoSS), 15 July, www.rwssp-mwrigoss.org/sites/default/files/Microsoft%20Word%20%20NEEDS%20ASSESSMENT%20REPORT%20WITHOUT%20ANNEXES.pdf, accessed January 2010.

Soltan, M. E. (1999) Evaluation of groundwater quality in Dakhla Oasis (Egyptian Western Desert), *Environmental Monitoring and Assessment*, 57, 157–168.

Taylor, R. G. and Howard, K. W. F. (1996) Groundwater recharge in the Victoria Nile basin of East Africa: Support for the soil moisture balance approach using stable isotope tracers and flow modelling, *Journal of Hydrology*, 180, 31–53.

Tindimugaya, C. (2008) Groundwater Flow and Storage in Weathered Crystalline Rock Aquifer Systems of Uganda: Evidence from Environmental Tracers and Aquifer Responses to Hydraulic Stress, unpublished PhD thesis, University of London, UK.

Tindimugaya, C. (2010) *Assessment of Groundwater Availability and Its Current and Potential Use and Impacts in Uganda*, February, report prepared for the International Water Management Institute, Colombo, Sri Lanka.

Tuinhof, A, Foster, S., van Steenbergen, F., Talbi, A. and Wishart, M. (2011) *Appropriate Groundwater Management Policy for Sub-Saharan Africa in Face of Demographic Pressure and Climatic Variability*, Strategic Overview Series Number 5, GW-MATE, World Bank, Washington, DC.

Tweed, O. S., Weaver, R. T. and Cartwright, I. (2005) Distinguishing groundwater flow paths in different fractured-rock aquifers using groundwater chemistry: Dendonong Ranges, southeast Australia, *Hydrogeology Journal*, 13, 771–789.

UNEP (United Nations Environmental Programme) (2010) *Africa Water Atlas*, UNEP, Nairobi, Kenya.

UNFPA (2011) *Population Issues – 1999, Demographic Trends by Region*, www.unfpa.org/6billion/populationissues/demographic.htm, accessed December 2011.

11

Wetlands of the Nile Basin

Distribution, functions and contribution to livelihoods

Lisa-Maria Rebelo and Matthew P. McCartney

Key messages

- Wetlands occur extensively across the Nile Basin and support the livelihoods of millions of people. Despite their importance, there are big gaps in the knowledge about the current status of these ecosystems, and how populations in the Nile use them. A better understanding is needed on the ecosystem services provided by the different types of wetlands in the Nile, and how these contribute to local livelihoods.
- While many of the Nile's wetlands are inextricably linked to agricultural production systems the basis for making decisions on the extent to which, and how, wetlands can be sustainably used for agriculture is weak.
- Due to these information gaps, the future contribution of wetlands to agriculture is poorly understood, and wetlands are often overlooked in the Nile Basin discourse on water and agriculture. While there is great potential for the further development of agriculture and fisheries, in particular in the wetlands of Sudan and Ethiopia, at the same time many wetlands in the basin are threatened by poor management practices and rising populations.
- In order to ensure that the future use of wetlands for agriculture will result in net benefits a much more strategic approach to wetland utilization is required; wetland management needs to be incorporated into basin management and, in addition, governance of wetlands should include a means of involving stakeholders from impacted or potentially impacted regions.

Introduction

The Nile is one of the longest rivers in the world, flowing through 10 countries, five of which are among the poorest in the world, with very low levels of socio-economic development (Awulachew *et al.*, 2010). Despite a wide range of productive ecosystems located within the basin, the Nile's land and water resources are not well utilized or managed and are degrading rapidly. While water development interventions and agricultural activities should be undertaken with caution within wetlands, to ensure the maintenance of ecosystem services, they offer a vast livelihood resource and development potential for agriculture and fisheries.

Wetlands and lakes cover approximately 10 per cent of the Nile Basin and play an important role in the hydrology of the Nile River system. Lakes and wetland storage are of particular importance in the White Nile Basin, where spill from the river and its tributaries into wetlands and subsequent evaporation are major components of the catchment water budget (Sutcliffe and Parks, 1999). The area covered is smaller, but wetlands are also important in the Blue Nile. Throughout the Nile Basin, patterns of flow and water chemistry are significantly modified by the complex movement of water within wetlands, which in turn affects the many ecosystem services upon which many millions of people living in the basin depend for their livelihoods.

This chapter provides an overview of the major wetlands in the Nile Basin, their important contribution to livelihoods, and the potential threats to the ecosystem services and functions they provide. Wetlands across the basin support agriculture (including livestock) and fisheries activities, and provide a critical dry-season resource, in particular in areas of low and erratic rainfall (i.e. much of the basin). Their importance to livelihoods will increase under future climate change scenarios and expanding populations, and with continuing pressure to improve food security in Nile Basin countries. Although many of the wetlands in the basin are not currently exploited to their full potential in terms of agriculture and fisheries activities, they are under threat. Better management of these ecosystems is vital, along with integration into water resources and basin management. If wetlands are not used sustainably, the functions that support agriculture as well as other food-security and ecosystem services, including water-related services, are undermined. Trade-offs between the various uses and users need to be better evaluated in order to guide management responses, and there is a pressing need for more systematic planning that takes into account trade-offs in the multiple services that wetlands provide.

Overview

The Nile River supports a range of wetland ecosystems distributed across the entire length of the basin (Figure 11.1), although there is a higher concentration in the upstream regions of both the Blue and White Nile. Defined by the Ramsar Convention (Article 1.1) as 'areas of marsh, fen, peatland or water, whether natural or artificial, permanent or temporary, with water that is static or flowing, fresh, brackish or salt, including areas of marine water the depth of which at low tide does not exceed six metres', wetlands are estimated to cover 18.3 million ha (i.e. about 5%) of the basin. These ecosystems are found at the source of the Blue Nile in Ethiopia (Lake Tana and associated floodplains) and along the tributaries of the Blue Nile and Atbara in Ethiopia (e.g. Finchaa, Didessa, Tekeze) and in Sudan (e.g. the Dinder). There are numerous wetlands at the source of the White Nile in the Equatorial Lakes region, in particular around Lakes Victoria, Albert and Kyoga. Further downstream, the Bahr el Jebel (the upper reach of the White Nile) enters the Sudan plains forming the vast Sudd wetland. Large wetlands are also found to the west of the Sudd, fed by the Bahr el Ghazal before it joins the Bahr el Jebel outflow from the Sudd at Lake No. Located on tributaries of the While Nile are the Baro-Akobo wetlands in Ethiopia and the Machar Marshes in Sudan. Upstream of Khartoum where the Blue and White Nile rivers merge, the Nile flows through the desert with only small fringing wetlands observed until it forms the Nile Delta at the Mediterranean Sea.

Within the basin 14 sites have been nominated by the individual countries as Ramsar Wetland Sites of International Importance (Table 11.1). These include one site in the Democratic Republic of Congo (DRC), 11 in Uganda and two in Sudan, covering a total area of 7.9 million ha. The only basin countries which are currently not signatories to the Ramsar convention are Ethiopia and Eritrea.

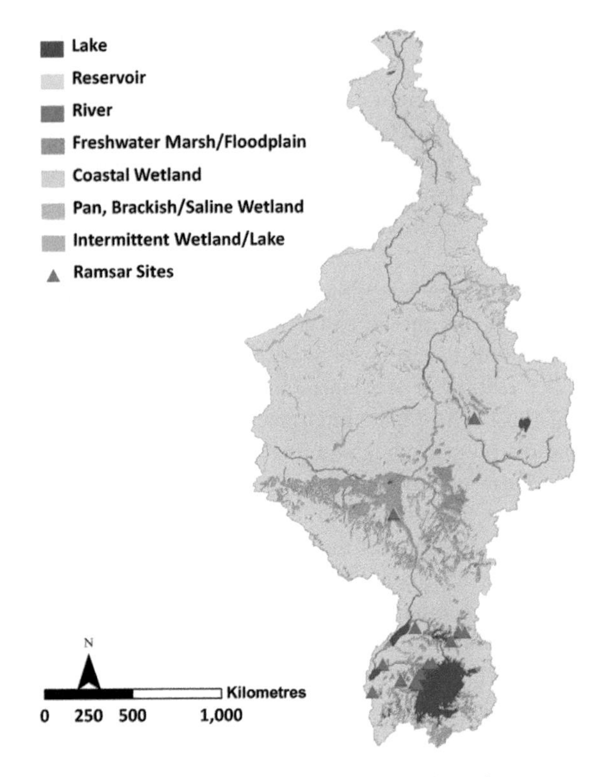

Figure 11.1 Spatial distribution and areal extent of wetlands within the Nile Basin

Source: Data are derived from the Global Lakes and Wetlands Database (Lehner and Döll, 2004) and country-based Africover data sets (FAO, 2002)

Table 11.1 Ramsar Wetland Sites of International Importance located within the Nile Basin

Country	*Site*	*Wetland area (ha)*	*Dominant type*
DRC	Virunga National Park	800,000	Permanent freshwater lakes
Uganda	Lake George	15,000	Permanent freshwater lakes
	Lake Nabugabo	3600	Permanent freshwater lakes
	Lake Bisina	54,229	Permanent freshwater lakes
	Lake Mburo-Navikali	26,834	Permanent freshwater lakes
	Lake Nakuwa	91,150	Permanent freshwater marshes or pools
	Lake Opeta	68,912	Permanent freshwater marshes or pools
	Lutembe Bay	98	Permanent freshwater marshes or pools
	Mabamba Bay	2424	Permanent freshwater marshes or pools
	Nabajjuzi	1753	Permanent freshwater marshes or pools
	Murchison Falls – Albert Delta	17,293	Permanent freshwater marshes or pools
	Sango Bay – Musambwa Island – Kagera	55,110	Seasonal/intermittent freshwater lakes
Sudan	Dinder National Park	1,084,600	Seasonal/intermittent freshwater lakes/ rivers
	Sudd	5,700,000	Permanent/seasonal rivers

Wetland ecosystem services

Ecosystem services are defined by the Millennium Ecosystem Assessment (MEA, 2005) as 'the benefits people obtain from ecosystems'. Different wetlands across the Nile Basin perform different functions thereby providing different ecosystem services, depending on the interactions between their physical, biological and chemical components, as well as their surrounding catchments. Ecosystem services are typically categorized into four groups; provisioning, regulating, supporting and aesthetics (Figure 11.2). The physical benefits which people derive from wetlands include provisioning services such as domestic water supply, fisheries, livestock grazing, cultivation, grasses for thatching, and wild plants for food, crafts and medicinal use. Wetlands in the Nile Basin play an important role in sustaining the livelihoods of many millions of people through the provision of numerous ecosystem services, including food. In many places these ecosystems are closely linked to cropping and livestock management. In arid and semi-arid regions with seasonal rainfall patterns the capacity of wetlands to retain moisture for long periods, sometimes throughout the year and even during droughts, means that they are of particular importance for small-scale agriculture, both cultivation and grazing. Such wetlands often provide the only year-round source of water for domestic use.

Other wetland ecosystem services are often not explicitly recognized by communities, but include a wide range of regulating services such as flood attenuation, maintenance of dry-season river flows, groundwater recharge, water purification, climate regulation and erosion control, as well as a range of supporting services such as nutrient cycling and soil formation. In addition, people also gain non-physical benefits from the cultural services, including spiritual enrichment, cognitive development and aesthetic experience. At many sites, the different types of service may be closely linked.

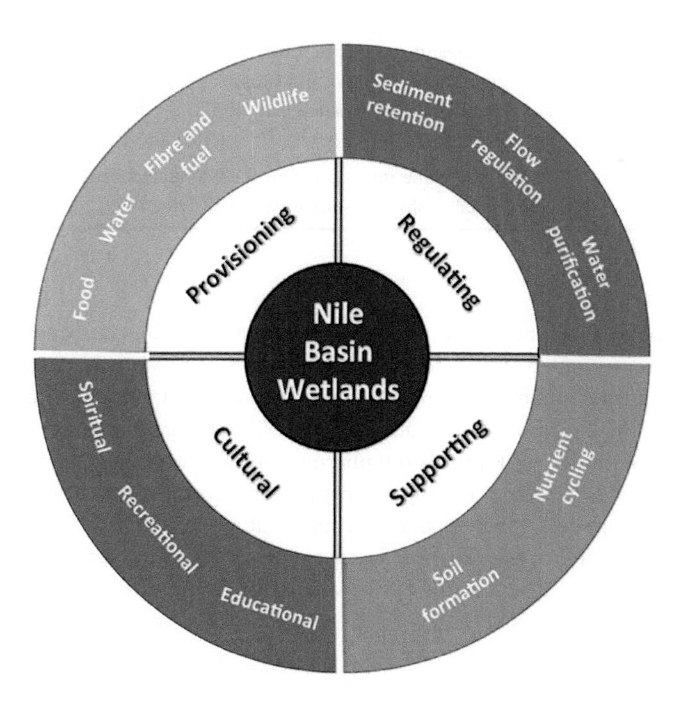

Figure 11.2 Wetland ecosystem services

Contributions to water resources

Due to their role in the provisioning of water, regulating flows and improving water quality, wetlands are increasingly perceived as an important component of water infrastructure (Emerton and Bos, 2004). The flow regulation functions of the various Nile wetlands contribute to the hydrology of the whole basin, although the magnitude of the contribution of an individual wetland depends on its type, location within the catchment and the presence/absence of upstream water resources infrastructure (Table 11.2). Because of their dependence on water and their importance in the hydrological cycle, it is essential that wetlands are considered as a key component in strategies for Integrated Water Resources Management (IWRM).

Table 11.2 Hydrological functions of major wetlands in the Nile Basin

Wetlands	*Hydrological functions*
Wetlands of Uganda	Most of the individual wetlands link to other wetlands through a complex network of permanent and seasonal streams, rivers, and lakes, making them an essential part of the entire drainage system of the country (UN-WWAP and DWD, 2005)
Headwater wetlands of the Baro Akobo	Regulate flow in the Baro Akobo River while believed to play an important role in maintaining downstream dry-season river flows
Lake Albert	Critical link between the White Nile and its headwaters; without the flow regulation of this lake the White Nile would be reduced to a seasonal stream and could play no significant role in maintaining the base flow of the main Nile (Talbot and Williams, 2009)
Sudd, Machar Marshes and wetlands of the Bahr Ghazal	Significantly attenuate flows of the White Nile and its tributaries reducing flood peaks and supporting dry-season river flows, thereby minimizing the seasonal variation in the flow of the White Nile (Sutcliffe and Widgery, 1997; Sutcliffe and Parks, 1999)
Nile Delta	Limits saline intrusion from the Mediterranean Sea, thereby protecting coastal freshwater sources (Baha El Din, 1999)

Wetlands can be very effective at improving water quality and, consequently, can be very important in the treatment of polluted water. This function of wetlands is achieved through processes of sedimentation, filtration, physical and chemical immobilization, microbial interactions and uptake by vegetation (Kadlec and Knight, 1996). In Uganda, sewage from 40 per cent of the residents of the city of Kampala (numbering approximately 500,000) is discharged into the 5.3 km^2 Nakivubo wetland. During the passage of the effluent through the wetland, the papyrus vegetation absorbs nutrients and the concentration of pollutants is reduced before the water enters Lake Victoria, the principal water source for the city (Kansiime and Nalubega, 1999). The water purification services of this wetland are estimated to be worth about US$1 million per year (Emerton, 2005).

Case studies

The Sudd wetland

While wetlands are found in all the Nile Basin countries, the largest and most important to the hydraulics of the downstream river is the Sudd in South Sudan. Derived from an Arabic word meaning obstacle or blockage of river channels, the Sudd is located between 6°0'–9°8' N and 30°10'–31°8' E (Figure 11.3). Its width varies from 10 to 40 km and its length is approximately 650 km. It is the largest freshwater wetland in the Nile Basin, one of the largest floodplains in Africa and one of the largest tropical wetlands in the world. Covering the area between the town of Mongalla in the south and Malakal in the north, the area of permanent swamps stretches over approximately 30,000 km^2 with the lateral extent of seasonal flooding varying considerably depending on the inflow conditions and season. In periods of high flood and rainfall such as in 1917–1918, 1932–1933, 1961–1964 and 1988–1989, the floodplain remained flooded well into the dry season, while during periods of low flood and rainfall such as in 1921, 1923 and 1984, the floodplain shrinks to the extent that even the permanent swamps dry up.

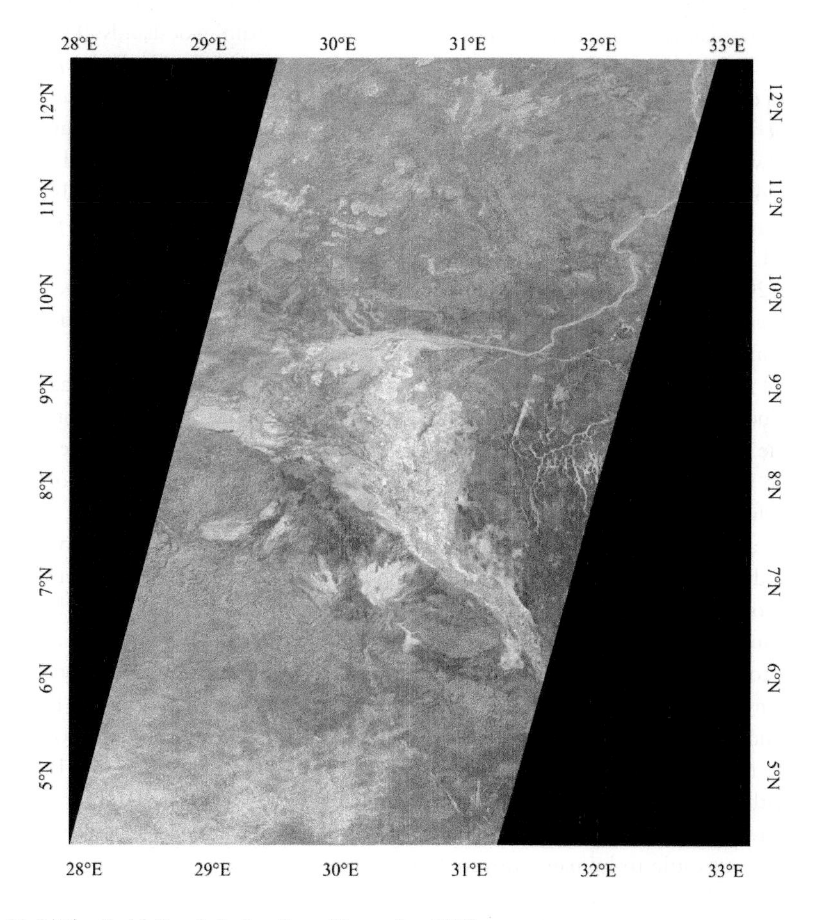

Figure 11.3 The Sudd, South Sudan, June–December 2007

Source: ALOS PALSAR data © JAXA/METI

The Comprehensive Peace Agreement, signed in 2005, ended 22 years of civil war in Sudan, and subsequently (in 2006) 57,000 km^2 of the floodplains of the Sudd were designated as a Ramsar Wetland Site of International Importance. Prior to this designation, three protected areas already existed within the Sudd region. These cover a total area of over 10,000 km^2 and include Zeraf Game Reserve (9,700 km^2, established in 1939), Shambe National Park (620km^2, established in 1985) and Fanyikang Game Reserve (480km^2, established in 1939).

The Sudd is part of the Bahr el Jebel system (the upper reach of the White Nile in Sudan), which originates in the African Lakes Plateau. Seasonal inundation drives the hydrologic, geomorphological and ecological processes and the annual flood pulse is essential to the functioning of the wetland. Flow in the Bahr el Jebel controls not only the hydrology but also many of the other biophysical characteristics of the Sudd (Sutcliffe, 2009). The inundated area of the Sudd varies both within and between years, as inflow and rainfall vary. Annually, the maximum extent of flooding occurs after the rainy season (i.e. between October and December). The rise of Lake Victoria is estimated to have resulted in a trebling of the area of permanent swamps after 1964, a smaller increase in the area of seasonal flooding and a decrease in the inter-annual variability of the flooded area (Sutcliffe and Parks, 1999).

The Sudd wetland comprises a complex maze of various ecosystems. Habitats within the region grade from open water and submerged vegetation to floating fringe vegetation, seasonally flooded grasslands, rain-fed grasslands and, finally, floodplain woodlands (Hickley and Bailey, 1987). The core area of the permanent swamps is dominated by *Cyperus papyrus*, with communities of *Phragmites communis* and *Vossia cuspidata*, bordered by stands of *Typha domingensis*. *Eichornia crassipes* (water hyacinth) found along the open channels. Surrounding the permanent swamps are vast floodplains which consist of seasonally river-flooded grasslands (referred to locally as '*toic*'). These are estimated to cover an area of approximately 16,000 km^2 (WWF, 2010) and are dominated by species of *Oryza longistaminata* and *Echinocloa pyramidalis*. An estimated 20,000 km^2 of rain-fed *Hyparrhenia rufa* grasslands surround the river floodplain (Robertson, 2001). Beyond these are the floodplain woodlands which are dominated by *Acacia seyal*, and *Balanites aegypticaca*. This diverse range of habitats supports a rich array of aquatic and terrestrial fauna including over 400 bird and 100 mammal species (Rzóska, 1974).

The Sudd and the surrounding areas are used extensively by the Dinka, Nuer and Shilluk tribes. The Sudd provides a source of water and essential dry-season grazing land for livestock, the backbone of the Nilotes' economy. The Nilotic pastoralists use a transhumance system to optimize the seasonal flooding and drying cycle, moving with their large herds of cattle in response to the annual regime of the Bahr el Jebel and rainfall. Three of the Sudd vegetation communities are used extensively for livestock grazing (Denny, 1991): river-flooded grassland, the most productive for year-round grazing because the dead grass has a high protein content; seasonally flooded grassland, which includes rain-flooded grasslands, seasonally inundated grasslands, and rain-fed wetlands on seasonally waterlogged clay soil, all three of which are heavily used by livestock; and floodplain scrub forest, at higher elevations on well-drained soils around the floodplains. Before the onset of the civil war the number of livestock using the floodplains of the wetland during the dry season was estimated to be 700,000 (Howell and Lock, 1988). There are no recent counts of livestock populations, but many Internally Displaced Persons are returning with their cattle, and head of livestock are likely to have increased. Recent estimates suggest the livestock population is 1 million head (BirdLife International, 2008), resulting in one of the highest cattle to human ratios in Africa (Okeny, 2007).

While livestock have historically been central to Sudan's economy, their contribution declined from 20 per cent of the GDP in 1999 to 3.2 per cent in 2005 (Fahey, 2007). In addition to the increase in oil exports, the decreasing contribution of livestock and related products is attributed to supply constraints (inadequate capacity at the port, deterioration in the road infrastructure), conflict in livestock-rich areas (in particular the Sudd region), and higher domestic demand (IMF, 2006). The Sudd region has a high potential for livestock, which could markedly contribute to the new economy of South Sudan. Increasing Sudd livestock productivity would also greatly improve the livelihoods of Sudd residents. Livestock production in South Sudan currently faces major challenges including limited access to water during the dry season, high levels of poverty and disease, and a rapidly growing population. Livestock is the most important source of income in the rural areas; however, it is also a potential contributor to water scarcity during the dry season. Although this topic requires further research, initial recommendations for improvements in the Sudd include: water storage with small ponds and larger reservoirs, access to, and development of, bore holes, promotion of productive range ecosystems with efficient livestock management, and establishment of organized livestock markets. In addition, agricultural and livestock training centres would educate herders, develop ranching systems, assist in a comprehensive livestock census within the context of the Sudd area to help in planning and management, provide veterinary services along with human health care services and build an awareness campaign for peace-building activities.

Fishing is the second most important occupation of the inhabitants of the wetlands, in particular for the Shilluk and Nuer tribes, and is typically conducted seasonally alternately with crop production and livestock-rearing. The Sudd is one of the only water bodies of the Nile which is currently not overfished, and the potential yield (based on a surface area of 30,000–40,000 km^2) has been estimated at 75,000 tonnes per year (Witte *et al.*, 2009). However, no direct stock assessment studies have ever been conducted for the Sudd fisheries. Many fish species migrate from the surrounding rivers to the nutrient-rich floodplains to feed and breed during the seasonal floods (Welcomme, 1979). While South Sudan has vast aquatic and fisheries resources with over 130 fish species reported, the full potential of these has yet to be exploited. This is mainly due to the lack of processing and storage facilities and inadequate transportation infrastructure both of which have limited the development of commercial fisheries.

Machar Marshes

A large expanse of wetlands comprising lakes and floodplains is found in the eastern part of Sudan and western Ethiopia, east of the White Nile and north of the Sobat rivers. Located between 8°27'–9°58' N and 32°11'–34°9' E, the wetland system extends at least 200 km from north to south and 180 km from east to west (Hughes and Hughes, 1992). The wetlands are fed by a combination of local precipitation, the torrents originating in the Ethiopian Highlands, and spillover from the Baro, the Akobo and the Sobat. Both the Baro and Akobo rivers spill during periods of high flows into the adjoining wetlands while the Baro spills north across the Ethiopia-Sudan border towards the Machar Marshes (Sutcliffe and Parks, 1999). Hughes and Hughes (1992) estimate the total area of the wetlands at around 9000 km^2, 5000 of which are located in Sudan and 4000 in Ethiopia, while the wetland along with the area of grassland which floods annually has been estimated to be between 6000 and 20,000 km^2 (JIT, 1954).

The Machar Marshes are one of the least studied wetland systems in the Nile Basin, and there is little information available in the literature describing the vegetation characteristics, the seasonal patterns of inundation, or livelihood activities within the wetland. It is noted by Sutcliffe and Parks (1999) that the hydrological regime of the Sobat is complicated by the influence of the wetlands, and the relative remoteness of these has meant that hydrological measurements and study have been less advanced than in the other tributaries. Hughes and Hughes (1992) describe the Marshes as extensive grassy floodplains and permanent swamps dominated either by papyrus along the watercourses or by *Phragmites* and *Typha* away from them. It is suggested that the area experiences a high variability in the timing and intensity of flooding, and that this may have an impact on the establishment of permanent wetlands dominated by vegetation such as papyrus sedge, *Phragmites* and *Typha* (Hassan *et al.*, 2009).

The floodplains are used for livestock grazing in the dry season, while hunting and fishing take place within the wetland. However, Hughes and Hughes (1992) note that the Marshes are little utilized due to the very low population density in the region, and more up-to-date information is not available. In Ethiopia, the Baro and Akobo wetlands provide direct benefits to more than half the population of the region through the provision of water, fisheries, construction materials and medicinal plants as well providing areas of grazing and cultivation.

The Nile Delta

The Nile Delta in northern Egypt is an extensive wetland system, comprising lakes, freshwater and saline wetlands and intertidal areas. Covering an area of approximately 22,000 km^2 encompassing 240km of the coastline (west to east) and 175km in length (north to south), it is one of the largest river deltas in the world (Hughes and Hughes, 1992). Formed as the Nile enters the Mediterranean Sea, the completion of the first Aswan Dam (between 1912 and 1934) dampened the annual flood pulse in the delta. The completion of the second, the Aswan High Dam (AHD), stopped flooding completely and most of the former seasonally or permanently flooded habitats have subsequently been converted to agriculture. Originally intended to produce clean energy and to conserve and protect agriculture (increasing cultivable land by 30%) by controlling the annual Nile flood, it had a dramatic negative effect on the sediment flux to the delta (Hamza, 2009). The delta is now composed of two branches, Rosetta and Damietta and has traditionally been one of the most important agricultural areas of Egypt (Dumont, 2009).

The delta is a very rich agricultural region, and before construction of the AHD recession farming had been practised on the floodplain for over 5000 years. Since the completion of the AHD the area is farmed year-round, causing the loss of much of the wetland habitat of the delta and lower Nile River floodplain. In addition, as the delta no longer receives an annual supply of nutrients and sediments from upstream due to the dam, the floodplain soils have become poorer and large amounts of fertilizers are now used. Although once known for the large papyrus (*Cyperus papyrus*) wetlands, due to the reduction in flooding these have largely disappeared and the remaining wetland consists of lakes and lagoons along the seaward side of the delta. Intensified by the construction of the AHD and other dams and barrages along the upper and lower Nile, and the extensive regulation of the Nile's waters, the delta is in an acute stage of subsidence (Stanley and Warne, 1994). The outer margins are eroding and salinity levels of some of the coastal lands are rising as a result of sea water infiltration to the groundwater (Hughes and Hughes, 1992; Baha El Din, 1999).

Fisheries and agriculture in the Nile Delta are well developed. Covering an area of approximately 22,000 km^2, the delta accounts for two-thirds of Egypt's agriculture. Although the delta comprises only 2.8 per cent of the country's area, it is home to 63 per cent of Egypt's population of 80 million, and is the most populated, cultivated and industrialized part of the country (Hamza, 2009). Due to the reduction of siltation as a result of the AHD farmers now have to use approximately 106 tonnes of artificial fertilizer as a substitute for the nutrients which no longer reach the floodplain, and salinity and waterlogging problems have developed due to over-irrigation (El-Shabrawy, 2009).

While fisheries have long been an important source of food and income to the inhabitants of the Nile Delta, the AHD has also had an impact on fish populations. Prior to its construction, the migrations of various fish species were dependent on the annual flood water. Following the construction of the AHD, the fish catch in the Mediterranean declined from 22,618 tonnes in 1968 to 10,300 tonnes in 1972, recovering to 13,450 tonnes in the 1980s (Biswas, 1992). Sardines, for example, used to breed in the Nile estuary but have now almost disappeared, and marine fish that used to seasonally migrate into the delta lakes have been virtually eliminated. Their place has, however, been taken by freshwater species (El-Shabrawy, 2009). In addition, with Lake Nasser the AHD created a completely new source of fish, which was producing 32,000 tonnes by 1982, thereby compensating for the initial loss of the Mediterranean catch (Biswas, 1992). More recently, exploitation of the delta fisheries has occurred at a level that is not sustainable (Dumont and El-Shabrawy, 2008) and, as a result, recent emphasis has focused on farmed fisheries, and aquaculture is currently booming in the delta. The expansion of aquaculture has resulted in the significant increase of total fisheries production in Egypt. The relative importance of Egyptian aquaculture to total fisheries production has increased from 16 to 56 per cent of total fisheries production between 1997 and 2007.

Other Nile wetlands

While the previous sections have focused on three of the large wetlands of the Nile, many other, equally significant, wetlands are also found across the basin. Located in the Ethiopian Highlands, Lake Tana is the largest freshwater lake in Ethiopia and is the source of the Blue Nile. The lake, covering an area of 3,156 km^2, is shallow, with an average depth of 8 m, and is the third largest lake in the Nile Basin. The lake is bordered by seasonal floodplains such as the Dembea in the north, the Fogera in the east, and the Kunzila in the west. Permanent *Cyperus papyrus* wetlands fringe much of the lake, forming the largest lake-wetland complex in the country. Lake Tana and adjacent wetlands support directly and indirectly the livelihoods of a population of over 500,000, and constitute the country's largest rice production area (Vijverberg *et al.*, 2009). During the rainy season the wetlands are connected with the lake, and act as nurseries for most of the fish populations in the lake, as well as serving as breeding grounds for waterfowl and mammals (Vijverberg *et al.*, 2009).

Wetlands are also located along the tributaries of the Blue Nile, such as the floodplain between the Dinder and Rahad rivers in Sudan, which is composed of a series of wetlands and pools which are part of the drainage systems of the two rivers. Like many wetlands in the basin, the Dinder wetlands on the Blue Nile in Sudan are an important source of water and nutritious grasses to livestock, in particular during the most severe period of the dry season.

Lake Victoria is the source of the White Nile and, with a surface area of approximately 68,800 km^2, is the second largest lake in the world. The White Nile outflow from Lake Victoria controls the levels of Nile base flow into Sudan and Egypt. The water balance of the lake is controlled mainly by precipitation over, and evaporation from, the lake, which vary greatly from year to year according to cloud cover and surface radiation balance (Lehman, 2009). Since 1956, outflow from the lake has been regulated by the Nalubaale Dam constructed for hydroelectric power generation at Owen Falls. Although it once supported species-rich fish communities, the introduction of exotic species and the commercial exploitation of these have drastically changed the biodiversity of the ecosystem. Lake Victoria is the second major commercial fishery in the basin, with the livelihoods of 1.2 million people directly or indirectly dependent on the lake fishery (Matsuishi *et al.*, 2006). In 2003, the estimated annual catch was worth over US$540 million in terms of fish landings, with a further US$240 million earned in fish exports (Balirwa, 2007). There is high population density around the lake, which serves a variety of important socio-economic purposes. It plays a vital role in providing drinking water for the major urban areas and lake shore communities, provides cheap animal protein to the surrounding populations, as well as supplying water to the lake basin, which is the most agriculturally productive and industrialized region in Uganda (Baecher *et al.*, 2000).

Located downstream of Lake Victoria, Lake Kyoga is a shallow lake surrounded by wetlands. During periods of low water, the lake splits into a series of satellite lakes with swamps of *Cyperus Papyrus* forming barriers between them when water levels rise. The lake and surrounding wetlands provide an important supply of water for domestic uses and livestock to the local population. Fishing is the main livelihood activity in the region, and subsistence and small-scale crop production is also undertaken. Between the late 1960s and the early 1980s, until the Victoria Nile perch fishery expanded, Lake Kyoga fish production was higher than that of Lake Victoria, reaching a peak in 1983 with 180,000 tonnes (75% of national production). Lake Kyoga discharges into Lake Albert, where large seasonal wetlands are found. Smaller seasonal wetlands are found along the banks of the Nile from Lake Albert to the border with Sudan. Wetlands are also found on tributaries to the White Nile; in the Baro Akobo catchment in south-western Ethiopia (i.e. the Illubabor region) with approximately 5 per cent of the land area covered by seasonal headwater wetlands. Extensive floodplains are found along the course of the Bahr el Ghazal, a tributary of the White Nile located to the west of the Sudd. The Bahr el Ghazal wetlands are estimated to cover an area of around 9000 km^2, with flora, fauna and livelihood activities similar to those of the Sudd (Hughes and Hughes, 1992). The river has a negligible impact on downstream Nile flow as only 2–3 per cent of the river flow reaches the White Nile as the remainder of the river inflow of the Bahr el Ghazal Basin (12 billion m^3 yr^{-1}) is evaporated before reaching the Nile (Sutcliffe and Parks, 1999; Mohamed *et al.*, 2006).

Threats to Nile wetlands

Nile Basin wetlands are vulnerable to a range of factors including water resource infrastructures, conversion to agriculture, increasing populations and overexploitation of resources, invasive species, extraction of minerals and oil, and climate change. Many hydrological interventions already exist or are planned across the Nile Basin in order to increase economic benefits and food security. However, these interventions will not be without negative consequences and both the costs and benefits need to be carefully evaluated. One likely consequence of increased flow regulation is reduced downstream flooding and dampening of the seasonal flood pulse, both of which will have an impact on wetlands. Uganda, Ethiopia and Sudan all have ambitious plans for dams along both the main stem of the White Nile and its tributaries, and the construction of these is already underway in some locations. In addition, it is not yet clear whether construction of the Jonglei canal (Figure 11.1), a major threat to the Sudd, will be resumed, with the aim of reducing evaporative losses from the Sudd and increasing water for irrigation downstream. The pastoral economy of the Sudd is dependent on the annual flooding which depends on relatively steady outflows from Lakes Victoria and Albert and the seasonal flows of the torrents above Mongalla. Any alteration of the natural flow would affect the regime of the Bahr el Jebel and the Sudd and would disrupt the economy of the area (Sutcliffe and Parks, 1999). If completed, the canal is also likely to have a significant impact on Nile hydrology, groundwater recharge, sedimentation and water quality; it is also likely to result in the loss of biodiversity, livestock grazing areas and fish habitats, and to interfere with the seasonal migration patterns of both cattle and wildlife, all of which will have an effect on the livelihoods of the local populations (WWF, 2010).

Dams on the Blue Nile and its tributaries also need careful consideration; under normal conditions, the Baro River only breaks its northern banks and provides water to the Machar Marshes when in peak flow. If water is extracted or stored upstream of this overflow so that the peak flows are reduced to a level that does not permit spillage northwards, the area of seasonally flooded marshes will be significantly reduced with serious effects on livestock and wildlife dependent on that wetland (Baecher *et al.*, 2000). Any dam built for storage and hydropower generation on the upper reaches of the Akobo or Baro rivers should therefore be reviewed for its ability as a regulator and flexible control structure with the goal to alter the rise of the Baro as little as possible, as this is the system that sustains the southern Machar Marshes (Baecher *et al.*, 2000).

Dams and the artificial wetlands that they create have brought some significant benefits to the region. For example, the construction of the AHD has significantly increased the amount of land available for agriculture, lengthened the agricultural year and provided hydroelectricity, thus benefiting many millions of people in Egypt (Biswas, 1992). However, dams almost always also bring about negative impacts, particularly for wetlands and the people who depend on the ecosystem services that wetlands provide. For example, the construction of the AHD has affected both the quantity and quality of discharge, and the limited Nile nutrient-rich sediments and water reaching the delta have negatively affected both the agricultural activities and the functioning of the ecosystems in the coastal area adjacent to the delta (Hamza, 2009).

While many wetlands in the Nile Basin support agriculture, trade-offs associated with the conversion of wetlands for agricultural use need to be carefully evaluated to ensure that the ecosystem services supported are not undermined. With the population of the Nile Basin predicted to grow to 300 million by 2010 and 550 million by 2030, increased pressure and competition for, and overexploitation of, increasingly scarce resources are to be expected. The need for appropriate management of wetland resources to ensure their sustainable use is therefore a matter of urgency. The conversion of wetlands to agriculture in Uganda has occurred extensively, affecting the hydrologic functioning of these wetlands (Baecher *et al.*, 2000). Conversion of wetlands for agricultural use is also a widespread practice in Ethiopia, in particular in the Baro Akobo wetlands (Teferi *et al.*, 2010). Across the country, wetlands are being lost or altered by unregulated overutilization including extraction of water for agricultural intensification, urbanization, dam construction, and pollution (Abunie, 2003). In this region, escalating population and the resultant need for increased food have resulted in increased agriculture in the wetlands and their subsequent degradation. The Nile Basin's most polluted wetlands are those of the Nile Delta, where irrigation drainage water, untreated or partially treated urban wastes and industrial effluents have reduced water quality, destroyed several forms of aquatic life, reduced the productivity of the fisheries and contaminated the fish catch (UNEP, 2006). In Sudan, the Sudd has come under considerable pressure during the past few years. Hunting has been uncontrolled during the civil war and with the signing of the CPA in 2005 the inflow of large numbers of refugees as well as their cattle has put pressure on the natural resources due to competition for grazing land, deforestation and infrastructural development.

In contrast to the Sudd, the degradation of Lake Victoria has been occurring for several decades. The fish community has been transformed from its native state of high species richness to a much simpler, largely introduced, fauna that appear to be unstable under prevailing exploitation regimes; nutrient enrichment and climate warming have contributed to deoxygenation of deep water habitats and promoted the rise of Cyanobacteria and other changes in the lower food web, and the lake now behaves as a light-limited, nutrient-saturated ecosystem that is becoming biologically sensitive to both the radiation balance and the water budget (Lehman, 2009). Overfishing has also had its effect on Nile perch in both Lake Victoria and

Lake Kyoga where the populations have lessened significantly. The aquatic diversity and fisheries of many other wetlands in the basin including the Sudd, Lake Kyoga and the Nile Delta are also vulnerable to invasive species such as the water hyacinth (*Eichhornia crassipes*).

Wetlands often contain rich reserves of oil and other mineral deposits, and those in the Nile Basin are no exception (De Wit, 2008). The Nile Delta area is currently Egypt's main source of hydrocarbons and natural gas, and the chemical industries located in the delta are a major source of hazardous waste (Hamza, 2009). In Sudan, recent discovery and exploitation of oil reserves in the Sudd threatens the diversity of the wildlife, aquatic macrophytes and floodplains, as well as the hydrology of the intricate ecosystem. Several blocks have already been allocated to oil companies and exploration drilling is under way in the permanent swamps. Concerns surrounding the exploration and extraction of oil include disruption of water flow patterns as a result of seismic testing and diking; wetland and floodplain fragmentation due to access roads and oil exploration sites; and contamination due to oil spills and human waste.

Some impacts of climate change are already being observed globally, and the *Fourth Assessment Report* of the Intergovernmental Panel on Climate Change (IPCC) confirmed that 'warming of the climate system is unequivocal, as is now evident from observations of increases in global average air and ocean temperatures, widespread melting of snow and ice, and rising global average sea level' (IPCC, 2007). Other impacts, because of inertia in the climate system and complex feedback mechanisms, will only become apparent in the future. However, the knock-on effects of changes in the climate, stemming largely from changes in rainfall and evaporation, will cause changes in many other natural systems, including wetlands.

Currently, there is considerable uncertainty about the exact impact of climate change in the Nile Basin. Results from global climate models (GCMs) are contradictory; some show increases in rainfall whilst others show decreases. A recent study of 17 GCMs indicated that precipitation changes between –15 and +14 per cent, which, compounded by the high climatic sensitivity of the basin, translated into changes in annual flow of the Blue Nile at the Sudan border of between –60 and +45 per cent (Elshamy *et al.*, 2008). Kim *et al.* (2008) found a generally increasing trend in both precipitation and run-off in the northern part of the Blue Nile Basin. To date, no studies have been conducted into the secondary impacts of climate change arising from changes in temperature and rainfall (e.g. changes in irrigation demand), which are also likely to affect run-off and river flows and hence wetlands.

Rising sea levels, both climate-related and due to other factors, will have an impact on coastal wetlands. Eustatic rise alone is estimated to potentially result in the loss of 22 per cent of the world's coastal wetlands by 2080 (Nicholls *et al.*, 1999). Rising sea levels would weaken the Nile Delta's protective sand belt, with serious consequences for essential groundwater, inland freshwater fisheries and the large expanses of intensively cultivated agricultural land (IIED, 2007). Future scenarios based on anticipated conditions of fluvial input, delta subsidence and acceleration of eustatic rise have been used to estimate land loss in the Nile Delta, with worst-case scenarios indicating a habitable land loss of 24 per cent by 2100 (Milliman *et al.*, 1989).

The lack of certainty in trends in rainfall, run-off and sea-level rise will greatly complicate future wetland management. It is likely that in some places in the basin increased rainfall and run-off will cause increases in flow into wetlands and vice versa, and even small rises in sea level will have large impacts on the delta. Further research is needed to improve quantitative understanding of the impacts of climate change on basin hydrology and hence on wetlands.

Conclusions and recommendations

Wetlands of various types occur across the length of the Nile Basin. These ecosystems are a vast livelihood resource, and either have the potential to contribute to, or already contribute in diverse ways to, the livelihoods of millions of people. Levels of wetland resources use vary considerably across the basin; while some ecosystems, such as Lake Victoria may currently be exploited to their full potential, others, such as the Sudd, can be further developed.

Livestock, fisheries and aquaculture are fundamental needs in the daily lives of people along the Nile, but they have been neglected topics in the water discourse. Livestock are essential to many groups in the Nile Basin; they establish the wealth of a family and ability to marry, and indicate the social standing of several groups within the Nile Basin countries. In many places throughout the basin, pastoralists are wholly dependent on wetlands to maintain their cattle and other animals. The Sudd region has a high potential for livestock, which could markedly contribute to the new economy of South Sudan; increasing livestock productivity in the Sudd would also greatly improve the livelihoods of Sudd residents.

Fish is a major source of protein across the basin, and there is scope for improvement in various Nile wetlands. Egypt has the best developed fisheries industry, and in some areas, overexploitation is suspected to occur. Fish landings in the delta lakes, for example, now overshoot the limits of sustainability (Dumont and El-Shabrawy, 2008). Fisheries contribute significantly to the GDP of Egypt and Uganda. In Uganda, fisheries also play a very important role for subsistence as well as for a commercial livelihood. While Lake Victoria is the largest and most important economically, other large lakes, including George, Edward, Albert and Kyoga, and associated wetlands also contribute substantially to the annual national catch. The Sudd and other wetlands contain huge untapped potential for fisheries. Fish farming has begun to play an increasingly important role in Egypt, where over 90 per cent of the basin's aquaculture is currently practised. However, there are opportunities elsewhere for the development of aquaculture.

The contribution of Nile wetlands to the livelihoods of local populations, as well as to the economies of the basin countries, is clear. Despite their importance, there are big gaps in the knowledge about the current status of these ecosystems, and how populations in the Nile use them. More information and a better understanding are needed on the ecosystem services provided by the different types of wetlands in the Nile, and how these contribute to local livelihoods. The values of many of these services are currently unknown, as are their interactions. As a result, it is difficult to assess trade-offs between the various competing uses of the wetlands, and thus the management responses required to balance the need to increase food security while at the same time ensuring that the ecosystem services which support these activities are sustained. While many of the Nile's wetlands are inextricably linked to agricultural production systems, rapidly increasing populations in conjunction with efforts to increase food security are escalating pressure to expand agriculture within them. The environmental impact of wetland agriculture can have profound social and economic repercussions for people dependent on ecosystem services other than those provided directly by agriculture; if wetlands are not used sustainably, the functions which support agriculture, as well as other food security and ecosystem services, including water-related services, are undermined (McCartney *et al.*, 2010). The basis for making decisions on the extent to which, and how, wetlands can be sustainably used for agriculture is weak, and there is a lack of information available describing the best agricultural practices to be applied within different types of wetlands within the Nile Basin and elsewhere. There is currently a pressing need for more systematic planning that takes into account trade-offs in the multiple services that wetlands provide, and a lack of understanding on how to establish appropriate management arrangements that will adequately safeguard important ecosystem services.

Due to these information gaps, the future contribution of wetlands to agriculture is poorly understood and wetlands are often overlooked in the Nile Basin discourse on water and agriculture. While there is great potential for the further development of agriculture and fisheries within these wetlands, in particular in Sudan and Ethiopia, at the same time many wetlands in the basin are threatened by poor management practices and rising populations. Although there is potential for more agriculture within these areas, there is a need for a much better understanding of how to practice agriculture sustainably. Very few governments in the Nile Basin countries have specific wetland policies, or national strategies/policies pertaining to either water or agriculture, that make explicit reference to wetland agriculture. As a result, wetlands are influenced by the policies of many different sectors. If future wetland agriculture is to bring about net benefits a much more strategic approach to wetland utilization is required.

Wetlands can be considered natural hydraulic infrastructure, bestowing many water resource benefits, which need to be carefully considered in planning and management of wetlands (McCartney *et al.*, 2010). As any activities which affect water use and diversions of water from wetland areas have important basin-wide and downstream implications, wetland management needs to be incorporated into basin management. In addition, governance of wetlands should include a means of involving stakeholders from impacted or potentially impacted regions. A policy framework for sustainable wetlands management should include two key factors: first, the maintenance of ecological integrity of wetlands should be clearly incorporated in policies dealing with larger landscapes (e.g., river basins, provinces, etc.); second, it should incorporate a mechanism that empowers local people to manage and control wetlands in their own landscape.

Looking to the future, increased wetland use in the Nile could either lead to prosperity, or be a flashpoint for conflict. Consequently, it is essential that future wetland management should significantly improve on what has occurred in the past, and is integrated in a systematic manner into the development plans and strategies for water and natural resources in the basin.

References

Abunie, L. (2003) The distribution and status of Ethiopian wetlands: an overview, in *Wetlands of Ethiopia, Proceedings of a Seminar on the Resources and Status of Ethiopia's Wetlands*, Y. Abebe and K. Geheb (eds), IUCN, Gland, Switzerland.

Awulachew, S., Rebelo, L.-M. and Molden, D. (2010) The Nile Basin: tapping the unmet agricultural potential of Nile waters, *Water International*, 35, 5, 623–654.

Baecher, G., Anderson, R., Britton, B., Brooks, K. and Gaudet, J. (2000) *The Nile Basin – Environmental Transboundary Opportunities and Constraint Analysis*, draft report prepared for United States Agency for International Development, October, http://rmportal.net/library/content/tools/environmental-policy-and-institutional-strengthening-epiq-iqc/epiq-environmental-policy-and-institutional-strengthening-cd-vol-1/epiq-cd-1-tech-area-biodiversity-conservation/the-nile-basin-2013-environmental-transboundary-opportunities-and-constraints-analysis, accessed 28 April 2012.

Baha El Din, S. M. (1999) *Directory of Important Bird Areas in Egypt*, Palm Press, Cairo, Egypt.

Balirwa, J. S. (2007) Ecological, environmental and socioeconomic aspects of the Lake Victoria's introduced Nile perch fishery in relation to the native fisheries and the species culture potential: lessons to learn, *African Journal of Ecology*, 45, 120–129.

BirdLife International. (2008) *BirdLife's Online World Bird Database: The Site for Bird Conservation*, version 2.1, BirdLife International, Cambridge, UK, www.birdlife.org, accessed 30 April 2009.

Biswas, A. K. (1992) The Aswan High Dam revisited, *Ecodecision*, September, 67–69.

Denny, P. (1991) Africa, in *Wetlands*, M. Finlayson and H. Moser (eds), pp115–148, International Waterfowl and Wetlands Research Bureau, London, UK.

De Wit, M. (2008) Mineral resources: wealth at a frightening price, *Quest*, 4, 2, 50–53, www.aeon.uct.ac.za/content/pdf/join%20us/p5053.viewpoint42.pdf, accessed 28 April 2012.

Dumont, H. J. (2009) *The Nile*, Monographiae Biologicae 89, Springer, Dordrecht, The Netherlands.

Dumont, H. J. and El-Shabrawy, G. I. (2008) Lake Borullus of the Nile Delta: a short history and an uncertain future, *Ambio*, 36, 677–682.

El-Shabrawy, G. I. (2009) Lake Nasser-Nubia, in *The Nile*, H. J. Dumont (ed.), Monographiae Biologicae 89, 125–155, Springer, Dordrecht, The Netherlands.

Elshamy, M. E., Seierstad, I. A. and Sorteberg, A. (2008) Impacts of climate change on Blue Nile flows using bias-corrected GCM scenarios, *Hydrology and Earth Systems Sciences Discussions*, 5, 1407–1439.

Emerton, L. (2005) *Values and Rewards: Counting and Capturing Ecosystem Water Services for Sustainable Development*, IUCN Water, Nature and Economics Technical Paper no 1, Ecosystems and Livelihoods Group, Asia, IUCN, Gland, Switzerland, p93,.

Emerton, L. and Bos, E. (2004) *Value: Counting Ecosystems as an Economic Part of Water*, IUCN, Gland, Switzerland, p88.

Fahey, D. (2007) *The Political Economy of Livestock and Pastoralism in Sudan,* IGAD LPI Working Paper 6–8, 59, IGAD (Intergovernmental Authority on Development), Rome, Italy.

FAO (Food and Agriculture Organization of the United Nations) (2002) *Africover – Eastern Africa Module: Land Cover Mapping Based on Satellite Remote Sensing*, FAO, Rome, Italy.

Hamza, W. (2009) 'The Nile delta', in *The Nile*, H. J. Dumont (ed.), Monographiae Biologicae 89, 75–94, Springer, Dordrecht, The Netherlands.

Hassan, M., Petersen, G., Dubeau, P., Beale, T. and Kombaz, T. (2009) *The Wetlands of the Nile Basin: Baseline Inventory and Mapping*, Nile Basin Initiative, Nile Transboundary Environmental Action Project, NBI-NTEAP, Khartoum, Sudan.

Hickley, P. and Bailey, R. G. (1987) Food and feeding relationships of fish in the Sudd swamps (River Nile, southern Sudan), *Journal of Fish Biology*, 30, 2, 147–160.

Howell, P. and Lock, M. (eds) (1988) *The Jonglei Canal: Impact and Opportunity*, Cambridge University Press, Cambridge, UK.

Hughes, R. H. and Hughes, J. S. (1992) *A Directory of African Wetlands*, International Union for the Conservation of Nature, Gland, Switzerland, United Nations Environment Programme, Nairobi, Kenya and World Conservation Monitoring Centre, Cambridge, UK.

IIED (International Institute for Environment and Development) (2007) *Water Ecosystem Services and Poverty Reduction Under Climate Change*, Issue Paper, IIED, London, UK.

IMF (International Monetary Fund) (2006) *Sudan: 2006 Article IV Consultation and Staff-Monitored Programme – Staff Report; Staff Statement; Public Information Notice on the Executive Board Discussion; and Statement by the Executive Director for Sudan*, IMF Country Report no 06/182, May, IMF, Washington, DC.

IPCC (Intergovernmental Panel on Climate Change) (2007) Summary for policymakers, in *Climate Change 2007: The Physical Science Basis. Contribution of Working Group I to the Fourth Assessment Report of the Intergovernmental Panel on Climate Change*, S. Solomon, D. Qin, M. Manning, Z. Chen, M. Marquis, K. B. Averyt, M. Tignor and H. L. Miller (eds) Cambridge University Press, Cambridge, UK.

JIT (Jonglei Investigation Team) (1954) *The Equatorial Nile Project and Its Effects on the Anglo-Egyptian Sudan*, vols I–IV, Sudan Government, Khartoum, Sudan.

Kadlec, R. H. and Knight, R. L. (1996) *Treatment Wetlands,* Lewis Publishers, CRC Press, Boca Raton, FL.

Kansiime, F. and Nalubega, M. (1999) *Wastewater Treatment by a Natural Wetland: The Nakivubo Swamp, Uganda*, A.A. Balkema, Rotterdam, The Netherlands.

Kim, U., Kaluarachchi, J. J. and Smakhtin, V. U. (2008) *Climate Change Impacts on Hydrology and Water Resources of the Upper Blue Nile River Basin, Ethiopia*, Research Report 126, International Water Management Institute (IWMI), Colombo, Sri Lanka.

Lehman, J. T. (2009) Lake Victoria, in *The Nile*, H. J. Dumont (ed.), Monographiae Biologicae 89, 215–241, Springer, Dordrecht, The Netherlands.

Lehner, B. and Döll, P. (2004) Development and validation of a global database of lakes, reservoirs and wetlands, *Journal of Hydrology*, 296, 1–4, 1–22.

Matsuishi, T., Muhoozi, L., Mkumbo, O., Budeba, Y., Njiru, M., Asila, A., Othina, A. and Cowx, I. G. (2006) Are the exploitation pressures on the Nile perch fisheries resources of Lake Victoria a cause for concern? *Fisheries Management and Ecology*, 13, 53–71.

McCartney, M., Rebelo, L.-M., Senaratna Sellamuttu, S. and de Silva, S. (2010) *Wetlands, Agriculture and Poverty Reduction*, IWMI Research Report 137, International Water Management Institute, Colombo, Sri Lanka, p39.

MEA (Millennium Ecosystem Assessment) (2005) *Ecosystems and Human Well-being: Wetlands and Water Synthesis*, World Resources Institute, Washington, DC.

Milliman, J. D., Broadus, J. M. and Gable, F. (1989) Environmental and economic implications of rising sea-level and subsiding deltas: the Nile and Bengal examples, *Ambio*, 18, 340–345.

Mohamed, Y. A., Savenije, H. G., Bastiaanssen, W. M. and van den Hurk, B. M. (2006) New lessons on the Sudd hydrology learned from remote sensing and climate modeling, *Hydrology and Earth System Sciences*, 10, 507–518.

Nicholls, R. J., Hoozemans, F. M. J. and Marchard, M. (1999) Increasing flood risk and wetland losses due to global sea-level rise: regional and global analysis, *Global Environmental Change*, 9, 69–87.

Okeny, A. (2007) *Southern Sudan: Launch of a Project to Improve Animal and Fish Production*, news release 2007/269/AFR, World Bank, Washington, DC.

Robertson, P. (2001) Sudan, in *Important Bird Areas in Africa and Associated Islands: Priority Sites for Conservation*, L. Fishpool and M. Evans (eds), BirdLife, Cambridge, UK, 877–890.

Rzóska, J. (1974) The upper Nile swamps, a tropical wetland study, *Freshwater Biology*, 4, 1–30.

Stanley, D. J. and Warne, A. G. (1994) Worldwide initiation of Holocene marine deltas by deceleration of sea-level rise, *Science*, 265, 228–231.

Sutcliffe, J. (2009) The hydrology of the Nile Basin, in *The Nile*, H. J. Dumont (ed.), Monographiae Biologicae 89, 335–364, Springer, Dordrecht, The Netherlands.

Sutcliffe, J. and Parks, Y. (1999) *The Hydrology of the Nile*, IAHS Special Publication no 5, IAHS Press, Wallingford, UK.

Sutcliffe, J. V. and Widgery, N. J. (1997) The problems of sustainable water resources management in Sudan, in *Sustainability of Water Resources under Increasing Uncertainty*, Proc. Rabat Symp. SI, April, 259–265, IAHS Publication no 240, Rabat, Morocco.

Talbot, M. R. and Williams, M. A. J. (2009) Cenozoic evolution of the Nile Basin, in *The Nile*, H. J. Dumont (ed.), Monographiae Biologicae 89, 37–60, Springer, Dordrecht, The Netherlands.

Teferi, E., Uhlenbrook, S., Bewket, W., Wenninger, J. and Simane, B. (2010) The use of remote sensing to quantify wetland loss in the Choke Mountain range, Upper Blue Nile basin, Ethiopia, *Hydrology and Earth Systems Sciences Discussions*, 7, 6243–6284.

UNEP (United Nations Environment Programme). (2006) *Africa Environment Outlook 2: Our Environment, Our Wealth*, UNEP, Nairobi, Kenya.

UN-WWAP and DWD (UN-World Water Assessment Programme and Directorate of Water Development, Uganda) (2005) *National Water Development Report: Uganda*, Second UN World Water Development Report, UNESCO, Paris, France, http://unesdoc.unesco.org/images/0014/001467/146760e.pdf, accessed on 12 January 2011.

Vijverberg, J., Sibbing, F. A. and Dejen, E. (2009) Lake Tana: Source of the Blue Nile, in *The Nile*, H. J. Dumont (ed.), Monographiae Biologicae 89, 163–192, Springer, Dordrecht, The Netherlands.

Welcomme, R. L. (1979) *Fisheries Ecology of Floodplain Rivers*, Longman, London, UK.

Witte, F., van Oijen, M. and Sibbing, F. (2009) Fish fauna of the Nile, in *The Nile*, H. J. Dumont (ed.), Monographiae Biologicae 89, 647–676, Springer, Dordrecht, The Netherlands.

WWF (World Wildlife Fund) (2010) *Saharan Flooded Grasslands*, AT0905, www.worldwildlife.org/wildworld/profiles/terrestrial/at/at0905_full.html, accessed June 2010.

12
Nile water governance

Ana Elisa Cascão

Key messages

- The governance regime of the Nile waters has been changing significantly in the past decade, with new transboundary settings being initiated, based on the principle of multilateral cooperation. The establishment of the Nile Basin Initiative and the negotiations for a new agreement in the basin (the Cooperative Framework Agreement) are the major highlights of the cooperation process in the Nile Basin. Successes, pitfalls and failures of a decade of hydropolitical cooperation are analysed in this chapter.
- The attempt to institutionalize the process of management and allocation of the Nile water resources over the last decade shows how the Nile is a politicized and securitized basin. The current outcome is a mix of progressive cooperative processes (such as the identification and implementation of projects with regional benefits) and an enduring diplomatic and legal deadlock between upstream and downstream riparians.
- The long-term future of cooperation in the Nile Basin depends, to a large extent, on the final outcome of the diplomatic negotiations between the Nile riparians, namely the adoption (or rejection) of the new Cooperative Framework Agreement. Based on the current political context, this chapter identifies four alternative emerging scenarios: 'one Nile' (all-inclusive Nile Basin Commission); 'two-speed Nile' (a Nile Basin Commission without all the riparians); 'cooperation-as-usual' (multilateral cooperation, but without a multilateral agreement); and 'end of multilateral cooperation' (partial cooperation or no-cooperation).

Introduction

This chapter aims to provide a comprehensive analysis of the past and the current water governance regime in the Nile Basin, and provides an updated analysis on the institutional set-up and progress of the ongoing cooperation process. The main goal is to consider what worked, what did not work and what the emerging institutional options at play in the Nile Basin Initiative (NBI) are.

The chapter is divided into three sections. The first section considers previous attempts at cooperation in the Nile Basin and its limited successes. It aims to understand why previous efforts at cooperation have not worked, and how the NBI is different from previous cooperative efforts. The second section analyses the ongoing two-track approach to transboundary cooperation that has been adopted in the basin since the 1990s. This part provides a critical, in-depth analysis of the two tracks of the cooperative process: the Nile Basin Initiative (NBI) and the negotiations for a Cooperative Framework Agreement (CFA). The progress and achievements in terms of implementation of the two cooperative tracks will be analysed. Finally, the third part looks forward to identifying and analysing the emerging scenarios for transboundary water cooperation in the Nile Basin, as well as the likelihood of each of the scenarios in light of the current developments (i.e. 2010).

Past Nile water governance

Background on the hydropolitics of the Nile Basin

The Nile is the longest river (6825 km) in the world and its basin is one of the only three basins in the world with ten or more riparian states. These states are: Burundi, Democratic Republic of Congo, Egypt, Eritrea, Ethiopia, Kenya, Rwanda, Sudan, Tanzania and Uganda. The Nile is a unique basin in many respects. Hydrologically, it is a complex body of water with its many tributaries grouped into two main sub-basins – the Eastern Nile and the Equatorial Lakes – which, in turn, are made up of several sub-basins. Economically, the region is characterized by underdeveloped economies and low levels of interregional trade and integration. Politically, the basin is a conflict-prone region and has been the stage for many armed conflicts. It was only recently that the region achieved a degree of political stability.

In hydropolitical terms, the basin has been characterized, historically, by the existence of low-level conflict (mainly diplomatic), opposing the two downstream riparians and main users of Nile water (Egypt and Sudan) and the upstream riparians, the main contributors to the Nile flows. The major point of contention revolves around water agreements, signed in 1929 and 1959, which afforded Egypt and Sudan, but not the other riparians, specific volumetric water allocations (Okidi, 1994; Dellapenna, 2002). Beginning in the 1990s, the Nile riparians have collectively initiated a transboundary cooperation process, but the hydropolitical complexities of the past remain visible.

A main feature of the Nile Basin hydropolitics has been, and remains, the existence of strong power asymmetries between upstream and downstream states (particularly Egypt). Asymmetries exist in terms of material, bargaining and ideational power (Zeitoun and Warner, 2006). In material power terms, all of Nile riparians lag far behind Egypt in terms of their GDP, economic diversification, external political support and access to international funding (Allan, 1999; Nicol, 2003). In terms of their bargaining power, upstream riparians have, so far, been much weaker than their downstream counterparts; these actors have had a comparatively weaker capacity to influence regional and global political and water agendas and also the basin's legal negotiations. In ideational terms, a wide gap exists between the capacities of upstream and downstream riparians to produce and disseminate knowledge, to sanction discourse and to define the red lines of cooperation (Cascão, 2008a, 2009). These asymmetries have been a crucial element in the maintenance of both the controversial 1959 Agreement and the positions of hegemony enjoyed by Egypt and Sudan in regional hydropolitics. Likewise, the asymmetries have also influenced the progress of the different cooperation attempts in the basin: power asymmetries have been internalized by institutions such as Hydromet, Undugu and TeccoNile. Eventually, the current cooperation process is of a different nature and might contribute to the levelling of the playing field between riparians, a potential discussed further, below.

Previous attempts of cooperation in the Nile Basin

Hydropolitical cooperation in the Nile Basin is not a recent phenomenon (see Figure 12.1). The goal of this chapter is not to discuss historical hydropolitics of the basin but to look at the most recent hydropolitical developments. Prior to the 1990s the Nile riparians, with the encouragement and financial support of international institutions and donors, have established institutions and platforms, their goal being to promote transboundary water cooperation in the region. Figure 12.1 summarizes the main milestones of hydropolitical relations in the Nile Basin region.

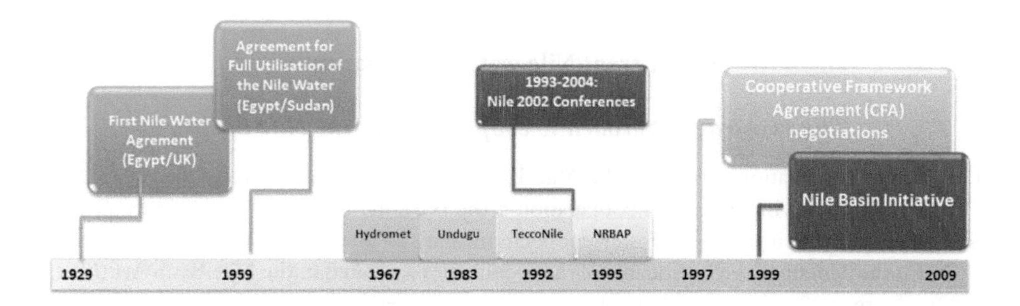

Figure 12.1 Timeline of hydropolitical relations in the Nile River

In general terms, the previous attempts towards cooperation in the Nile Basin have failed to establish a basin-wide type of cooperation, for two main reasons. First, none of the previous initiatives was all-inclusive. Hydromet, Undugu and TeccoNile (in 1967, 1983 and 1992, respectively) could gather together several Nile riparians, but not all, given the refusal, by key upstream riparians including Ethiopia, Kenya and Tanzania, to become members. Although these states have variously acted as observers to the initiatives, they have considered them (i) to be under the direct control of the downstream riparians and of benefit only to downstreamers' interests; and (ii) to remain silent on the main and most controversial issue in the basin: the unfair legal allocation of the Nile waters and the need for renegotiation of past agreements (Tamrat, 1995; Collins, 2000; Arsano, 2004). Unable to bring all of the riparians together, there was little chance that the initiatives could promote any measure of basin-wide hydropolitical cooperation.

Second, none of the previous initiatives successfully promoted a basin-wide or integrated river basin perspective because of their limited scope. They were mostly concerned with technical issues, but not with addressing issues of infrastructure, investment or economic development. Neither did these initiatives address one central hydropolitical dilemma of the basin: the legal issues. As such, the interests of upstream riparian countries were frustrated; to these parties this was the most important element of a potential basin-wide cooperation process and, as such, there was little hope that previous initiatives would lead to a broad involvement of all riparians.

However, developments at the beginning of the 1990s planted the seeds for some change in terms of perceptions and ambitions of all Nile riparian states. On the one hand, the Nile 2002

Conferences substantially increased the basis for dialogue (Shady *et al.*, 1994; Dinar and Alemu, 2000). Looking back at the conference's proceedings, it is clear that several crucial and innovative issues were already being publicly debated (MoWR, 1997). On the other hand, the TeccoNile was responsible for initiating and facilitating the design of the Nile River Basin Action Plan (NRBAP), which led to the establishment of the NBI in 1999 (Dombrowski, 2003). The NRBAP was the first document in the history of Nile cooperation to address issues of economic development and the equitable utilization of water resources. It was also under TeccoNile auspices that the Nile riparians initiated the D3 Project, which aimed to address legal and institutional issues (Amare, 1997; Brunnee and Toope, 2002). Therefore, the Nile 2002 Conferences and TeccoNile may be viewed as the preliminary steps made towards a multilateral, basin-wide cooperation.

The current Nile water governance

The two-track approach to cooperation in the Nile Basin

The current cooperation process in the Nile Basin may be defined as a two-track approach. The first, and most visible, track of cooperation is the NBI, the *transitional cooperative mechanism* established in 1999 in order to 'foster cooperation and sustainable development of the Nile River for the benefit of the inhabitants of those countries', as stated in the Nile Basin Act (NBI, 2002). The second and less visible track comprised multilateral negotiations for a Cooperative Framework Agreement (CFA). The eventual adoption of the CFA by the Nile riparian states will result in the establishment of a *permanent river basin commission*, which will replace the transitional mechanism (NBI, 2002). The progress of both tracks is analysed in this chapter after a preliminary introduction of the current cooperation process.

Current cooperation on the Nile Basin – goal and programmes

The main goal of the cooperation in the Nile Basin is 'to achieve sustainable socio-economic development through the equitable utilization of, and benefit from, the common Nile Basin water resources' (NBI, 1999). In order to achieve the goal, the NBI Strategic Action Programme entailed two complementary programmes: (1) the Shared Vision Programmes (SVPs), encompassing grant-based activities to foster trust and cooperation and build an enabling environment for investment; and (2) the Subsidiary Action Programmes (SAPs), with projects aimed at identifying cooperative opportunities to realize investments and tangible benefits through activities in the Eastern Nile and the Nile Equatorial Lakes regions. The correlation between the two programmes is represented in Figure 12.2.

Evolution of the institutional set-up of the NBI

The institutional set-up of the NBI has evolved since its inception. In 1999, the NBI's institutional set-up comprised three main bodies – the Nile-COM (the decision-making body), the Nile-TAC (already formed during the TeccoNile) and the Nile-SEC established in 1999. Figure 12.3 shows the specific mandates of the three bodies. Ten years later, the NBI's expanded institutional set-up has expanded although the three main bodies remain exactly with the same mandates (see Figure 12.4). By 2009, several other bodies had been incorporated, such as Eastern Nile Subsidiary Action Programme Technical (ENSAPT), Nile Equatorial Lakes Technical Advisory Committee (NELTAC), Eastern Nile Technical Regional Office

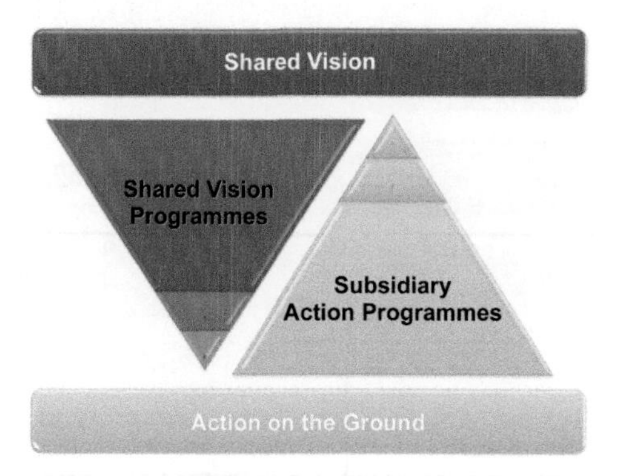

Figure 12.2 Correlation between Shared Vision Programmes and Subsidiary Action Programmes

(ENTRO), and Nile Equatorial Lakes Subsidiary Action Programme Coordination Unit (NELSAP-CU) and NBI National Offices. These institutional changes are an outcome of the application of the principle of subsidiarity – with decision-making processes taking shape at the sub-basin level, although being overseen by the Nile Council of Ministers (Nile-COM).

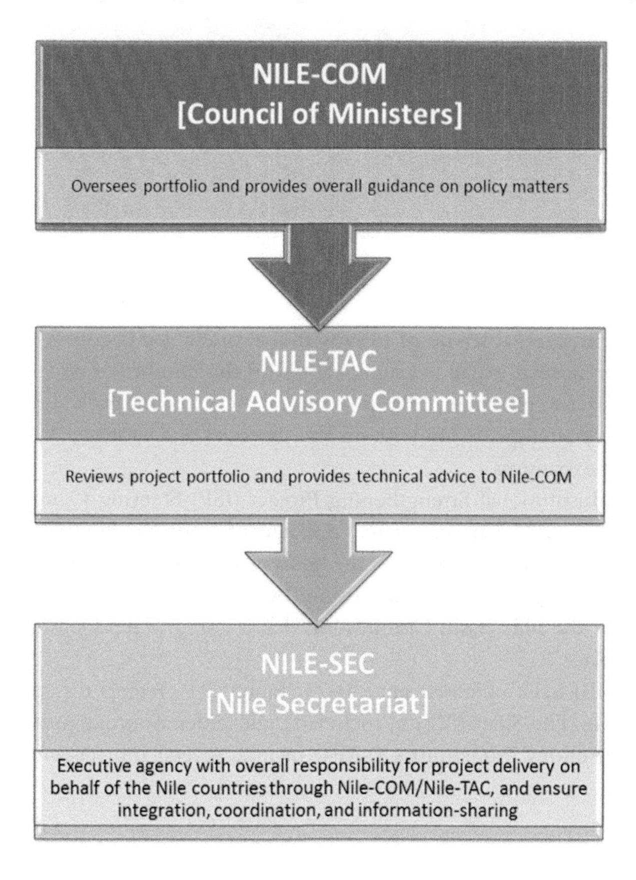

Figure 12.3 Nile Basin Initiative institutional set-up in 1999

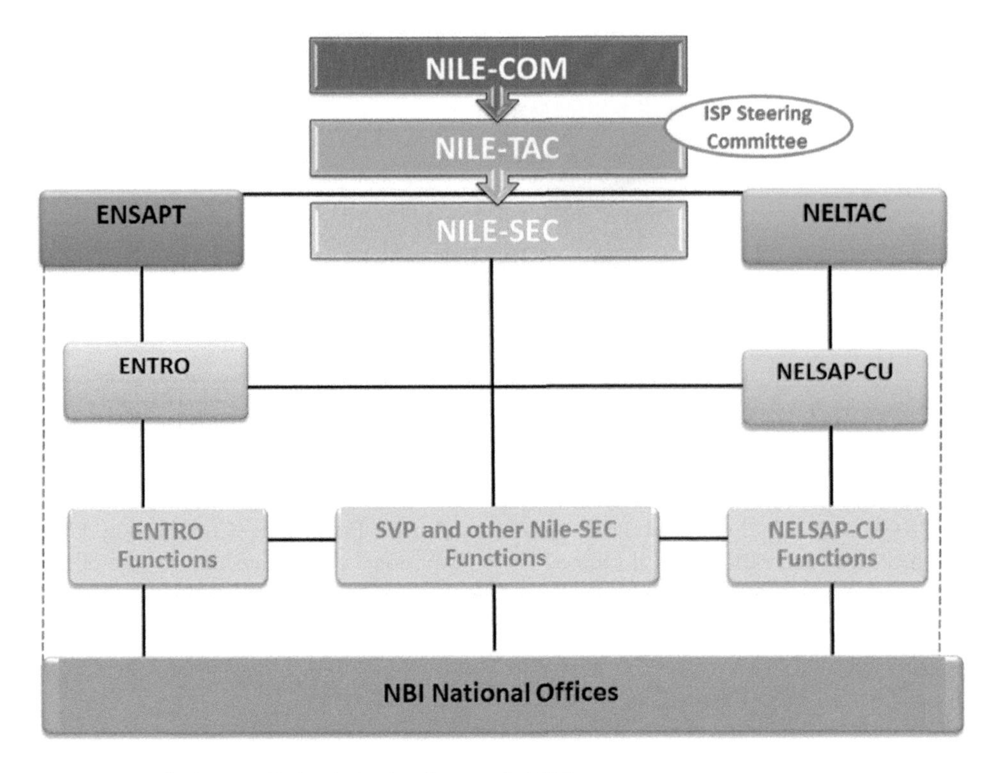

Figure 12.4 Nile Basin Initiative institutional set-up in 2009

In comparing the institutional set-up of 1999 with that of 2009, it becomes clear that (i) the NBI's operational structure has expanded significantly, (ii) the number of nodes and the organization's complexity have increased, and (iii) several new technocratic levels have been established. This chapter attempts to understand the successes and obstacles faced during this decade of institutional expansion.

More recently, the Institutional Strengthening Project (ISP) Steering Committee was also established. The ISP consists of an integrated package of institutional strengthening to be implemented by the NBI institutions. One of the goals of ISP is to pave the way for the establishment of the would-be permanent commission, by identifying and analysing alternative institutional models for the Nile Basin Commission that is going to replace the NBI once the new agreement is in place.

But how has the NBI been operated since 1999? Table 12.1 presents the specific structure of the NBI programmes. The Shared Vision included eight different programmes at the basin-level. The Subsidiary Actions Programmes (SAPs) include several projects envisioned at the sub-basin level.

Table 12.1 Structure of the Nile Basin Initiative Strategic Action Program

Shared Vision Programmes (SVPs)	1. Applied training
	2. Socio-economic development and benefit-sharing
	3. Confidence-building and stakeholder involvement
	4. Transboundary environmental action
	5. Water resources planning and management
	6. Regional power trade
	7. Efficient water use for agriculture
	8. Coordination of shared vision projects
Subsidiary Action Programmes (SAPs)	*Eastern Nile Subsidiary Action Programme (ENSAP)*
	1. Eastern Nile planning model
	2. Flood preparedness and early warning
	3. Watershed management
	4. Irrigation and drainage
	5. Ethiopia–Sudan transmission interconnection
	6. Eastern Nile power trade programme study
	7. Baro–Akobo–Sobat multi-purpose development
	8. Joint multi-purpose programme
	Nile Equatorial Lakes Subsidiary Action Programme (NELSAP)
	1. Kagera Transboundary Integrated Water Resources Management (TIWRM)
	2. Sio Malaba Malakisi TIWRM
	3. Mara TIWRM
	4. Lakes Edward and Albert Fisheries (LEAF) project
	5. Regional trade and agricultural production
	6. Regional power transmission line interconnection
	7. Rusumo falls hydroelectric and multi-purpose development

External support to the transboundary water cooperation in the Nile Basin

The actors 'internal' to the basin (i.e. those from its riparian states) are not the only actors that have influenced the establishment and evolution of Nile Basin cooperation. External actors such as bilateral and multilateral donors have also been an important part of the process. From the very beginning, the World Bank, United Nations Development Programme (UNDP) and Canadian International Development Agency (CIDA) were major financial supporters of the NBI, and their early support was crucial to the NBI process (Nicol *et al.*, 2001). In 2001, the International Consortium for Cooperation on the Nile (ICCON) meeting in Geneva brought together several institutions to support the NBI programmes, which pledged US$140 million to support the launch of cooperation institutions and the Shared Vision Programmes (World Bank, 2001). By that time, several other bilateral and multilateral partners had joined the Nile donor community.

In 2003, the Nile Basin Trust Fund (NBTF), a funding mechanism that helps to administrate and harmonize donor partner support, was established. The NBTF is managed by the World Bank and gathers financial contributions from several donor partners. There are ten NBTF donors: the World Bank, European Commission, Canada, Denmark, Finland, France, Netherlands, Norway, Sweden and United Kingdom. The non-NBTF donors are UNDP,

African Development Bank, Germany, FAO, Global Environmental Facility (GEF), Japan, Switzerland and the United States. This chapter also includes a detailed discussion of the donors' roles in, and contributions to, the evolution of the NBI and its programmes and projects.

Assessment of Shared Vision Programmes and Subsidiary Action Programmes

The goals of SVPs were building confidence, institutional and technical capacity, and creating an enabling environment for the investment projects. These SVPs were initiated between 2003 and 2005, and phased out in 2009 (NBI, 2009a). The SAPs have been designed to implement the shared vision on the ground through concrete water-related projects. In order to manage the SAPs, two institutions were established: ENTRO and NELSAP-CU. The main roles of these institutions are to identify and carry out feasibility studies and prepare for implementation of a portfolio of investment projects, including hydropower, irrigation and multi-purpose projects. Here we analyse what worked well in the context of the SVPs and SAPs – the main achievements, as well as some of the limitations – as well as identifying areas for improvement.

Shared Vision Programmes: main achievements and limitations

Increased dialogue, trust and confidence among riparians and stakeholders

Increased and improved dialogue between riparians is usually considered a major achievement of the NBI and the SVPs. Ten years ago, communication among the different riparians was rare and often engendered conflict. But over the last decade, national decision makers have frequently met in diverse regional forums, and other stakeholders (technicians, academics, legal experts, civil society) have become increasingly present and influential in the decision-making process. But trust and confidence are difficult parameters to measure. On the one hand, riparian states are now keener to maintain dialogue on critical issues which shows their increased confidence. Yet their historic grievances remain decisive factors in the context of negotiations and national media outputs. Sustainable trust and confidence between riparians may only be corroborated in the medium and long term.

Enhanced institutional and technical capacity

In institutional terms, prior to the establishment of the NBI, the transboundary water folder was mainly or exclusively in the hands of national authorities. With the creation of the Nile-SEC, ENTRO and NELSAP-CU, the basin benefits from the presence of a high-quality team of experts, working on several transboundary programmes. In terms of technical capacity, prior to the NBI there existed a large expertise gap between downstream and upstream riparians. Currently, however, this gap seems to be progressively decreasing. Upstream riparians are now building or reinforcing their respective internal capacities. Finally, the NBI institutions have become a platform for exchange of knowledge and eventually for the creation of a new type of expertise informed by a basin-wide approach.

Towards a basin-wide approach to the management of the Nile water resources

One of the SVP's main goals was the creation of an enabling environment for the joint management and development of the Nile water resources. This would engender a move away from narrow national-based approaches to a basin-wide approach. The four SVP sectoral projects were considered essential for the promotion of this approach in four key sectors (environment, water planning, power trade and agriculture). The outcomes of these projects have been mixed but positive examples include preparation of regional baseline assessments, basin-wide guidelines, and the launching of the ground-breaking Nile Basin Decision Support System (expected by 2012). Another example of a developing shared vision is the recognition by all the riparians of the crucial economic and political value of regional power grids and trade.

Shared Vision Programmes: areas for improvement

Understanding and operationalizing the benefit-sharing paradigm

The establishment of the NBI represented the gaining of currency of a fashionable concept circulating in the global water community: the benefit-sharing paradigm. The aim is to shift away from controversial water-sharing agreements towards a more comprehensive understanding of transboundary water cooperation and its potential to generate multiple benefits (Sadoff and Grey, 2002, 2005). The Nile Basin became a testing-ground for the paradigm, but not without criticism. The paradigm is still considered by several of the Nile stakeholders as theoretical, ambiguous, and lacking in real examples. It is also considered that it remains difficult for decision makers to understand fully the range of benefits and how they could be traded among the riparians. Representatives from upstream riparians insist that in the absence of joint investment projects on the ground, it is difficult to talk about sharing benefits.

Coordination between Shared Vision Programmes, and between Shared Vision Programmes and Subsidiary Action Programmes

From the very beginning of NBI's activities, it became clear that the programmes could not coexist as a web of programmes working separately from one another, or from the SAPs, and that coordination was required. The task of coordinating a set of multi-country, cross-cutting, multi-sectoral, multi-donor programmes was, from the outset, a monumental task. The outcome puts in evidence that this coordination has somehow failed for two main reasons: (i) the nature, objectives and languages of the programmes were not harmonized; and (ii) communication and liaison have not always occurred (and, in particular, the crucial liaison between SVPs and SAPs had always been weak).

Subsidiary Action Programmes: main achievements and limitations

Towards a strategic sub-basin-wide approach

One of the main elements of the NBI's institutional design is its organizing principle of subsidiarity (i.e. 'action on the ground needs to be planned [and implemented] at the lowest appropriate level', according to the Policy Guidelines for the Strategic Action Program; NBI, 1999). This principle is at the core of the two sub-basin programmes and their projects. Since their inception, ENTRO and NELSAP-CU have promoted ground-breaking work in the application of this principle. Both have advanced with a strategic approach in order to identify investment projects in their respective sub-basins, for example the Blue Nile Basin or the Kagera Basin. This has allowed a more effective identification of (i) the appropriate planning level, (ii) the potential benefits, costs and impacts, and (iii) the comparison of benefits and impacts between different projects and possible trade-offs.

Towards professional sub-organizations and a transboundary knowledge base

One of the main achievements of the SAPs has been the institutionalization of two regional organizations in order to deal with NBI's investment projects. These institutions are now professional bodies, with well-defined missions, and recruitment is merit-based (although country-balance remains an important factor). A major contribution of the SAPs is the launching of an embryonic transboundary knowledge base (i.e. the joint development of data, knowledge and analytical tools). Examples of this are the Cooperative Regional Assessments (CRAs). The Joint Multipurpose (JMP) in the Eastern Nile, for example, is considered to be the first regional attempt to introduce a 'one river system inventory', which includes other conceptualizations such as 'no borders perspective' (NBI, 2007a; Blackmore and Whittington, 2008). Such studies contribute to the emergence of a new perspective based on long-term, multilateral and multi-sector planning. This may be a first step towards a genuine transboundary decision-making process.

Successful identification of investment projects

The main role of the two SAPs has been, from the very beginning, the identification (pre-feasibility and feasibility studies) and preparation of a portfolio of investment projects. And not, contrary to what many think, to implement the projects. Bearing this important detail in mind, it is possible to say that the performances of both ENTRO and NELSAP-CU have been broadly successful. There is a general consensus that the identification phase in both sub-basins had been successful, but that some concerns had been raised concerning the next phase: the implementation of the identified projects.

Some cases of action on the ground

As mentioned above, from the outset, the role of the SAPs was not to implement projects on the ground. Nonetheless, some of the fast-track projects identified have already begun to be implemented. Examples from the Eastern Nile Basin include: the Ethiopia–Sudan Transmission Interconnection, the Watershed Management and the Irrigation and Drainage projects. Their implementations were made possible because of a combination of favourable factors:

- the projects were identified earlier;
- by their nature, these projects were less political and so there were fewer obstacles to their implementation; and
- funds were successfully mobilized by the regional coordinators.

Although these are cases of action on the ground, several critical voices still consider that this is still a long way removed from initial expectations.

Subsidiary Action Programmes: areas for improvement

Not as much implementation as expected?

The issue of expectations of cooperation is problematic. Interviews with NBI officers highlight that cooperation is a lengthy process, and large-scale investment projects will take time to be implemented and tangible benefits will be forthcoming only in the long term. Other national and regional experts highlight particular concerns that the NBI is not delivering significant results and has not, so far, generated concrete benefits. This difference in opinions is related in part to the fact that the NBI did not clearly define its expected short-, medium- and long-term outcomes in 1999. As a result, the NBI has generated high, and now frustrated, expectations among the riparians. According to the updated NBI schedule (NBI, 2009b), the regional multi-purpose projects will take 10–20 years in their implementation, and 20–30 years for tangible benefits to be derived.

Projects implemented so far are mainly nationally identified and implemented

In a recent presentation given by NBI (NBI, 2009c), a differentiation between the NBI investment projects was advanced. Four types of projects were identified:

Type 1 nationally identified and nationally implemented (consultative projects);
Type 2 regionally identified and prepared, but nationally implemented (cooperative projects);
Type 3 regionally identified, regionally implemented (cooperative projects);
Type 4 beyond the river, towards regional integration.

With this typology in mind, it is possible to observe that the SAP investment projects already under implementation are mainly Type 1 (irrigation and drainage projects) or Type 2 (e.g. Ethiopia–Sudan transmission interconnection) projects. Type 1 and 2 projects are mainly national-based projects that have benefited from the NBI to get access to international funding. They are examples of a narrow type of cooperation, which fails even to bestow multilateral benefits. By contrast, there are still no examples of Type 3 or Type 4 projects being already implemented. Type 3 and 4 projects are, by their nature, transformational: they have the potential to be genuine regional projects which include joint studies, consultation, implementation, benefits and eventually even joint ownership.

In summary, the projects implemented so far lack a real transboundary character. This is intimately related to the fact that the NBI is still a transitional arrangement lacking a cooperative institutional and legal framework that could back large-scale regional projects.

Large investment projects are pending

A lingering question is: can the identified investment projects, especially the large-scale multi-purpose projects, move ahead without the establishment of a permanent Nile Basin Commission? Most of the experts in the Nile Basin consider it very unlikely, for several reasons. First, because implementing regional projects without a clear regional framework is a complex task, as it becomes more complicated to guarantee the long-term commitment of individual countries. Second, because without such a framework, it would be more difficult to mobilize large-scale funding from external donors. And third, because if the Nile Basin Commission (NBC) will not be established within the coming years, it is unclear what would happen to the NBI and the SAPs, that might even lose its *raison d'être*. This issue will be discussed in detail in examining the emerging scenarios later in this chapter.

Funding the Nile Basin cooperation process

Who finances what in the Nile cooperation process?

Over the last decade cooperation in the basin has been financially supported by several multilateral and bilateral donors, known commonly as the Nile development partners. Their commitment to the process has been an element crucial to the launching of the NBI and its performance. The Nile Basin had indeed been one of the darlings of the donor community, both in terms of the number of donors involved and in the magnitude of financial contributions. In particular, since the ICCON meeting in 2001, the NBI has benefited from grants and loans from multiple donors (Jägerskog *et al.*, 2007).

Several of the development partners have supported the NBI through the Trust Fund (NBTF), a funding mechanism established in 2003. Its main role has been to administer the funds pledged by the NBTF donors and harmonize donor partner support. The diagrams below display information about the allocation of funds (World Bank, 2009). Figure 12.5 shows the extents of the donors' financial commitment towards the NBTF. Figure 12.6 shows how these funds have been distributed among the different NBI components. However, it is important to bear in mind that the NBTF is not the only source of funding for the NBI. Multilateral and bilateral donors also finance NBI activities outside the NBTF framework. This includes the financial contributions of non-NBTF donors. Additionally, NBTF donors have also financed projects through bilateral or direct protocol agreements.

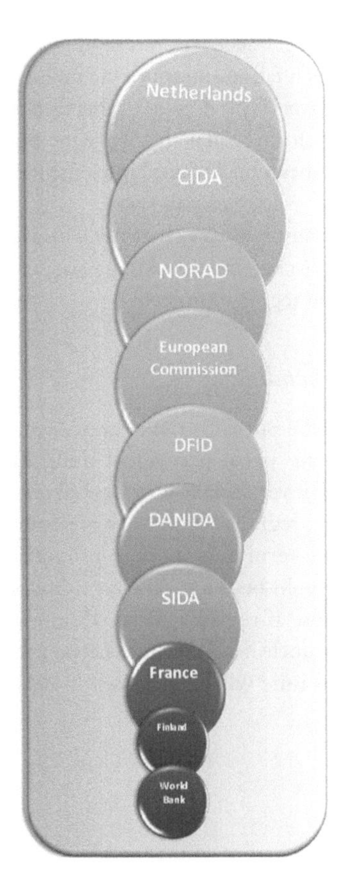

Legend:

1-5 $ million
5-15 $ million
15-25 $ million
25-30 $ million
30-40$ million

Figure 12.5 Commitments to the Nile Basin Trust Fund by the 10 partners (US$ millions)

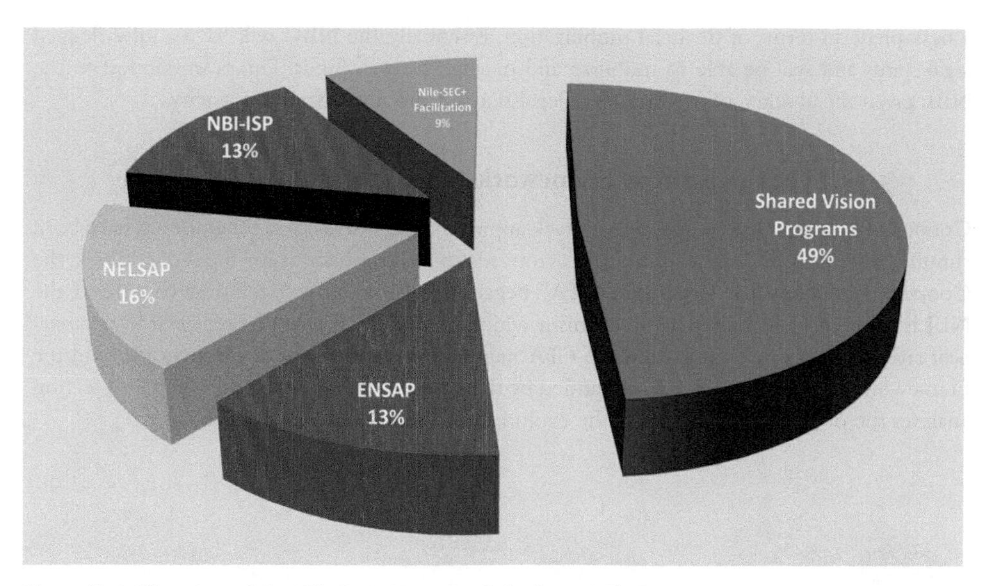

Figure 12.6 Allocation of the Nile Basin Trust Fund funds per Nile Basin Initiative component (as in March 2009)

Ownership of the cooperation process

The financial support of development partners had played a crucial role in the evolution of the basin's cooperation process. Contrary to other examples of transboundary cooperation, cooperation in the Nile Basin has remained to a great extent dependent on the external financial contributions. Because of its over-reliance on donors, the NBI has also been frequently portrayed by several critical voices as excessively donor-driven. Several of the regional experts consider that, ultimately, it is the donors who retain the upper hand in the Nile cooperation process; they influence the cooperative agenda, the design and pace of the programmes. Many basin actors consider their influence in basin hydropolitics to represent unwarranted interference and, since recent events surrounding the CFA negotiations (see next section), they do not see donors as neutral facilitators.

If some consider the donor involvement as a 'necessary evil' for the first stages of cooperation, the large majority of regional experts consider it already to be high time that donors took a step back and grant the riparians ownership of the process. Most of these experts clustered around a shared belief, that ownership of a process is currently one element vital for the internalization and intensification of a cooperation process.

Over-reliance on donor support can undermine the long-term ownership of a transboundary cooperation process (Nicol *et al.*, 2001). Accordingly, once the cooperative institutions and the 'shared vision' have been established, riparian states and the national and regional institutions should take ownership of the process. This does not mean that the overall financial burden should be transferred to the riparian states; instead it means that the process should adopt the form of a genuine partnership between donors and national authorities.

This is certainly a politically sensitive topic, and one that will become more significant if the permanent commission is to be established. With the NBC, the cooperation process will enter a new phase in terms of financial mobilization. Essentially, the NBC will have a fully-fledged legal status and will be able to mobilize and manage its own funds. This is in contrast to the NBI: given the absence of its own, clear legal status it did not have that capacity.

The Cooperative Framework Agreement negotiations

Cooperation in the Nile Basin is a two-track approach. The NBI and its programmes have been running since 1999, as analysed in the previous sections. Negotiations for the second track, the Cooperative Framework Agreement (CFA), began in 1997. The two are intimately related: the NBI is a *transitional* cooperative mechanism which, in due course, will be replaced by a *permanent* river basin organization when the CFA negotiations have come to a conclusion, and the Framework Agreement adopted and ratified by the riparian states (see Figure 12.7). This section analyses the origins and, principally, the evolution of the negotiations for the CFA.

Figure 12.7 Relationship between the Nile Basin Initiative and the Cooperative Framework Agreement

Background to the Cooperative Framework Agreement

Historically, one of the main obstacles for regional cooperation in the Nile Basin has been the existence of historical legal agreements over water, specifically, the 1929 Agreement signed between Egypt and Great Britain (as colonial power in the basin), and the 1959 Agreement between Egypt and Sudan (Caponera, 1993; Dellapenna, 2002). For the purposes of its analysis it should be remembered that these agreements (i) were partial in scope (they did not include all of the Nile riparians), (ii) granted specific volumetric allocations to Egypt and Sudan alone, (iii) are binding for signatories but are not legally recognized by the other riparians, (iv) have historically been the major stumbling block in the hydropolitical relations between the two downstream riparians and the upstream neighbouring states, and (v) still carry an enormous influence over the ongoing cooperation process.

One of the major failures of the past cooperative attempts in the basin has revolved around the fact that the sensitive issue of the previous agreements has not been approached. The downstream riparians claim that multilateral cooperation is possible without having to address the past water agreements and what they consider their 'historic and acquired rights' (Amer, 1997; Hefny and Amer, 2005). Upstream riparians consider that no cooperation is possible without a revision of past agreements and endorsement of a new multilateral agreement (Okidi, 1994; Amare, 1997, 2000).

Therefore, when, at the beginning of the 1990s, technical experts and international donors began to plan for the establishment of a new cooperative initiative, the upstream riparians stipulated that negotiations for a new legal framework *must* be included in the cooperation process; without this condition the new initiative would, once again, turn out to be a non-all-inclusive organization (Arsano and Tamrat, 2005). As a result, in 1997, a Panel of Experts (PoE) including technical and legal experts from all riparians was established and the D3-Project, concerned with multilateral legal negotiations, was initiated. Following these developments the upstream riparians agreed to become full members of the future initiative, the NBI.

1997–2007: stiff negotiations

The goal of the negotiations was to identify cooperation options and to identify main operational and legal principles relevant to the basin. In 2001, the mandate of the PoE was extended to include concrete negotiations for a legal and institutional framework. A Transition Committee was established and its goal was to bring the draft framework to the next stage of negotiations. In 2003, a permanent Negotiating Committee was launched and it has worked as a part of the CFA negotiations ever since.

Against a background of continuous and fundamental divergence of expectations between upstream and downstream riparians over legal agreements, the CFA negotiations were expected to be a long, complex and thorny process. Contrary to the NBI process that mainly involved 'soft' cooperative issues, the CFA process was at the crossroads of some of the most 'hard' and sensitive issues in the Nile Basin: sovereignty, national interests and security, international law principles and water rights. Therefore, it is not surprising that the individual riparian states have tended to assume positions based around national interest, sidelining the basin-wide perspective.

However, contrary to the usual belief, the CFA negotiations have not been about water-sharing per se. Instead, they have been about the delineation of (i) general legal principles regulating the cooperation process (legal framework) and (ii) the institutional structure to be adopted by the future river basin organization (institutional framework). The negotiations for the legal framework were particularly stiff because the countries could not immediately agree on several of the commonly accepted legal principles. But during ten years of negotiations (1997–2007) each of the articles to be included in the final Draft Agreement has been negotiated, bargained and agreed upon, except the one pertaining to the status of the existing agreements (NBI, 2007b). Indeed, in June 2007, when the CFA negotiations were concluded, the countries had agreed on the contents of 38 of the 39 articles. Article 14b, on the status of the previous past Nile agreements, was the only one not backed by consensus. The lack of agreement on Article 14b evidences that the fundamental predicament of Nile cooperation remains the stumbling block. Historic water agreements remain the primary obstacle to sustainable cooperation in the Nile Basin.

2007–2009: A situation of deadlock and increasing political pressure

After 2006, the media in the Nile region reinforced the idea that the negotiators were about to finalize the CFA draft (IPR, 2006; Ethiopian Herald, 2007). Such hopes were frustrated by the Nile-COM meetings of 2006 and 2007 (New Vision, 2006; Addis Fortune, 2007). Meanwhile, and due to lack of agreement between the different parties, a proposal emerged to rephrase Article 14b to include the ambiguous term 'water security' in order to accommodate and harmonize the differing claims of riparians (Cascão, 2008b, c). Ambiguity is often a feature of water legal negotiations and can help to accommodate conflicting interests and resolve enduring deadlocks (Fischhendler, 2008).

The final draft agreement, concluded in June 2007, included the ambiguous Article 14b on 'water security', which states that: 'the Nile Basin States therefore agree, in a spirit of cooperation, to work together to ensure that all states achieve and sustain *water security* and *not to significantly affect the water security* of any other Nile Basin State' (East African Business Week, 2007; emphasis added). The seven upstream riparians, forming one single block for the first time in Nile history, accepted this formulation as an acceptable departure, potentially helpful in accomplishing the conclusion of the CFA. However, Egypt and Sudan had reservations and proposed an alternative formulation: 'to work together to ensure that all states achieve and sustain water security and *not to adversely affect the water security and current uses and rights* of any other Nile Basin State' (East African Business Week, 2007; emphasis added). The reformulation was not accepted by the upstream riparians; they considered that it would perpetuate the old agreements (New Vision, 2007, 2008). In May 2009, during a Nile-COM meeting in Kinshasa, all of the upstream riparians decided that they would not wait any longer. They decided, if necessary, to go ahead with the signature of the CFA without the downstream riparians (East African, 2009). According to the CFA itself, the ratification of a two-thirds majority is the requirement to establish the NBC.

The unanimous decision of the upstream riparians to move forward with the CFA ratification without Egypt and Sudan represented a unique decision in the hydropolitical history of the Nile Basin; it contributed to the resurrection of old grievances but also revealed a new vocal attitude and determination of upstream riparians. The political pressure within, and from outside, the basin has intensified.

2010–2011: The signature of the CFA

At the end of 2009, during the Nile-COM meeting in Sharm El-Sheikh, the upstream riparians decided they would wait no longer and announced that signing of the new agreement would take place on 14 May 2010 after which the Agreement would remain open for signature for a year.

By May 2010, five upstream riparians had signed the CFA – Ethiopia, Kenya, Rwanda, Tanzania and Uganda (East African, 2010). Burundi and Congo were also expected to sign the CFA on the 14 of May 2010, but the two countries have not done so to date, despite the fact that they had promised to do so in the near future. The regional media reported that the reasons for Congo and, in particular, Burundi for not signing the CFA were intimately related to political pressure from the Egyptian side (Daily News Egypt, 2010).

For Egypt, the stakes were extremely high – at any cost, it wanted to prevent Burundi and Congo from signing the CFA, which needs only one more signature to reach the two-thirds majority. Since 14 May, Burundi has been at the top of Egypt's agenda on foreign policy. The small upstream country, earlier totally ignored by Egypt, had the 'honour' of exchanging high-level visits with the downstream neighbour (Al-Masry Al-Youm, 2010b). The outcome of these visits was that Burundi acknowledged that no agreement should be signed against Egyptian interests (Al-Masry Al-Youm, 2010b).

In February 2011, the hydropolitical relations in the Nile Basin experienced a rapid change with the fall of Mubarak's regime in Egypt. That is when Burundi, taking advantage of Egypt's power vacuum, decided to ignore its previous promises and became the sixth country signing the CFA thus opening the Agreement to ratification (Bloomberg, 2011). By doing so, Burundi might have opened a new chapter in the hydropolitical relations in the Nile Basin.

A decade of cooperation in the Nile Basin: what did and did not work?

The analysis above demonstrated that cooperation in the basin over the last decade has been a very dynamic process, and one that has shown signs of both achievement and presence of limitations. Optimistically, the establishment of the NBI symbolized an initial departure from the previous scenario in the basin, one characterized by conflict of interests, towards a more cooperative picture. The NBI had successfully promoted dialogue between the riparians and in increasing levels of trust, at least among technical experts. The NBI was also successful in mobilizing the donors' extensive financial support despite its status as merely a transitional mechanism. To a certain extent, the NBI successfully fostered a basin-wide perspective, and managed to introduce complementary water governance concepts (e.g. subsidiarity). The NBI's different programmes have delivered several important outputs. Finally, in hydropolitical terms, it is also possible to observe that the NBI/CFA cooperation tracks have at least provided a new platform for the upstream riparians, which are generally weaker than their downstream neighbours, to influence the regional water agenda.

But there remain plenty of questions unanswered: after ten years, is the NBI's output greater than the sum of its parts? Are the existing facts on the ground enough to show that the NBI is already contributing to the formation of a new hydropolitical regime in the basin? Finally, in hydropolitical terms, did the NBI really contribute to level the playing field and reduce the asymmetries existing between the long-standing hegemonic riparians downstream and their less powerful neighbours upstream? One might answer that there is still a long way to go until the NBI can reach its initial ambitious goals, and that complete cooperation remains an ideal scenario. In fact, several factors still hinder a change of the hydropolitical regime, towards one of effective cooperation and, particularly, towards the 'equitable utilization' of the Nile water resources. First, it has yet to be proven that a cooperative modus operandi is already in place at high-level national political echelons (i.e. if policy-makers at the ministries and presidential/ministerial cabinets have moved away from perspectives based strictly around national interests towards a basin-wide perspective). Analysis of the CFA negotiations shows that this is not yet the case. Nonetheless, this also raises a question about the extent to which the NBI and its technicians have been able to really influence their national governments. Second, the NBI/CFA might have helped to create a better balance between the downstream and upstream riparians in terms of their ability to influence the regional agenda, but it is still Egypt that calls the shots; these states are still the ones with the capacity to establish the red lines and influence the timing of negotiations and ultimately of project implementation. Third, a major obstacle to sustainable cooperation certainly relates to the absence of the adoption of the legal and institutional framework. The absence of ratification of the CFA protracts the transitional (and uncertain) character of Nile cooperation, and there are several potential risks that are already emerging:

- no Nile Basin Commission can be established – and the Commission is considered by many to be the only way for a sustainable cooperation in the basin;
- the frustration of the upstream riparians, for whom the NBI is the way to have access to investment for facts on the ground, will only increase;
- the NBI's shared vision, as implemented over the last decade, may prove fragile, and riparians may return to, or continue with, unilateral water development;
- the willingness of international donors to support the cooperative process and finance the investment projects may fade; and
- in the worst-case scenario, the NBI may lose its *raison d'être* and collapse as have the previous basin's cooperative efforts.

What next? Emerging hydropolitical scenarios in the Nile River Basin

Alternative emerging scenarios for the Nile Basin Cooperation

The medium- and long-term future of cooperation process in the Nile Basin depends, to a large extent, on the final outcomes of the CFA process. It will be ratification and adoption of the CFA, or its lack, that will represent a tipping point in the basin's hydropolitical relations. Based on the current political context, this section identifies four alternative emerging scenarios:

- **Scenario 1: 'One Nile'** (all-inclusive cooperation, via Nile Basin Cooperation). In the short-, medium-term, the CFA will be signed and ratified by all the nine Nile riparians. As a result, a river basin organization will be established and replace the transitional NBI mechanism. All of the Nile riparian states will be members of the Commission.
- **Scenario 2: 'Two-speed Nile'** (partially inclusive cooperation, with Nile Basin Cooperation). Only some of the riparians will sign and ratify the CFA. With a two-thirds majority, the NBC may be established. All signatories to the CFA will be members of the NBC, and the remaining riparians may opt for the status of observers, remain out, and/or join later.
- **Scenario 3: Cooperation-as-usual** (multilateral cooperation, without Cooperative Framework Agreement). This assumes that the CFA will not be signed in the short term. As such, the Commission would not be established but alternative cooperative arrangements will still be in place. This is an overarching scenario and, below, the different options that it contains will be analysed.
- **Scenario 4: End of multilateral approach** (partial cooperation or no cooperation). This assumes that the CFA will not be signed and that a multilateral approach, including all riparians, becomes no longer viable. This scenario can also include different options, ranging from unilateralism to types of partial, rather than all-inclusive, cooperation.

Scenario 1: 'One Nile'

This scenario is considered by many as the best-case scenario: having all the Nile riparians on board of a Nile Basin Commission, a permanent river basin organization (RBO) to be established once the countries sign and ratify a consensual legal and institutional agreement. In this scenario, all Nile riparians would be involved in the selection of the institutional and financial arrangements for the new institution. Dialogue between stakeholders of the different countries will take place soon and selection by the Nile-COM of the most suitable institutional arrangement for the Nile Basin will follow. Once the NBC is established, the cooperative investment projects previously identified by the NBI could start being implemented on the ground.

Advantages

An all-inclusive river basin commission would come in line with the NBI's 'One Nile' strategic vision, having the basin as the unit to promote regional socio-economic development, but simultaneously strengthening and empowering the sub-basin organizations to implement the investment projects. The NBC, contrary to the NBI, would have an international legal personality, and this would mean that the commission itself could be fully responsible for financial mobilization and allocation to the different investment projects. This would entail two advantages. On the one hand, riparian states would increase their own ownership of the political process and would be less donor-driven. The new commission will also be in a better position to mobilize funding and investment among non-traditional external partners. On the other hand, the establishment of the all-inclusive commission would also ease the relations with the current development partners. The traditional donors, however, are usually in favour of doing it through a commission that includes all the Nile riparians, and not only part of them (see Scenario 2).

Challenges

Despite being considered the most desirable and most sustainable scenario for the Nile River Basin, this would-be Nile Commission is expected to face some challenges. The main one is the potential politicization of the new institution; knowing that regional hydropolitical complexities and conflict of interests between upstream and downstream riparians will not disappear overnight, the risk of the NBC to become extremely politicized is high. The other associated risk is that, in order to avoid politicization, the devolution of authority by the countries to the commission will be very limited and, as such, the commission might end up simply as a technical body but without sufficient political weight. One of the grey areas related to the future cooperation is to know exactly what will be the positionality of the NBC vis-à-vis the ongoing and planned national developments.

Likelihood

At the beginning of 2011, this scenario does not appear to be the most likely. As mentioned in the previous section, only six out of nine countries have signed the CFA, but the two downstream riparians – Egypt and Sudan – maintain their reservations on the agreement. Indeed, since 2010, Egypt and Sudan have frozen their participation in some of the NBI activities in retaliation for the decision of the upstream riparians to sign the CFA. Although not impossible, Egypt and Sudan are not expected to reverse their position in the short term; or at least not before the CFA is ratified by the six countries and becomes a valid international agreement.

Scenario 2: 'Two-speed Nile'

This scenario assumes that only some of the riparians will sign and ratify the CFA, but not all of them (at least in a first stage). According to the agreement itself, with the ratification by two-thirds of the countries (i.e. 6 out of 9), the NBC can be established. In such a scenario, all those riparians that ratify the CFA will be members of the Commission, and the remaining riparians may opt for the status of observers, remain outside, and/or join later.

Advantages

The idea of having a two-speed process had often been used as a way to solve institutional or political deadlocks and simultaneously to incentivize the acceleration of further cooperation and integration among those countries that want to move faster (e.g. in the European Union process during the 1990s). This is not a rare situation in transboundary river basins – this had been the case in the Senegal Basin, and it is also occurring in the Zambezi and the Mekong, for example.

Challenges

Although this scenario might be considered the most likely at this point in time (beginning of 2011) taking into account the recent political developments, it appears as very challenging for several different reasons. First, in institutional terms this scenario was not an option until recently (prior to May 2009, when upstream riparians decided to move forward with the CFA adoption). The design of a partly-inclusive river basin organization was not included in the institutional design studies. It remains unclear what the status and role of non-signatory riparian states could be and the possibility of their joining later.

Second, this scenario is very complex in operational and financial terms. The main challenge is to understand how the commission could operate without having on board two of its main countries – Egypt and Sudan are the main users of the Nile water resources. This raises several questions. How can projects be studied and implemented in a comprehensive manner if all concerned riparians will not be involved? How could the Commission get regional projects approved and financed without involving Egypt and Sudan, in particular, knowing that international financial institutions usually require consent from downstream riparians for projects to be financed?

Last but not least, the 'two-speed scenario' is politically very challenging, as it comes to break the idea of comprehensive multilateral cooperation as adopted since the mid–1990s. It represents a clear departure from a status quo; it is understood as a move towards further cooperation by upstream riparians, but the downstream riparians see it as threat to their water security. Many in the international community see this option as a sub-optimal case scenario.

Likelihood

The likelihood of this scenario appears to be high. Although many doubted in 2009 that upstream countries would even sign the CFA without the consent of Egypt and Sudan, they have done it. Despite all the political pressure since then, the CFA now has six signatories, although none have ratified it as yet. Ratification can be a long process and there might be a lot of political back calculations in the meantime, but if the six countries ratify the agreement, then the NBC will be established.

Scenario 3: Cooperation as usual

Scenario 3 assumes that the Cooperative Framework Agreement (CFA) will not be adopted as soon as expected and, as such, the establishment of the Nile Basin Commission (NBC) will be postponed. Still, this scenario considers that cooperation between riparians will be kept in the business-as-usual manner. If we think in the short or medium term, two different sub-scenarios can be identified:

Maintenance of the Nile Basin Initiative

The NBI is a transitional arrangement for cooperation between the Nile riparians. It was not predetermined when exactly this arrangement would be replaced by the permanent commission. As such, the NBI can continue to operate according to its current limited mandate. This is a likely scenario in the short and medium term, but not in the long term. First, the institution might become redundant and inefficient, in particular, after the existing funds are over. Second, not having a permanent legal status, the NBI will find it difficult to move forward with the implementation of the major cooperative projects. Third, constant delays to adopt the CFA and establish the NBC might contribute to a confidence crisis within the NBI, with predictable frustration or fatigue of the countries as well as of the external partners.

Nile Basin Initiative renamed or revamped

This sub-scenario comes somehow as a continuation of the first one. Assuming that no NBC is established and the NBI becomes redundant, countries and donors might opt for a cosmetic operation. A possibility is to rename the initiative (e.g. Nile Water Committee or Nile Climate Platform) although keeping similar characteristics. In this case, the 'new' cooperative initiative would be operating with a transitional arrangement, but having regional projects to be financed by several sources. This is not a very likely scenario and not something that is being considered in the public discussion, but it is not totally impossible.

Scenario 4: End of multilateral approach

This scenario assumes that the CFA will not be signed and that a multilateral approach to the management of the Nile water resources, including all riparians, becomes no longer viable. This scenario can include different options, ranging from unilateralism to types of partial, rather than all-inclusive, cooperation. It is important to highlight that these scenarios are not only likely in the future; indeed they have been a dominant facet of the hydropolitical developments in the Nile Basin during the last decade.

Partial cooperation

Although multilateral cooperation had been promoted as the optimal approach to transboundary water management, the fact is that many of the developments in the Nile Basin are still being done at the bilateral and/or trilateral level. For example, projects being developed and implemented by the Lake Victoria Basin Commission (LBVC), usually involve two or three of the Lake Victoria riparians. Also in the Eastern Nile Basin, one of the most successful cooperative projects had been the power-transmission line built between Ethiopia and Sudan. Approval, funding, and operationalization of these projects appear easier than the multilateral ones. A continuation of this modus operandi (partial cooperation) in the future is likely, but it involves risks. First, it moves away from a more comprehensive management of the water resources and decreases the scope of potential cooperation benefits (economic, environmental and political). The basin-wide gains are potentially lost. Second, this partial cooperation had been successful in the past mainly because the projects were of small or medium scale. The same type of cooperation might not be the ideal for large-scale projects, as fund-raising will not be an easy process, as examples in the past have shown. Third, legal dilemmas in the basin will remain and downstream riparians will likely continue to block funds from major financial institutions.

Unilateral developments

Although not usually acknowledged by regional and NBI experts, unilateral water developments have never stopped in the period since the NBI was established. The Toshka Project in Egypt, Merowe Dam in Sudan, Tekezze Dam in Ethiopia and Bujagali Dam in Uganda are examples of unilateral projects already or almost completed. Often, these projects have not included any consultation with the neighbouring countries and have failed to provide benefits that go beyond the national borders. Again, recently, the Ethiopian government announced the plans to build another large-scale dam in the Blue Nile. The scenario of unilateral development may continue and increase in scope in the next decades. National plans/projects in the waiting lists of all the Nile riparians often feature several projects – including hydropower as well as irrigation. The Nile region is also calling the attention of many international investors in land for commercial uses. Countries such as Ethiopia, Kenya, Sudan and Uganda are already leasing large tracts of land (and water) to these foreign investors.

The 'unilateral scenario' is considered by many as the worst-case scenario, especially if countries would be in a position to substantially develop these projects. And what the reality on the ground is showing nowadays is that countries can do it through new business models. China, Gulf countries, Arab development banks, private constructors, and others usually do not require consent from other riparian states to finance or implement projects.

From the perspective of the national authorities this is an evident option, in particular if the cooperation process will not result in the medium term. But the impacts of this option are manifold in terms of water flows, environmental and socio-economic impacts, as well as in terms of relations with neighbouring countries and the political stability in the region.

Conclusion: the future of the Nile cooperation – not a 'black or white' scenario

The analysis of the four hydropolitical scenarios demonstrates that the future of cooperation in the Nile Basin is not 'black or white': the choice is not between full cooperation on the one hand and non-cooperation on the other. On the contrary, there exists a large grey area, and the different emerging scenarios involve their own complexities. The goal has been not to identify the most desirable scenario, but to understand each of the scenarios, according to current political circumstances. Some preliminary conclusions may be drawn:

- The momentum for the establishment of the NBC has been created over the last couple of years, but it remains unclear as to whether or not the riparians (either all of them or a few) will seize the momentum.
- The acceleration of the multilateral cooperation process is dependent on the political and hydropolitical back calculations of the individual Nile riparian states.
- Multilateral basin-wide cooperation may be at risk if riparians opt for plenty of large-scale unilateral water developments.
- Last but not least, there has been a general (although shy) optimism, shared among the NBI officials and the actors of the donor community, that multilateral cooperation will soon experience a breakthrough through the adoption of the CFA and the establishment of the NBC. However, this optimism is not always shared by the high-level representatives of national authorities. Ultimately, they are the main political decision makers in the Nile Basin cooperation process.

In conclusion, the jury is still out on whether the Nile Basin is gradually moving towards a new water governance regime marked by multilateral cooperation and joint management of the transboundary resources, or whether partial cooperation and unilateralism will dominate the decades to come.

References

Addis Fortune (2007) Commission or initiative: Nile countries cannot decide, *Addis Fortune*, Ethiopia, 25 March.

Allan, J. A. (1999) *The Nile Basin: Evolving Approaches to Nile Waters Management*, Occasional Paper 20, SOAS Water Issues Group, University of London, London.

Al-Masry Al-Youm (2010a) Egypt's top spy chief and finance minister in East Africa to discuss water, *Al-Masry Al-Youm*, 23 June.

Al-Masry Al-Youm (2010b) Burundi will not take stand inimical to Egypt's interests, *Al-Masry Al-Youm*, 14 November.

Amare, G. (1997) The imperative need for negotiations on the utilization of the Nile waters to avert potential crises, in *Proceedings of the 5th Nile 2002 Conferences, Addis Ababa 24–28 February 1997*, pp287–297, Ethiopian Ministry of Water Resources, Addis Ababa, Ethiopia.

Amare, G. (2000) Nile waters – hydrological cooperation vs. hydropolitics, in *Proceedings of the 8th Nile 2002 Conferences, Addis Ababa 26–29 June 2000*, pp573–580, Ethiopian Ministry of Water Resources, Addis Ababa, Ethiopia.

Amer, S. E. (1997) Cooperation in the Nile Basin: Appropriate legal and institutional framework, in *Proceedings of the 5th Nile 2002 Conferences, Addis Ababa 24–28 February 1997*, pp325–336, Ethiopian Ministry of Water Resources, Addis Ababa, Ethiopia.

Arsano, Y. (2004) Ethiopia and the Nile: Dilemma of National and Regional Hydro-politics, PhD thesis, Center for Security Studies, Swiss Federal Institute of Technology, Zurich, Switzerland.

Arsano, Y. and Tamrat, I. (2005) Ethiopia and the eastern Nile Basin, *Aquatic Sciences*, 67, 1, 15–27.

Blackmore, D. and Whittington, D. (2008) *Opportunities for Cooperative Water Resources Development on the Eastern Nile: Risks and Rewards*, an independent report of the scoping study team prepared to the eastern Nile Council of Ministers, World Bank, Washington, DC.

Bloomberg (2011) Burundi Government signs Accord on use of Nile River water, *Bloomberg*, 28 February.

Brunnee, J. and Toope, S. J. (2002) The changing Nile Basin regime: Does law matter? *Harvard International Law Journal*, 43, 1, 105–159.

Caponera, D. A. (1993) Legal aspects of transboundary river basins in the Middle East: the Al Asi (Orontes), The Jordan and The Nile, *Natural Resources Journal*, 33, 3, 629–664.

Cascão, A. E. (2008a) Ethiopia – Challenges to Egyptian hegemony in the Nile Basin, *Water Policy*, 10, S2, 13–28.

Cascão, A. E. (2008b) New Nile Treaty – A threat to the Egyptian hegemony on the Nile, in *Chroniques Egyptiennes 2007*, H. Aouardji and H. Legeay (eds), Centre d'Études et de Documentation, Économiques, Juridiques et Sociales, Cairo, 2008, 139–162.

Cascão, A. E. (2008c) Ambiguity as strategy in transboundary river negotiations: The case of the Nile River Basin, Paper presented at II Nile Basin Development Forum, 17–19 November 2008, Khartoum, Sudan, Nile Basin Initiative/Sudanese Ministry of Irrigation and Water Resources.

Cascão, A. E. (2009) Changing power relations in the Nile River Basin: unilateralism vs. cooperation? *Water Alternatives*, 2, 2, 245–268.

Collins, R. O. (2000) In search of the Nile waters, 1900–2000, in *The Nile: Histories, Cultures, Myths*, H. Erlich and I. Gershoni (eds), pp245–267, Lynne Rienner Publishers, Boulder, CO.

Daily News Egypt (2010) Congo and Burundi expected to sign new Nile water treaty, *Daily News Egypt*, 15 June..

Dellapenna, J. W. (2002) The Nile as a legal and political structure, in *Conflict and Cooperation Related to International Water Resources: Historical Perspectives*, IHP-VI Technical Documents in Hydrology 62, S. Castelein and A. Otte (eds), pp35–47, UNESCO, Paris, France.

Dinar, A. and Alemu, S. (2000) The process of negotiation over international water dispute: the case of the Nile Basin, *International Negotiation*, 5, 2, 331–356.

Dombrowski, I. (2003) Water accords in the Middle East peace process: moving towards cooperation? in *Security and Environment in the Mediterranean – Conceptualising Security and Environmental Conflict*, H. G. Brauch, P. H. Liotta, A. Marquina, P. Rogers and M. el Sayed (eds), pp729–744, Springer, Heidelberg, Germany.

East African (2009) Egypt, Sudan renege on new Nile Pact, *The East African*, 22 June.

East African (2010) Kenya: Govt signs Nile Basin Agreement, *The East African*, 19 May.

East African Business Week (2007) River Nile agreements – no change for poorer downstream countries, *East African Business Week*, Uganda, 20 August.

Ethiopian Herald (2007) Ethiopia: NBI Permanent Cooperative Framework progressing, *Ethiopian Herald*, 23 February.

Fischhendler, I. (2008) Ambiguity in transboundary environmental dispute resolution: the Israeli–Jordanian Water Agreement, *Journal of Peace Research*, 45, 1, 91–109.

Hefny, M. and Amer, S. E. (2005) Egypt and the Nile Basin, *Aquatic Sciences*, 67, 1, 42–50.

IPR (Info-Prod Research) (2006) Egypt: Keenness on reaching agreement with Nile Basin countries, *Info-Prod Research*, 2 April.

Jägerskog, A., Granit, J., Risberg, A. and Yu, W. (2007) *Transboundary Water Management as a Regional Public Good, Financing Development – An Example from the Nile Basin*, Report 20, Stockholm International Water Institute, Stockholm, Sweden.

MoWR (Ministry of Water Resources, Ethiopia) (1997) *Proceedings of the 5th Nile 2002 Conferences, Addis Ababa 24–28 February 1997*, Ethiopian Ministry of Water Resources, Addis Ababa, Ethiopia.

NBI (Nile Basin Initiative) (1999) *Policy Guidelines for the Nile River Basin Strategic Action Program*, prepared by the NBI Secretariat in cooperation with the World Bank, Nile Basin Secretariat, Kampala, Uganda.

NBI (2002) *Nile Basin Act*, faolex.fao.org/docs/pdf/uga80648.pdf, accessed 29 March 2009.

NBI (2007a) First draft conceptual design of the Nile Basin decision support system reviewed by Keyholders, *Nile News – A Newsletter of the Nile Basin Initiative*, 4, 4, October–December, pp6, 9.

NBI (2007b) *Ministers Agree a Cooperative Framework for the Nile Basin*, 26 June, www.nilebasin.org/index.php?option=com_content&task=view&id=50&Itemid=84, accessed10 July 2007.

NBI (2009a) *Growing Cooperation through Joint Actions*, NBI Annual Report, January–December, 2008, Nile-SEC, Entebbe, Uganda.

NBI (2009b) NBI presentation to the session 'Towards achieving institutional sustainability of river basin organizations', World Water Week 2009, 16–22 August, Stockholm, Sweden.

NBI (2009c) Nile Basin Initiative: building a cooperative future through regional investments, NBI presentation to the session 'National Goals and Regional Cooperation in Transboundary Waters: Incentives and Barriers for Basin-wide Partnerships', World Water Week 2009, 16–22 August 2009, Stockholm, Sweden.

New Vision (2006) Africa drying up, *The New Vision*, 13 December.

New Vision (2007) Country blocked from tapping Nile waters, *The New Vision*, 30 July.

New Vision (2008) River Nile treaty talks hit deadlock, *The New Vision*, 9 November.

Nicol, A. (2003) *The Nile: Moving Beyond Cooperation*, UNESCO Technical Documents in Hydrology, PC-CP Series 16, UNESCO, Paris, France.

Nicol, A., van Steenbergen, F., Sunman, H., Turton, A. R., Slaymaker, T., Allan, J. A., de Graaf, M. and van Harten, M. (2001) *Transboundary Water Management as an International Public Good*, Ministry of Foreign Affairs, Stockholm.

Okidi, C. O. (1994) History of the Nile Basin and Lake Victoria basins through treaties, in *The Nile: Sharing a Scarce Resource – A Historical and Technical Review of Water Management and of Economical and Legal Issues*, P. P. Howell and J. A. Allan (eds), pp321–350, Cambridge University Press, Cambridge, UK.

Sadoff, C. W. and Grey, D. (2002) Beyond the river: the benefits of cooperation on international rivers, *Water Policy*, 4, 5, 389–403.

Sadoff, C. W. and Grey, D. (2005) Cooperation on international rivers: a continuum for securing and sharing benefits, *Water International*, 30, 4, 1–8.

Shady, A. M., Adam, A. M. and Ali, M. K. (1994) The Nile 2002: the vision toward cooperation in the Nile Basin, *Water International*, 19, 2, 77–81.

Tamrat, I. (1995) Constraints and opportunities for basin-wide cooperation in the Nile: a legal perspective, in *Water in the Middles East: Legal, political and commercial implications*, J. A. Allan and C. Mallat (eds), pp177–188, I. B. Tauris, London.

World Bank (2001) *Donor community supports poverty reduction, prosperity and peace through the Nile Basin Initiative*, 28 June, http://go.worldbank.org/CAPQ532BP0, accessed 30 March 2007.

World Bank (2009) *Nile Basin Trust Fund Annual Report 2009*, prepared by the World Bank for the Sixth NBTF Committee Meeting, World Bank, Washington, DC.

Zeitoun, M. and Warner, J. (2006) Hydro-hegemony: a framework for analysis of transboundary water conflicts, *Water Policy*, 8, 5, 435–460.

13

Institutions and policy in the Blue Nile Basin

Understanding challenges and opportunities for improved land and water management

Amare Haileslassie, Fitsum Hagos, Seleshi B. Awulachew, Don Peden, Abdalla A. Ahmed, Solomon Gebreselassie, Tesfaye Tafesse, Everisto Mapedza and Aditi Mukherji

Key messages

- In the past decades, both upstream and downstream countries of the Blue Nile Basin (BNB) had developed and adopted several policies and strategies related to land and water management. Yet there are important policy and institutional gaps that impeded adoption of improved land and water management strategies. An example of these gaps is the lack of upstream–downstream linkage and incentive-based policy enforcement mechanisms.
- In spite of long-standing efforts in improving land and water management in the BNB, achievements have been negligible to date. This is accounted for by land and water management policy and institutional gaps mentioned above. Addressing these gaps only at local level may impact the basin communities at large. Therefore, institutional arrangements need to be built across different scales (nested from local to international) that build trust, facilitate the exchange of information and enable effective monitoring required for successful water resources management (e.g. dam operation, cost and benefit sharing, demand management, etc.).
- Payment for environmental services (PES) is a potential incentive-based policy enforcement mechanism for improved land and water management and conflict resolution between upstream and downstream users both at the local scale and in the BNB at large. This potential must be comprehended to bring about a win-win scenario in upstream and downstream parts of the BNB.
- Financing improved land and water management practices is an expensive venture and mostly within a long-term period of returns. A fully farmer-financed PES scheme may not be financially feasible (at least in the short term). Therefore, options for user and state co-financing must be sought.

Introduction

Overview

Lives and livelihoods in the BNB are strongly linked with crop production and livestock management and, therefore, with land and water. Over 95 per cent of the food-producing sector in upstream areas (i.e. Ethiopia) is based on rain-fed agriculture. In Sudan, downstream, the Blue Nile supplies water for major irrigation development and also for livestock production (Haileslassie *et al.*, 2009). Agriculture is a system hierarchy stretching across plot, farm, watershed and basin. For such a hierarchy operating within the same hydrological system, such as the BNB, water flows create intra- and inter-system linkages, and therefore changes in one part of a basin will affect water availability and attendant livelihoods and ecosystem services (provision, regulation, support and cultural) in other parts.

In the BNB, threats to these co-dependent livelihoods arise from new dimensions like population growth and associated need for agricultural intensification (Haileslassie *et al.*, 2009). In this respect, a question arises as to how the current policy and institutions, at local and basin scales, enhance complementary associations between these co-dependent livelihoods.

Purposes and organization of this chapter

The purposes of this chapter are to:

- Explore the set-up and gaps of land and water management policy and institutions at different scales of the BNB.
- Identify determinants and intensity of adoption for improved land and water management practices and their implications for institutions and policy interventions.
- Assesses mechanisms for basin- and local-level upstream/downstream community cooperation through, for example, benefit-sharing by taking payment for environmental services as an example.

This chapter reports on challenges and opportunities of institutions and policy for improved land and water management in the BNB. It considers different spatial scales ranging from international and national via region, to watershed and community. Below we present the overall analytical framework, before addressing institutional set-ups and gaps, adoptions of improved land and water management technologies, payment for environmental services and benefic-sharing. The last section presents the overall conclusion, key lessons learnt and the policy implications thereof.

Analytical framework and methodology

In terms of analytical framework, the chapter follows a nested approach: from the local perception through to the international. It considers policy and institution interventions and its upstream–downstream impacts at the community, sub-catchment, basin and international levels, as appropriate. Each level of analysis involves different physical dynamics, stakeholders, policies and institutions, and therefore options for interventions. Where relevant, it also looks at the interactions between these levels. This chapter is synthesized based on different case studies representing different spatial scales in the BNB. Detailed methodologies for the respective level of studies are elaborated by Alemayehu *et al.* (2008), Mapedza *et al.* (2008), Gebreselassie *et al.* (2009) and Hagos *et al.* (2011).

Land and water management institutions and policy in the BNB: their set-up and gaps

In Ethiopia (upstream) and Sudan (downstream) parts of the BNB institutional arrangements related to land and water are broadly categorized into three different tiers: federal (national), regional (state) and local-level organizations. More recently, in Ethiopia, basin-level organizations have also come into the picture. Formal institutions are structured at federal and regional levels. Regional states adopt federal land and water institutions as they are, or, as in some cases, they develop region-specific institutions based on the general provisions given at the federal level. Informal institutions are locally instituted and may lack linkages with the formal institutions and among themselves. In this study, we focus on the assessment of federal land and water management institutions as they apply to regional, sub-basin and local scales. We focused only on those institutions and policy related to water resources, agriculture and environmental protection.

Land and water-related organizations

Bandaragoda (2000) defined institutions as established rules, norms, practices and organizations that provide a structure to human actions related to water management. The framework of Bandaragoda (2000) also presents the overall institutional framework in three broad categories: policies, laws and administration. Here we used this category to explore institutional performances of the BNB by (i) elaborating organizational attributes, (ii) developing a list of essential organizational design criteria and comparing these against its current state, and (iii) identifying missing key policy elements and instruments.

Organizational set-up, their attributes and coordination in the BNB

There are at least three federal and other subsidiary agencies and the same number, if not more, of NGOs, of regional bureaus/authorities working in the areas of land, water and environmental protection in Ethiopia (Haileslassie *et al.*, 2009). A comparable organizational structure is reported for Sudan (Hussein *et al.*, 2009). In Ethiopia, the Ministry of Water Resources (MoWR), Ministry of Agriculture and Rural Development (MoARD) and Ethiopian Environmental Protection Authority (EPA) are key actors, while in Sudan the Ministry of Irrigation and Water Resources (MIWR), Ministry of Agriculture and Forests (MoAF), Ministry of Animal Resources and Fisheries (MoARF) and Higher Council for Environment and Natural Resources (HCENR) are reported as important organizations for land and water management. Water user associations (WUAs) and irrigation cooperatives (IC) are the most common local organizations engaged in water management (e.g. Gezira). The role of a WUA is commonly restricted to the distribution of water between members, rehabilitation and maintenance of canals, and addressing water-related conflicts.

The presence of clear institutional objectives in the BNB is fairly well established (Haileslassie *et al.*, 2009; Hagos *et al.*, 2011). There are organizations with clear mandates, duties and responsibilities, and given by-laws. The policies and laws in place have also clear objectives, and some have developed strategies and policy instruments to meet these objectives (Haileslassie *et al.*, 2009; Hussein *et al.*, 2009; Hagos *et al.*, 2011).

However, there are important problems noticed in the organizational setting that affect activities and actors and, therefore, outputs (Table 13.1). A careful look into the work portfolios of ministries indicates the presence of overlaps in mandates between MoWR, MoARD and EPA in

Table 13.1 Assessment of institutional design criteria against current organizational structure and operations in the case study area (Tana-Beles sub-basin)

Institutional design criteria	*Key issues*	*Focus institutions* MoWR	MoARD	EPA
Clear institutional objectives	Key objectives from among the many objectives?	*Inter alia* inventory and development of the country's surface water and groundwater resources; basin-level water management and benefit-sharing	Development and implementing of a strategy for food security, rural development, and natural resources protection; development of rural infrastructure and agricultural research	Formulation of policies, strategies, laws and standards to foster social and economic development and the safety of the environment
	Key constraints in meeting these objectives?	Overlap with EPA and MoWR; high manpower turnover; frequent restructuring; weak enforcement capacity; lack of hierarchy; upstream downstream not considered	Overlap with MoWR and EPA; high manpower turnover; frequent restructuring; weak enforcement capacity	Overlap with MoWR and MoARD; high manpower turnover; weak enforcement capacity
Interconnectedness between formal and informal institutions	Relation between formal and informal institutions;	Note the linkage matrix	Note the linkage matrix	Note the linkage matrix
	Cases where informal institutions replace formal institutions?	Water user association	EDIAR gives some micro credit	
Adaptiveness	The common forms of adaptive management	Evolutionary management	Evolutionary management	Evolutionary management
Scale	Spatial scale	Hydrological boundary	Administrative boundary	Administrative boundary
Compliance capacity	Dealing with violations of norms;	Not clear	Not clear	
	typical forms of enforcement?	Command-control	Command-control	Command-control

Note: EDIAR is an informal institution in Ethiopia mainly engaged in burial services

Source: Haileslassie *et al.*, 2009

upstream and MoIWR, MoEPD and MoARF in downstream (Haileslassie *et al.*, 2009; Hussein *et al.*, 2009; Hagos *et al.*, 2011). For instance, MoWR and MoARD, in upstream areas, have responsibilities related to water resources development; MoWR focuses on medium and large-scale works while MoARD focuses on small-scale irrigation and micro-watershed management. The broad areas of integrated natural resources management also fall into the mandates of these two ministries and the EPA (Haileslassie *et al.*, 2009; Hagos *et al.*, 2011).

It seems there is a further dilemma of split jurisdiction between federal- and regional-level organizations that may create problems in implementation and enforcement. For example, environmental impact assessment (EIA) and water pollution control in the upstream portion also fall under the jurisdiction of EPA and MoWR. There is already possible overlapping of responsibility between general and broad mandates of EPA and regional environmental bureaus or authority in the field of pollution control. If these organizations work separately, this would lead to a clear duplication of effort and waste of resources. Interestingly, linkages and information-sharing mechanisms in place do not ensure institutional harmony and efficient information and resource flows.

Table 13.2 shows an example of information flows and linkages between organizations operating in land and water management in the upstream part of the BNB. It is apparent that horizontal communications between ministries and bureaus belonging to different sectors is seldom common. There are hardly any formal information flows and linkages between sectors. Lack of an integrated information management system exacerbates this problem. Therefore, organization of ministries, bureaus and departments seems to follow 'disciplinary' orientation while problems in the sector call for an interdisciplinary and integrated approach. In Sudan, Hussein *et al.* (2009) also indicated that a lack of coordination and formal information flow was a major threat to organizations' performance in the downstream part of the basin.

Table 13.2 Map of information flow and linkages between major actors in upper parts of the Blue Nile Basin

	BoARD	*BoWRD*	*EPLAUA*	*AARI*	*SHWISA (NGO)*	*Water Aid (NGO)*	*MoARD*	*MoWR*	*EPA*	*EIAR*
BoARD		IFL	IFL	FFL	FFL	NFL	FFL	IFL	IFL	IFL
BoWRD	IFL		IFL	IFL	IFL	FFL	NFL	FFL	IFL	NFL
EPLAUA	IFL	IFL		IFL	IFL	NFL	NFL	NFL	FFL	IFL
AARI	FFL	IFL	IFL		NFL	NFL	IFL	NFL	NFL	FFL
SHWISA (NGO)	FFL	IFL	IFL	IFL		NFL	NFL	NFL	NFL	NFL
Water Aid (NGO)	NFL	FFL	NFL	NFL	NFL		NFL	IFL	NFL	NFL
MoARD	FFL	NFL	NFL	NFL	NFL	NFL		IFL	IFL	FFL
MoWR	NFL	FFL	NFL	NFL	NFL	IFL	IFL		IFL	IFL
EPA	NFL	NFL	FFL	NFL	NFL	NFL	IFL	IFL		IFL
EIAR	NFL	NFL	NFL	NFL	NFL	NFL	NFL	NFL	NFL	

Notes: Linkages: FFL, institutionalized flow and linkage; IFL, indirect flow and linkage; NFL, no flow and linkage. Actors: AARI, Amhara Agricultural Research Institute; BoARD, Bureau of Agriculture and Rural Development; BoWRD, Bureau of Water Resources Development; EIAR, Ethiopian Institute of Agricultural Research; EPLAUA, Environmental Protection Land Administration and Land Use Authority; EPA, Environmental Protection Authority; MoARD, Ministry of Agriculture and Rural Development; MoWR, Ministry of Water Resources

Source: Hagos *et al.*, 2011

In both upstream and downstream parts of the BNB, ministries of water are responsible for water resources that are transboundary in nature and not confined within a regional state, while regional counterparts are responsible for water resources within their jurisdictions. At the same time, for example in the downstream part, MIWR is responsible for managing schemes (e.g. Sennar Dam) in the BNB. An important point here is that the central ownership of these resources is incompatible with decentralized management that both countries are following.

What is more relevant is that organizations involved in land and water management in the upstream and downstream part of the BNB were marked by frequent restructuring and reorganization over the last few years and the process seems to be going on. For example, since the 1990s, there has been an institutional reform process in water sectors of Sudan (Hussein *et al.*, 2009). Adjusting organizational responsibilities and frequent redesigning of organizational structures have certainly produced uncertainties and made capacity-building difficult. To achieve the objectives of sustainable outcome, the gaps mentioned in BNB organizations' attributes and coordination need to be addressed.

Enforcement capacity of organizations

Enforcement capacity of an organization is one of the important indicators of organizational performance. The point here is to see how violations of accepted institutions were dealt with and typical forms of enforcement (Table 13.1).

Overall, emerging evidence suggests that regulations on water resources management, pollution control, land use rights, watershed development, etc. are not effective because of weak enforcement capacity in both upstream and downstream parts of the BNB. A similar observation is reported by NBI (2006). For example, while the Ethiopian and Sudanese water development and environmental protection policies and laws recognize the need to take proper EIAs in pursuing any water-related development interventions, traditional practices still dominate. This problem is identified as more serious in the downstream part of the BNB (NBI, 2006). EPA complains of inadequate staff and resources to do proper enforcement of these environmental provisions. The poor enforcement capacity of institutions can also be linked to the absence of an integrated system of information management at the country or sub-basin level. While the land and water organizations, both in Sudan and Ethiopia, are mandated to collect and store relevant data to support decision making, the data collection is at best inadequate and haphazard. Information-sharing and exchange between organizations to support timely policy decision making and to encourage cooperation between upstream downstream regions are generally appraised as weak (NBI, 2006). In light of this, various organizations keep and maintain a wide range of data to meet their purposes (NBI, 2006).

Institutional adaptiveness

We have described the various aspects of land and water management institutions in the BNB. In this regard it is interesting to assess how these institutions evolved and the type of adaptive management pursued (Table 13.2). Hagos *et al.* (2011) suggested that adaptive evolutionary management is the typical type of strategy followed in drafting structuring of these organizations.

Organizational efficacy is measured not only in fulfilling daily work mandates but also in developing forward-looking solutions to emerging issues. One related issue in this regard is the adaptive capacity of institutions to exogenous factors. In general, in both upstream and downstream of the BNB, there is hardly any indication that the emerging challenges are reflected upon and strategies to address emerging issues are designed (Haileslassie *et al.*, 2009; Hussein *et al.*, 2009). There are allusions in the policy documents that envisaged how water sector and broader development strategies in upstream and downstream parts of the BNB are expected to provide mechanisms to mitigate some, if not all, of the environmental challenges. However, these strategies assume that there is plenty of water potential to tap into from the sub-basins. Economic water scarcity is considered a greater challenge than physical water scarcity. Climate change scenarios and their impact on water resources are hardly taken into account in the development of these strategies. This will obviously put sustainability of development efforts in both upstream and downstream parts of the basin under question.

Appropriateness of scale

The Ethiopian and Sudanese water policies advocate integrated water resources development, where the planning unit should be a river basin. It seems, however, that there is confusion in the definition of the appropriate scale. For example, in Ethiopia regional bureaus and federal office are organized on the basis of administrative scale (i.e. regions or the country). On the other hand, relevant water resources policy and watershed management guidelines advocate that the basin or watershed be the basic planning unit for intervention. In the downstream part of the BNB, the Ministry of Water Resources and Irrigation (MoWRI) in Sudan has organs operating at the basin and at the same time at the state level. A critical constraint against effective river basin management is the commonly prevalent conflict between boundaries of river basins and those of political units (nations, regions, districts, etc.). The administrative boundaries also pose potential constraint in management of small watersheds that fall between two smaller administrative units or farmers association. This calls for establishing viable and acceptable institutional mechanisms for shared management of water resources in the BNB.

Assessment of policy framework, elements and instruments

The policy framework

An example of how BNB policy framework considerations impact on important policy elements is depicted in Table 13.3. In the upstream part, environmental policy lacks climate change; upstream–downstream linkage; role of educational activities and need for research (Table 13.3; FDRE, 1997). The environmental framework act (2001) in Sudan also does not explicitly recognize important issues like climate change, despite a compelling evidence of climate change. The enforcement of some policy elements mentioned in the policy documents is constrained by the low level of regional states' implementation capacity (Hagos *et al.*, 2011; Haileslassie *et al.*, 2009). This is a major point of concern to reduce impacts of upstream-region intervention on downstream (e.g. siltations of water infrastructures in the downstream).

One of the most important water-related policies, strategies, regulations or guidelines in Ethiopia is the water resources management policy (MoWR, 1999). Sudan developed the first national water policy in 1992 and revised it in 2000 (NBI, 2006). A number of important policy elements mentioned in Table 13.3 are reflected in both countries' policy documents: community participation, institutional changes, duty of care and general intent of the policy/law jurisdiction. For the environmental policy, the water resources policy also lacks important elements such as climate scenarios, upstream–downstream linkage, role of education and the need for research and investigation.

The Integrated Water Resources Management (IWRM) approach in both upstream and downstream water policies has relevant provisions: regarding the needs for water resources

Table 13.3 Examples of essential elements of water and land management policies in Blue Nile Basin

Element	*WRMP*	*EPE*	*LULA*	*WSG*
General intent of the policy/law	✓	✓	✓	✓
Jurisdiction – spatial and administrative scales	✓	✓	✓	✓
Responsibility (establishes or enables commitment)	✓	✓	✓	✓
Specific goals and objectives	✗	✗	✗	✗
Duty of care (ethical, legal responsibility, attitude, responsibility or commitment)	✓	✓	✓	✓
Hierarchy of responsibilities ('rights and obligations' of hierarchies)	✗	✓	✓	✓
Institutional changes (statements of an intended course of action/needed reform or legal change)	✓	✓	✓	✓
Climate change scenarios/demand management	✗	✗	✗	✗
Upstream–downstream linkages (e.g. watershed level)	✗	✗	✓	✓
Role of educational activities	✗	✗	✗	✗
Research and investigation	✗	✗	✗	✗
Community participation	✓	✓	✓	✓
Green and blue water/land use planning	✗	✗	✓	✗
Financing	✓	✗	✗	✗
Enforcement/regulation (self- versus third-party enforcement)	✗	✓	✓	✗
Mechanisms for dispute resolution	✗	✗	✓	✗

Notes: ✗, not clear/uncertain; ✓, clearly reflected; EPE, Environmental Policy of Ethiopia; LULA, Land Use and Land Administration Policy; WSG, Watershed Management Guideline; WRMP, Water Resources Management Policy/Regulation/Guideline.

Source: Hagos *et al.*, 2011

management to be compatible and integrated with other natural resources as well as river basin development plans. In practice, however, some of the policies are not coherent and coordination between sectors to realize such integration is loose (Hagos *et al.*, 2011 Hussein *et al.*, 2009). The states have a stronger power to administer land in their regions; however, administration of water (particularly of the international regions and those rivers crossing two or more regions) is an issue of the federal states, which manifests a lack of integrated approaches in practice. The weak status of integrated approaches can also be realized from a lack of land use planning and rainwater management in the policy element, which is an interface between different elements of integrated approaches (Table 13.3). This is particularly true for parts of the downstream where the key policy focus is blue water management (Hussein *et al.*, 2009).

Typology of essential policy instruments

There are different types of policy instruments and approaches to internalize externalities (Kerr *et al.*, 2007), which include regulatory limits, taxes on negative externalities, tradable environmental allowances, indirect incentives, payment for environmental services, etc. These instruments could be broadly classified into economic, market-based, and command-and-control instruments. For example, administrative and legal measures against offenders, technology standards, closure or relocation of any enterprise and permits in the case of hazardous waste or substances (as indicated in EPA) fall under the category of command-and-control instruments. Among the many incentive-based policy enforcement mechanisms only subsidies are mentioned in EPA.

The new proclamations on land use and land administration in the upstream have specific regulations on land use obligations of the land user. It lists a set of obligations of the land user not only to protect the land under his/her holding but also to conserve the surroundings of lands obtained as rent (CANRS, 2006, p21). Non-compliance is likely to lead to deprivation of use rights and penalty. This is mainly a command control type of instrument. As suggested in a number of empirical studies, security of tenure is a critical variable determining incentives to conserve land quality. For example, Gebreselassie *et al.* (2009) also suggested that farmers with registered plots were more likely to adopt conservation investments than those with non-registered plots. But these farmers' interest in the decision to invest in land and water management is highly correlated to farmers' asset holdings (Gebreselassie *et al.*, 2009), and this suggests the need for mechanisms to finance land and water management (Table 13.4).

Similarly, in Sudan, land tenure is a complicated issue. The overwhelming majority of farmers in the irrigated sub-sector are tenants without recognized rights over their landholdings. A tenant has no freedom in trading his tenancy. He cannot, for example, use his tenancy as a collateral security for bank loans. Nor has he the leisure of choosing the crops that suit him. The Gezira Scheme Act of 2005 tried to address these and other land-tenure issues by giving the farmers, among other things, the freedom of choosing the crops to grow and to gradually shift from land tenancy to landownership.

Incentive-based enforcement mechanisms are lacking in the water resources policy document in both upstream and downstream parts. Those mentioned (e.g. cost- and benefit-sharing) are not implemented. For example, the water policy of Ethiopia has specific stipulations

Table 13.4 Typology of policy instruments in environmental management

Policy instruments	*WSG*	*LULA*	*WRMP*	*EPE*	*Responsible*
Information and education	✓	✗	✗	✓	
Regulations/standards	✗	✓	✗	✓	EPA/EPLAUA
Incentive-based subsidies	✗	✓	✗	✓	EPA/EPLAUA
Taxes	✗	✗	✗	✓	
Charges/penalties	✗	✓	✗	✓	
Certification (property rights)	✗	✓	✓	✓	
Cost- and benefit-sharing	✗	✗	✓	✗	
MoWR cost recovery	✗	✗	✓	✗	MoWR
Public programmes (PSNP, FFW, CFW/free labour contribution, etc.)	✓	✗	✗	✗	MoARD/BoARD
Conflict resolution	✓	✓	✗	✗	EPLAUA/social courts

Notes: CFW, cash for work; EPA, Environmental Protection Authority; EPLAUA, Environmental Protection, Land Administration and Land Use Authority; FFW, food for work; IWSM, Integrated Watershed Management Policy; LULA, Land Use and Land Administration; MoARD, Ministry of Agriculture and Rural Development; MoWR, Ministry of Water Resources; PSNP, Productive Safety Net Program; WRMP, Water Resources Management Policy

Source: Hagos *et al.*, 2011

pertaining to tariff setting. It calls for rural tariff settings to be based on the objective of recovering operation and maintenance (O&M) costs while urban tariff structures are based on the basis of full cost recovery. Users from irrigation schemes are also required, at least, to pay to cover O&M costs (Table 13.4). The institutionalization of cost recovery schemes and tariff-setting is expected not only to generate funds for maintaining water points/schemes but also to change users' consumption behaviour (i.e. demand management).

One of the principal policy objectives of structural adjustment in Sudan is to be able to recover the cost of goods and services rendered (Hussein *et al.*, 2009). In line with this policy, the Irrigation Water Corporation, a parastatal within the MIWR, was established in the mid-1990s as a part of restructuring of the water sector to provide irrigation services to the national irrigation schemes. The corporation was supposed to levy irrigation fees for its services. Unfortunately, it could not collect enough fees to cover its operations. This led to empowering the water user associations to manage minor irrigation canals, collect irrigation fees and pay for the services rendered. But the achievement has been appraised as weak to date.

Overall, there is a tendency to focus on command-control type policies (Hagos *et al.*, 2011), but not on carefully devised incentive mechanisms for improved environmental management. Through proper incentives farmers could be motivated to conserve water, prevent soil loss and nutrient leakage, and, hence, reduce downstream externalities (e.g. payment for environmental services; Table 13.4). There is an argument that policy instruments building on command and control, like regulations and mandatory soil conservations schemes in the upstream part have limited or negative effects (Kerr *et al.*, 2007; Ekborn, 2007). There are suggestions for the increased use of positive incentives, like payment for environmental services to address land degradation problems in developing countries (Table 13.4; Ekborn, 2007). It could be argued that various forms of incentives have been provided to land users to conserve the land resources in Ethiopia and elsewhere in eastern Africa. However, most of the incentives were aimed at mitigating the effects of the direct causes of land degradation. The underlying causes of land degradation remained largely unaddressed. Hence, there is a need to carefully assess whether the proposed policy instruments address incentive problems of actors, form improved environmental management and whether those selected instruments must be realistic and their formulation must involve the community.

Determinants of adoption of improved land and water management practices in the BNB: policy and institutional implication for out-scaling of good practices?

States of land and water management today: Is adoption sufficient and diverse?

The major reason for the poor performance of agriculture in many countries of sub-Saharan Africa is the deterioration of the natural resource base. Soil erosion and resultant nutrient depletion are reported as two of the triggers of dwindling agricultural productivity in the BNB (Haileslassie *et al.*, 2005). The problem is severe, mainly, on the highlands where rain-fed agriculture constitutes the main source of livelihood of the people. There are also off-site impacts: sedimentation of wetlands, pollution of water and flooding of the downstream. This raises a concern on the sustainability of recent development initiatives for irrigation and hydropower development in the BNB.

As a countermeasure, various land and water management programmes have been undergoing for decades. A range of watershed management practices have been introduced at different landscapes; for example, these include physical soil conservation measures, water harvesting, and soil fertility management (MoARD, 2005). However, the trends hitherto show that these efforts have had limited success in addressing these problems. Among others, poor adoption and transitory use of conservation techniques are often mentioned as the major factors (Shiferaw and Holden, 1998).

From an upstream case study of BNB, Gebreselassie *et al.* (2009) demonstrated that farmers are focusing more on short-term gain than on long-term investment in land and water management (Table 13.5). Technologies with immediate productivity-enhancing effects take priority in farmers' decisions. The most widely used long-term improved soil conservation technologies were soil and stone bunds (Table 13.6). This suggests that there is a widespread use of a few technologies despite the recommendations based on agro-ecological and landscape suitability (MoARD, 2005). Some of the technologies introduced to the smaller watersheds in the BNB could not be diffused into the community practice. It is understood that wider adoption of these policy and institutional factors is limited.

Table 13.5 Proportion of sample farm households and farm plots by type of regular agronomic practices used in the Blue Nile Basin

	Upstream		*Downstream*		*Households*		*Farm plots*	
	Number	*%*	*Number*	*%*	*Number*	*%*	*Number*	*%*
Manuring	136	22.86	134	18.21	239	73.5	294	19.8
Composting	93	15.63	66	8.97	120	36.9	169	11.4
Counter ploughing	315	53.03	308	41.85	186	57.2	649	43.6
Strip cropping	21	3.54	59	8.02	65	20.0	96	6.5
Intercropping	54	9.09	58	7.89	90	27.7	131	8.8
Crop rotation with legumes	497	83.81	590	80.38	315	96.9	1194	80.3
Fallowing	6	1.01	13	1.77	11	3.4	19	1.3
Mulching and crop residue management	–	–	2	0.27	5	1.5	5	0.3
Relay cropping	–	–	1	0.14	1	0.3	1	0.1
Alley cropping	–	–	1	0.14	1	0.3	1	0.1
Use of Broad Bed Maker to drain water	8	1.65	1	0.14	3	0.9	9	0.6
Reduced tillage/no tillage	52	8.77	87	11.84	36	11.1	139	9.3
Inorganic fertilizer application	228	38.15	339	46.06	211	64.9	652	43.8

Source: Gebreselassie et al., 2009

Conserving land and water in the BNB: what limits adoption of improved land and water management practices?

The number of policy- and institution-related factors are mentioned as determinants of adoption of improved land and water management (Gebremedhin and Swinton, 2003). In this regard, an example of farmers' adoption of improved land and water management practices was studied upstream of the BNB by Gebreselassie *et al.*, (2009). Using econometric modelling

Table 13.6 Number of households and farm plots by type of long-term soil and water conservation structures used in the Blue Nile Basin

Type of structure	*Upstream*		*Downstream*		*Households*		*Farm plots*	
	Number	*%*	*Number*	*%*	*Number*	*%*	*Number*	*%*
Stone bund	146	50.52	92	34.85	114	44.0	238	43.0
Soil bunds	127	43.94	158	59.85	157	60.6	285	51.5
Bench terraces	5	1.73	–	–	4	1.5	5	0.9
Grass strips	1	0.35	–	–	1	0.4	1	0.2
Fanya Juu	8	2.77	–	–	5	1.9	8	1.5
Vegetative fence	–	–	2	0.76	1	0.4	2	0.4
Multi-storey gardening	–	–	6	2.27	5	1.9	6	1.1
Life check dam	–	–	4	1.52	4	1.5	4	0.7
Tree planting	2	0.69	2	0.76	4	1.5	4	0.7

Source: Gebreselassie *et al.*, 2009

tools, they demonstrated that land tenure security increases the probability of adoption significantly. Farmers with registered plots were more likely to adopt the conservation investments than those with the non-registered plots. Other empirical studies (e.g. Gebremedhin and Swinton, 2003) also show that security of tenure is a critical variable determining incentives to conserve land quality. A secured land-tenure right reinforces private incentives to make long-term investments in soil conservation.

Although access to market is perceived as one of the major determinants to farmers' adoption of land and water management technologies, Gebreselassie *et al.* (2009) suggested that this can be site-specific and depends on the return farmers are expecting from such investment. They suggested that households allot their labour to non-conservation activities in case returns from agriculture are not significantly higher than those from non-farm employment. This calls for incentive mechanisms emphasized in the preceding section. Particularly, market-based incentive mechanisms, such as eco-labelling and taxes and subsidies, can enhance farmers' adoption of improved land and water management techniques.

Plot characteristics such as plot area, slope, soil type and fertility are factors that significantly affect farmers' adoption decisions (Pender and Kerr, 1998; Pender and Gebremedhin, 2007; Gebreselassie *et al.*, 2009). Plot area has relatively the most vivid effect on the probability of farmers' decision to adopt land and water management techniques: with one unit increase in the area of plot, the probability of a farmers' decision to use land and water management practices increased 2.2 times. The most commonly adopted physical soil and water conservation practices in the area, stone bund and soil bund, occupy space and this reduces the actual area under crops. Thus farmers with larger plot areas are more likely to adopt these practices given the technological requirement for space. Slope of the land increases the adoption decision implying that flat land is less likely to be targeted for conservation. Shiferaw and Holden (1998) noted the importance of technology-specific attributes and land-quality differentials in shaping conservation decisions. Therefore, the findings of these case studies call for policy measures against land fragmentation (e.g. minimum plot size) and promotion of technology specific to land size and quality.

Factors that determine the decision to adopt improved land and water management technologies may not necessarily determine the intensity of use. The degree of intensification is a good indicator for the scale of adoption. Therefore, those variables that explain both adoption and intensification can give better ideas where policy and institutions related to improved land and water management should focus to increase adoption and intensification. In this regard, Gebreselassie *et al.* (2009) concluded that plot area, tenure security, walking distance to output markets and location in relation to access to extension services influence both farmers' decision and intensity of adoption.

Payment for environmental services in the BNB: prospects and limitations

Payment for environmental services (PES) is a paradigm to finance conservation programmes. PES implies that users of environmental services compensate people and organizations that provide them (Stefano, 2006; Wunder, 2005). PES principles within watersheds and basins imply that downstream farm households and other water users are 'willing to compensate' upstream ecosystem service providers. The institutional analyses for BNB have illustrated that PES as an alternative policy tool for improved land and water management has received little attention. The question here is whether PES can better motivate upstream and downstream stakeholders to manage their water and land for greater sustainability and benefits for all.

Willingness to pay: opportunities and challenges

The key to the successful implementation of PES schemes lies in the motivation and attitudes of individual farmers and government policies that would provide incentives to farmers to manage their natural resources efficiently. In this regard, an example of farmers' willingness to pay (WTP), in cash and labour for improved ecosystem services, was studied by Alemayehu *et al.*, (2008) in the upstream of the BNB (Koga and Gumera watersheds, Ethiopia). The authors reported the downstream users' willingness to compensate the upstream users for continuing land and water management. The upstream users were also willing to pay for land and water conservation and, in fact, rarely expect compensation for what they do, as minimizing the on-site costs of land degradation is critical for their livelihood. The authors reported a stronger magnitude of farmers' WTP in labour for improved land and water management compared with cash and a significantly higher mean willingness to pay (MWTP) by downstream users (Table 13.7). These differences in MWTP, between upstream and downstream, can be accounted for by the discrepancy of benefits that can be generated from such intervention (e.g. direct benefits from irrigation schemes, reduced flood damages, etc.) and also from the differences in resources holdings between the two groups, and PES is widely supported as one of the promising mechanism for transfer of resources.

Table 13.7 Farmers' willingness to pay for ecosystem services, in cash and labour units (Koga and Gumera watersheds, Blue Nile Basin, Ethiopia)

	Upstream		*Downstream*		*Total*	
	Willing	*Not willing*	*Willing*	*Not willing*	*Willing*	*Not willing*
WTP (number of respondents)	99	76	112	38	211	114
WTP (labour PD month^{-1})	169	6	147	3	316	9

Notes: PD, person-days; WTP, willingness to pay

Source: Alemayehu *et al.*, 2008

Farmers' willingness to pay in labour was twofold higher compared to their willingness to pay in cash. This implies that farmers are willing to invest in improved environmental services but that they are obstructed by the low level of income and lack of institution and policy that consider PES as an alternative policy instrument. Here, the major point of concern is also whether these farmers' contribution (either in cash or labour) is adequate for investment and maintenance costs of conservation structures and, if this is not the case, what the policy and institutional options to fill the gaps could be.

As indicated in Table 13.8, the average labour contributions for upstream and downstream farmers were 3.3 and 3.9 PD month^{-1}, respectively; whereas the average cash contributions of the upstream and downstream farmers were 10.4 and 13.1 Ethiopian birr (ETB) month^{-1}, respectively. The MoWR (2002) reported an estimated watershed management cost of 9216 ETB (US$760) ha^{-1}. Taking mean current landholding per household and inflation since the time of estimate into account, a farm householder may require about 13,104 ETB (US$1,365) ha^{-1} to implement improved land and water management on his plots. From this it is apparent that the general public in the two watersheds are willing to pay for cost of activities to restore ecosystem services, although this amount is substantially less than the estimated costs. This trend could be argued from the point of view of Stefanie *et al.* (2008), who illustrated that PES is based on the beneficiary-pays rather than the polluter-pays principle, and as such is attractive in settings where environmental service providers are poor, marginalized landholders or powerful groups of actors. The authors also make a distinction within PES between user-financed and PES in which the buyers are the users of the environmental services and government-financed PES in which the buyers are others (typically the government) acting on behalf of environmental service users. In view of these points it can be concluded that implementation of PES can be an opportunity in BNB but will require the coordinated effort of all stakeholders including the governments, and the upstream and downstream communities.

Table 13.8 Estimated mean willingness to pay for ecosystem services in cash and labour units (Koga and Gumera watersheds, Blue Nile Basin, Ethiopia)

MWTP	*n*	*Mean value*	*CI (95%)*	*p > t*
MWTP in ETB month^{-1} (upstream)	175	10.4	8.2–12.6	0.0029
MWTP in ETB month^{-1} (downstream)	150	13.1	11.8–14.5	
MWTP in labour PD month^{-1} (upstream)	175	3.3	3.15–3.40	0.0000
MWTP in labour PD month^{-1} (downstream)	150	3.9	3.69–4.01	

Notes: CI, confidence interval; ETB, Ethiopian birr, where US$1 = ETB 9.6; MWTP, mean willingness to pay; PD, person-days

Source: Alemayehu *et al.*, 2008

Overall conclusions and policy recommendations

This chapter explored the set-up and gaps of land and water management policy and institutions in the BNB. It identified determinants and intensity of adoption for improved land and water management practices and its implications for institutions and policy interventions and it assessed also mechanisms for basin- and local-level upstream and downstream community cooperation by taking payment for environmental services as an example.

Despite decades of efforts to improve land and water management in the BNB, achievements made are negligible to date. This is accounted for by the fact that farmers' conservation decision and intensity of use of improved land and water management are influenced by a number of policy and institutional factors. Some of these factors are related to access to resources while others are related to policy incentive (e.g. access to market, payment for environmental services, benefit-sharing, and property right), appropriateness of technology (e.g. lack of niche-level technology), the way organizations are arranged, and their weak enforcement capacity.

The question is whether addressing these policy and institutional issues only at local/country level would be effective at the basin level. The agrarian-based livelihood in the basin is operating within the same hydrological boundary. This also means policy measures that respond to local needs (e.g. poverty alleviation in upstream) may affect downstream users. Therefore, while addressing local- and regional-level policy and institutional issues, mechanisms for basin-level cooperation must be sought (e.g. virtual water trade to improve market access of farmers, PES, benefit-sharing, etc.).

The findings from the PES study substantiate the hypothesis of PES as a potential policy instrument for improved land and water management and conflict resolution between upstream and downstream users. This potential must be realized to bring about a win-win scenario in the upstream and downstream of a watershed and at large in the BNB. Above all, the low magnitude of farmers' bid can be a challenge for its realization and thus a sole user-financed PES scheme may not be feasible in short terms both at the local and the basin scale. Alternatively, a PES paid by the users and government-financed PES schemes can be a strategy. The modality for government support can be part of investment in irrigation infrastructure and can be also linked to the global target of increasing soil carbon through land rehabilitation and tree plantation.

One of the critical constraints, indicated in this chapter, against effective and common river basin management is that institutions and policy frameworks do not consider upstream or downstream users. No-win outcomes are likely to occur if the current scenario of unilateral acts continues to persist. Hence, it is incumbent upon co-basin countries to go beyond that and apply a positive outcome if they opt to share the benefits coming out of water. The first step in this direction would be to establish transboundary river-basin institutions which offer a platform for such an engagement. However, the virtue of establishing such an institutional architecture may not guarantee the success of cooperative action. Benefits, costs and information have to be continuously shared among the different stakeholders within the country and between countries in order to build trust and confidence. The latter is not an event, but rather a process that should be continuous and built on an iterative procedure.

References

Alemayehu, B., Hagos, F., Haileselassie, A., Mapedza, E., Gebreselasse, S., Bekele, S. and Peden, D. (2008) Payment for environmental service (PES) for improved land and water management: the case of Koga and Gumara watersheds of the BNB, Ethiopia, in *Proceedings of CPWF Second International Workshop November 2008, Addis Ababa, Ethiopia*, Challenge Program on Water and Food, CGIAR, Washington, DC

Bandaragoda, D. J. (2000) *A Framework for Institutional Analysis for Water Resources Management in a River Basin Context*, IWMI Working Paper 5, International Water Management Institute, Colombo, Sri Lanka

CANRS (Council of Amhara National Regional State) (2006) The Revised Amhara National Regional State

Rural Land Administration and Use, Proclamation No. 133/2006, Zikre Hig, 11th year, no 18, 29 May, CANRS, Bahir Dar, Ethiopia

Ekborn, A. (2007) Economic Analysis of Agricultural Production, Soil Capital and Land Use in Kenay, PhD thesis, Department of Economics, University of Gothenburg, Sweden

FDRE (Federal Democratic Republic of Ethiopia) (1997) *Environmental Policy of Ethiopia,* Environmental Protection Authority in collaboration with the Ministry of Economic Development and Cooperation, Addis Ababa, Ethiopia

Gebremedhin, B. and Swinton, S. M. (2003) Investment in soil conservation in Northern Ethiopia: the role of land tenure security and public programs, *Agricultural Economics*, 29, 69–84

Gebreselassie, S., Hagos, F., Haileslassie, A., Bekele S.A., Peden, D. and Tafesse, T. (2009) *Determinants of Adoption of Improved Land and Water Management Practices in the BNB: Exploring Strategies for Outscaling of Promising Technologies*, Proceedings of the 10th Conference of the Ethiopian Society of Soil Science (ESSS), 25–27 March 2009, EIAR, Addis Ababa, Ethiopia

Hagos, F., Haileslassie, A., Bekele, S., Mapedza, E. and Taffesse T. (2011) Land and water institutions in the BNB: setups and gaps for improved land and water management, *Review of Policy Research*, 28, 149–170

Haileslassie, A., Priess, J., Veldkamp, E., Teketay, D. and Lesschen, J. P. (2005) Assessment of soil nutrient depletion and its spatial variability on smallholders' mixed farming systems in Ethiopia using partial versus full nutrient balances, *Agriculture, Ecosystems and Environment*, 108, 1, 1–16

Haileslassie, A., Hagos, F., Mapedza, E., Sadoff, C., Bekele, S., Gebreselassie, S. and Peden, D. (2009) *Institutional Settings and Livelihood Strategies in the BNB: Implications for Upstream/Downstream Linkages*, IWMI Working Paper 132, International Water Management Institute, Colombo, Sri Lanka

Hussein, A., Abdelsalam, S. A., Khalil, A. and El Medani, A. (2009) *Assessment of Water and Land Policies and Institutions in the BNB, Sudan*, unpublished report from Improved Land and Water Management in The Ethiopian Highlands: Its Impact on Downstream Stakeholders Dependent on the Blue Nile project, International Water Management Institute (IWMI), Addis Ababa, Ethiopia

Kerr, J., Milne, G., Chhotray, V., Baumann, P. and James, A. J. (2007) Managing watershed externalities in India: Theory and practice, *Environmental, Development and Sustainability*, 9, 263–281

Mapedza, E., Haileselassie, A., Hagos, F., McCartney, M., Bekele, S. and Tafesse, T. (2008) Transboundary water governance institutional architecture: reflections from Ethiopia and Sudan, in *Proceedings of CPWF Second International Workshop November 2008, Addis Ababa, Ethiopia*, Challenge Program on Water and Food, CGIAR, Washington, DC

MoARD (Ministry of Agriculture and Rural Development) (2005) *Community Based Participatory Watershed Development: A Guideline*, Ministry of Agriculture and Rural Development, Addis Ababa, Ethiopia

MoWR (Ministry of Water Resources) (1999) *Water Resources Management Policy*, Ministry of Water Resources, Addis Ababa, Ethiopia

MoWR (2002) *Assessment and Monitoring of Erosion and Sedimentation Problems in Ethiopia*, final report V, MoWR/Hydrology Department, Addis Ababa, Ethiopia

NBI (Nile Basin Initiative) (2006) *Baseline and Needs Assessment of National Water Policies of the Nile Basin Countries, A Regional Synthesis*, Shared Vision Program, Water Resources Planning and Management Project, NBI, Addis Ababa, Ethiopia

Pender, J. and Gebremedhin, B. (2007) Determinants of agricultural and land management practices and impacts on crop production and household income in the highlands of Tigray, Ethiopia, *Journal of African Economies*, 17, 3, 395–450

Pender, J. and Kerr, J. (1998) Determinants of farmers' indigenous soil and water conservation investments in semi-arid India, *Agricultural Economics*, 19, 113–125

Shiferaw, S. and Holden, S. T. (1998) Resource degradation and adoption of land conserving technologies in the Ethiopian highlands: a case study in Andit Tid, North Shewa, *Agricultural Economics*, 18, 233–247

Stefanie, E., Stefano, P. and Sven, W. (2008) Designing payments for environmental services in theory and practice: an overview of the issues, *Ecological Economics*, 65, 663–674

Stefano, P. (2006) *Payments for Environmental Services: An Introduction*, Environment Department, World Bank, Washington, DC

Wunder, S. (2005) *Payments for Environmental Services: Some Nuts and Bolts*, Occasional Paper no 42, Center for International Forestry Research (CIFOR), Jakarta, Indonesia

14

Simulating current and future water resources development in the Blue Nile River Basin

Matthew P. McCartney, Tadesse Alemayehu, Zachary M. Easton and Seleshi B. Awulachew

Key messages

- Both Ethiopia and Sudan have plans to unilaterally develop the water resources of the Blue Nile for hydropower and irrigation. The extent to which these plans will actually be implemented is unclear. However, if both countries totally fulfil their stated objectives the following are estimated to occur:
  - the total reservoir storage in Ethiopia will increase from the current 11.6 to more than 167 billion cubic metres (i.e. more than 3 times the mean annual flow at the Ethiopia–Sudan border);
  - large-scale irrigation withdrawals in Sudan will increase from the current 8.5 to 13.8 billion m^3 yr^{-1};
  - large-scale irrigation withdrawals in Ethiopia will increase from the current 0.26 to 3.8 billion m^3 yr^{-1}; and
  - electricity generation in Ethiopia will increase from the current 1383 to 31,297 gigawatt hours (GWh) yr^{-1}.
- Increased water storage in dams and greater withdrawals will inevitably alter the flow regime of the river and its main tributaries. If full development occurs, the total flow at the Ethiopia–Sudan border is predicted to decrease from the current (near natural) 45.2 to 42.7 billion m^3 yr^{-1} and at Khartoum from the current 40.4 to 31.8 billion m^3 yr^{-1}. However, although there is a significant reduction in wet season flow at both locations, dry season flow will actually increase because of the greater upstream flow regulation. By increasing water availability in the dry season and reducing flooding in the wet season this increased regulation promises significant benefits for Sudan.
- There is great potential for increased water resources development in the Blue Nile. However, if Ethiopia and Sudan continue to implement development unilaterally, the benefits of water resources development are unlikely to be fully realized. It is therefore essential that the countries cooperate closely to (i) identify priority development options, (ii) improve irrigation efficiencies, (iii) mitigate any adverse impacts (e.g. to the environment)

and (iv) manage water resources in a way that brings benefits to all. To take full advantage of the water resources of the basin it is necessary that they are managed as a single system (i.e. without considering national borders) that, in turn, requires the establishment of much more effective institutional arrangements than those currently existing.

Introduction

The Blue Nile River is an important shared resource of Ethiopia and Sudan, and also – because it is the major contributor of water to the main Nile River – Egypt. However, tensions over the basin's water resources remain unresolved. Although the riparian countries have agreed to collaborate in principle, formal mechanisms to cooperatively develop the basin's water resources are limited. Currently, a Cooperative Framework Agreement (CFA) is being negotiated, but this process has been under way for more than a decade and no final agreement has yet been achieved (Cascão, 2009). Recently, five of the riparian countries, Ethiopia, Kenya, Rwanda, Tanzania and Uganda signed an agreement, but both Egypt and Sudan remain opposed to the current version.

Under the auspices of the Nile Basin Initiative (NBI) two primary programmes have been established: (i) the basin-wide Shared Vision Program, designed to build confidence and capacity across the basin, and (ii) the Subsidiary Action Program, which aims to initiate concrete investments and action at sub-basin level (Metawie, 2004). However, both programmes are developing slowly, and there are few tangible activities on the ground. As a result, all riparian countries continue to pursue unilateral plans for development.

The potential benefits of regional cooperation and integrated joint basin management are significant and well documented (Whittington *et al.*, 2005; Jägerskog *et al.*, 2007; Cascão, 2009). A prerequisite for such cooperation is the development of shared knowledge bases and appropriate analytical tools to support decision-making processes. Currently, knowledge of the basin is fragmented and inconsistent and there is limited sharing of data and information. There is also a lack of analytical tools to evaluate water resources and analyse the implications of different development options. These are major impediments to building consensus on appropriate development strategies and cooperative investments in the basin.

A number of computer models have been developed to assess various aspects of hydropower and irrigation potential within the Blue Nile and the wider Nile basins (Guariso and Whittington, 1987; Georgakakos, 2003; Block *et al.*, 2007; Elala, 2008). However, these models have focused primarily on the upper Blue Nile in Ethiopia and the development of hydraulic infrastructure on the main stem of the river. Relatively little attention has been paid to water diversions and development on the tributaries or future development in Sudan.

In this chapter we report the findings of research conducted to determine the impact of current and possible future water demand throughout the whole of the Blue Nile Basin (BNB). The Water Evaluation And Planning (WEAP) model was used to evaluate the water resource implications of existing and proposed irrigation and hydropower development in both Ethiopia and Sudan. The current situation and two future development scenarios were simulated; one representing a relatively near future (the *medium-term* scenario) and the other a more distant future (the *long-term* scenario). Since year-to-year variation is important for water management, 33 years of monthly time step flow data were used to simulate the natural hydrological variation in all the major tributaries. The water demands of the scenarios, incorporating all existing and planned development on both the main stem and the tributaries were superimposed on these time series. However, because the planned large reservoirs require considerable time to fill, a 20-year warm-up period was used and comparison between the scenarios was made over

13 years. Although necessarily based on many assumptions, the work illustrates how a relatively simple model, used in conjunction with data from both countries, can provide a credible basis for assessing possible future water resources development throughout the basin.

In the following section of this chapter, the natural characteristics and the current socio-economic situation in the basin as well as the planned water resources development are described. The following section describes the WEAP model and its configuration and application to the Blue Nile River Basin through development scenarios. Thereafter, the results are presented and discussed and finally some conclusions are drawn.

Water availability in the Blue Nile River Basin

Natural characteristics

The Blue Nile River (known as the Abay River in Ethiopia) rises in the Ethiopian Highlands in the region of West Gojam and flows northward into Lake Tana, which is located at an elevation of just under 1800 m (Figure 14.1). It leaves the southeastern corner of the lake, flowing first southeast, before looping back on itself, flowing west and then turning northwest, close to the border with Sudan. In the highlands, the basin is composed mainly of volcanic and Pre-Cambrian basement rocks with small areas of sedimentary rocks. The catchment is cut by deep ravines through which the major tributaries flow. The valley of the Blue Nile River itself is 1300 m deep in places. The primary tributaries in Ethiopia are the Bosheilo, Welaka, Jemma, Muger, Guder, Finchaa, Anger, Didessa and Dabus on the left bank, and the North Gojam, South Gojam, Wombera and Beles on the right bank.

The Blue Nile enters Sudan at an altitude of 490 masl. Just before crossing the frontier, the river enters a clay plain, through which it flows to Khartoum. The average slope of the river from the Ethiopian frontier to Khartoum is only 15 cm km^{-1}. Within Sudan, the Blue Nile receives water from two major tributaries draining from the north, the Dinder and the Rahad, both of which also originate in Ethiopia. At Khartoum, the Blue Nile joins the White Nile to form the main stem of the Nile River at an elevation of 400 masl. The catchment area of the Blue Nile at Khartoum is approximately 311,548 km^2.

Within the basin, rainfall varies significantly with altitude and is, to a large extent, controlled by movement of air masses associated with the Inter-Tropical Convergence Zone. There is considerable inter-annual variability, but within Sudan the mean annual rainfall over much of the basin is less than 500 mm, and it is as low as 140 mm at Khartoum. In Ethiopia, it increases from about 1000 mm near the Sudan border to between 1400 and 1800 mm over parts of the upper basin, and exceeds 2000 mm in some places in the south (Awulachew *et al.*, 2008). The summer months account for a large proportion of mean annual rainfall: roughly 70 per cent occurs between June and September. This proportion generally increases with latitude, rising to 93 per cent at Khartoum.

Potential evapotranspiration also varies considerably, and, like rainfall, is highly correlated with altitude. Throughout the Sudanese part of the basin, values (computed using the Penman–Monteith method; Monteith, 1981) generally exceed 2200 mm yr^{-1}, and even in the rainy season (July–October) rainfall rarely exceeds 50 per cent of potential evapotranspiration. Consequently, irrigation is essential for the growth of crops. In the Ethiopian Highlands potential evapotranspiration ranges from approximately 1300 to 1700 mm yr^{-1} and, in many places, is less than rainfall in the rainy season. Consequently, rain-fed cultivation, producing a single crop in the rainy season, is possible, though risky in low rainfall years.

The flow of the Blue Nile is characterized by extreme seasonal and inter-annual variability.

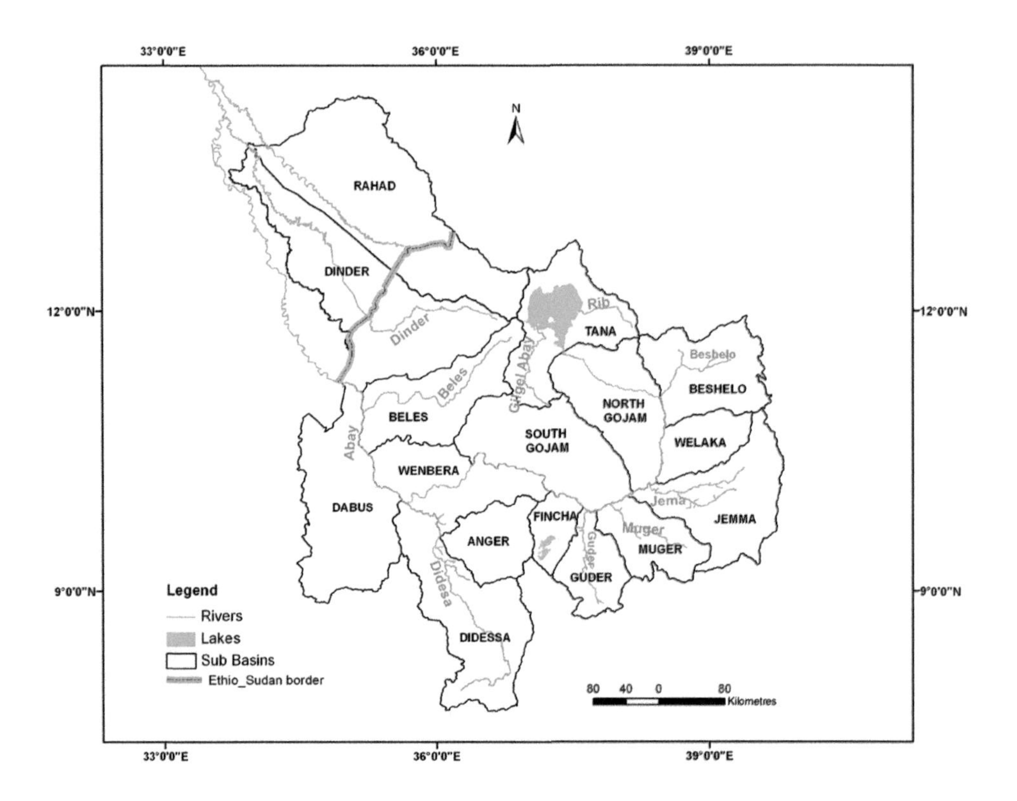

Figure 14.1 Map of the Blue Nile Basin showing the major tributaries and sub-basins
Source: Yilma and Awulachew, 2009

At Khartoum, annual flow varies from approximately 23 billion to 63 billion m^3 (Figure 14.2). Mean monthly flow also varies considerably at all locations along the river (Table 14.1; Figure 14.3). Typically, more than 80 per cent of the flow occurs during the flood season (July–October) while only 4 per cent of the flow occurs during the dry season (February–May) (Awulachew *et al.*, 2008).

Current water resources development

Currently, Ethiopia utilizes very little of the Blue Nile water, partly because of its inaccessibility, partly because the major centres of population lie outside of the basin and partly because, to date, there has been only limited development of hydraulic infrastructure on the river. To date, only two relatively minor hydraulic structures (i.e. Chara Chara weir and Finchaa dam) have been constructed in the Ethiopian part of the catchment (Table 14.2). These two dams were built primarily to provide hydropower. They regulate flow from Lake Tana and the Finchaa River, respectively. The combined capacity of the power stations they serve (218 MW) represented approximately 13 per cent of the total installed generating capacity of the country in 2009 (i.e. 1618 MW, of which 95% was hydropower). In 2010, a new power station on the Beles River came on line (see below) and the total installed capacity increased to 1994 MW.

a

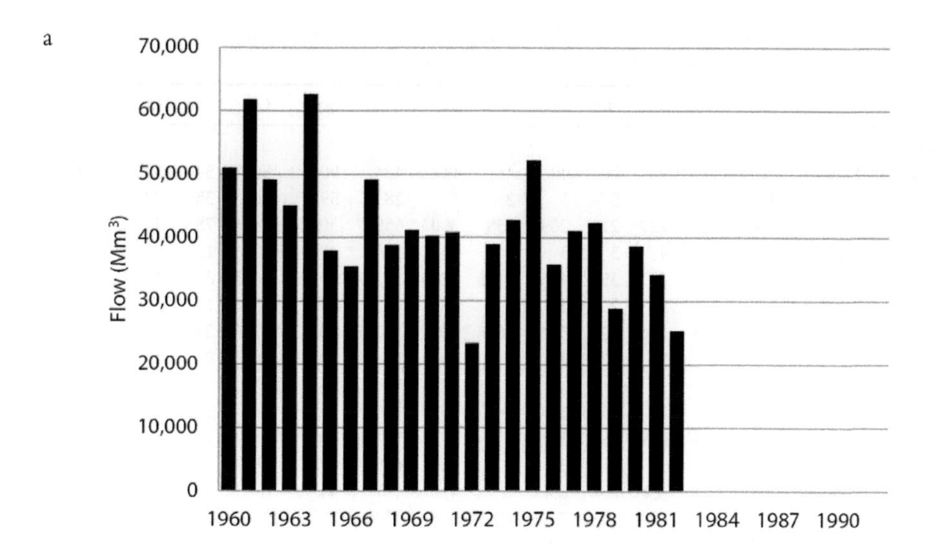

b

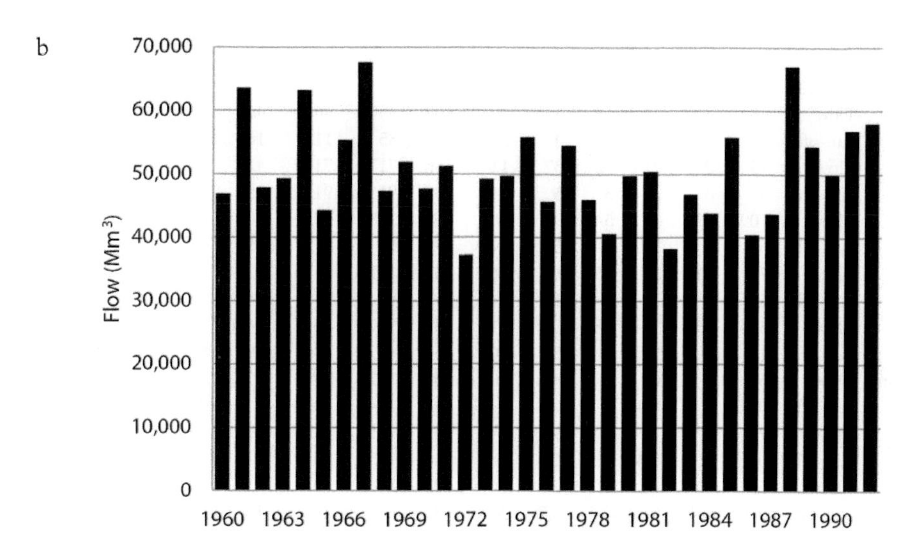

Figure 14.2 Annual flow of the Blue Nile measured at (a) Khartoum (1960–1982) and (b) the Ethiopia–Sudan border (1960–1992)

Source: Data obtained from the Global Data Runoff Centre and Ethiopian Ministry of Water Resources

Agriculture, which is the main occupation of the inhabitants in the basin, is primarily rain-fed with almost no irrigation. Although there is some informal small-scale irrigation, currently the only formal irrigation scheme in the Ethiopian part of the catchment is the Finchaa sugar cane plantation (8145 ha), which utilizes water after it has passed through the Finchaa hydropower plant (Table 14.2).

In contrast to Ethiopia, Sudan utilizes significant volumes of Blue Nile water for both irrigation and hydropower production. Two dams (i.e. Roseires and Sennar), constructed on the

Table 14.1 Mean monthly flow (million m³) and run-off (mm) measured at gauging stations located on the main stem and major tributaries of the Blue Nile River

Location		*Jan*	*Feb*	*Mar*	*Apr*	*May*	*June*	*July*	*Aug*	*Sept*	*Oct*	*Nov*	*Dec*	*Annual*
Main stem														
Lake Tana	Flow	203	127	94	70	49	45	114	434	906	861	541	332	3776
	run-off	13	8	6	5	3	2	7	28	59	56	35	22	247
Kessie	Flow	331	221	227	211	209	258	3003	6594	3080	1456	788	503	16,881
	run-off	5	3	4	3	3	4	46	100	47	22	12	8	257
Border	Flow	949	545	437	359	446	1175	6293	15,502	13,068	7045	3105	1709	50,632
	run-off	5	3	2	2	2	6	31	78	65	35	16	9	253
Khartoum	Flow	724	448	406	427	503	1084	4989	15,237	13,625	7130	2451	1257	48,281
	run-off	3	2	2	2	2	4	18	55	50	26	9	5	176
Major tributaries														
Besheilo	Flow	4	4	4	5	5	14	494	1303	527	74	19	9	2462
	run-off	0.3	0.3	0.3	0.4	0.4	1	37	98	40	6	1	0.6	186
Welaka	Flow	2	2	2	3	3	7	261	689	279	39	10	5	1302
	run-off	0.3	0.3	0.4	0.4	0.4	1	41	107	44	6	2	0.7	203
Jemma	Flow	6	5	6	7	7	18	662	1748	707	100	25	11	3301
	run-off	0.4	0.3	0.4	0.4	0.5	1	42	111	45	6	2	0.7	209
Muger	Flow	1	1	1	2	2	6	268	753	312	44	10	4	1402
	run-off	0.1	0.1	0.1	0.2	0.2	0.7	33	92	38	5	1	0.5	171
Guder	Flow	0	0	0	0	0	7	43	66	50	15	1	0	182
	run-off	0	0	0	0	0	1	6	9	7	2	0.1	0	26
Finchaa	Flow	45	29	21	18	16	20	108	347	464	409	220	91	1786
	run-off	11	7	5	4	4	5	26	85	113	100	54	22	437
Anger	Flow	44	25	21	22	37	114	386	717	716	436	141	75	2733
	run-off	6	3	3	3	5	14	49	91	91	55	18	10	346
Didessa	Flow	109	62	52	54	93	283	958	1782	1779	1084	352	186	6791
	run-off	6	3	3	3	5	14	49	91	91	55	18	10	346
Wombera	Flow	72	41	34	35	61	187	632	1176	1174	715	233	123	4483
	run-off	6	3	3	3	5	14	49	91	91	55	18	10	346
Dabus	Flow	306	155	114	88	94	214	534	917	1336	1460	1070	602	6888
	run-off	15	7	5	4	5	10	25	44	64	69	51	29	328
North Gojam	Flow	6	5	6	8	8	20	730	1927	779	110	27	13	3639
	run-off	0.4	0.4	0.4	0.5	0.5	1	51	134	54	8	2	1	253
South Gojam	Flow	7	6	7	9	9	24	855	2257	913	128	32	15	4262
	run-off	0.4	0.4	0.4	0.5	0.6	1.4	51	135	54	8	2	1	254
Beles	Flow	6	2	2	1	2	36	393	846	637	218	42	12	2195
	run-off	0.4	0.1	0.1	0	0.2	3	28	60	45	15	3	1	155
Rahad	Flow	0	0	0	0	0	1	132	342	354	201	26	1	1056
	run-off	0	0	0	0	0	0.1	16	41	43	24	3	0.1	128
Dinder	Flow	0	0	0	0	0	17	291	968	917	376	34	4	2609
	run-off	0	0	0	0	0	1	20	65	62	25	2	0.2	176

Source: BCEOM (1998), with slight modifications based on more recent feasibility studies of ENTRO (2007) and Sutcliffe and Parks (1999)

main river approximately 350 and 620 km southeast of Khartoum (Table 14.2), provide hydropower (primarily for Khartoum) as well as water for several large irrigation schemes. These include the Gezira irrigation scheme (882,000 ha), which is one of the largest in the world. As well as irrigating land immediately adjacent to the Blue Nile River, some water is diverted from the Blue Nile downstream of the Roseires reservoir to the Rahad River, where it is used to supplement the irrigation of the Rahad irrigation scheme (168,037 ha). The total

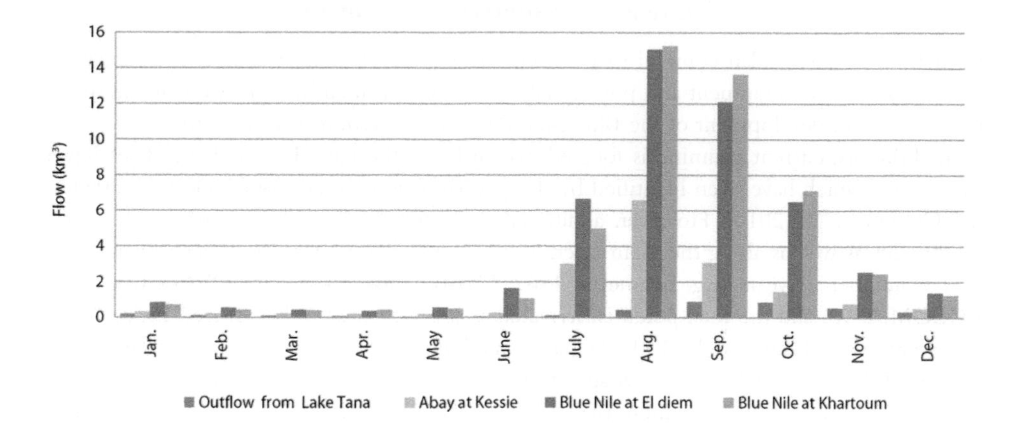

Figure 14.3 Mean monthly flow (million m³) at gauging stations located on the main stem of the Blue Nile

Source: Data provided by the Ministry of Water Resources, Ethiopia and the UNESCO Chair in Water Resources, Sudan

irrigated area in the Sudanese part of the Blue Nile is estimated to be 1,305,000 ha, consisting of a variety of crops including cotton, sugar cane and vegetables. The installed power capacity at the two dams is 295 MW, which represents 25 per cent of the country's total generating capacity (i.e. 1200 MW from both thermal and hydropower stations).

Table 14.2 Existing dams in the Blue Nile catchment

Dam	*Country*	*River*	*Storage (million m³)*	*Year dam was built*	*Purpose*
Chara Chara	Ethiopia	Abay	9100[a]	2000	Regulation of Lake Tana outflows for hydropower production at Tis Abay I and Tis Abay II power stations (installed capacity 84 MW) and, since 2010, for transfer of water to the Beles River hydropower station (installed capacity 460 MW)
Finchaa[b]	Ethiopia	Finchaa	2395	1971	Regulation for hydropower production (installed capacity 134 MW) and also for irrigating sugar cane plantations (8145 ha)
Roseires	Sudan	Blue Nile	3024	1964	Regulation for hydropower production (installed capacity 280 MW) and for supply to irrigation schemes (1,305,000 ha)
Sennar	Sudan	Blue Nile	930	1925	Regulation for hydropower production (installed capacity 15 MW) and for supply to irrigation schemes (1,093,502 ha)

Notes: [a] This is the active storage of Lake Tana that is controlled by the operation of the weir (i.e. lake levels between 1784 and 1787 masl). It represents 2.4 times the average annual outflow of the lake

[b] There is a small dam located on the Amerty River (storage 40 million m³), which diverts water from the Amerty River into the Finchaa reservoir

Future water resources development

Both Ethiopia and Sudan contend that utilization of the Nile water resources is essential for socio-economic development and poverty alleviation. Consequently, both countries are planning significant development of the Blue Nile River water resources in the future.

In Ethiopia, current planning is focused primarily on the Lake Tana and the Beles River catchments which have been identified by the government as an economic 'growth corridor' (McCartney *et al.*, 2010). However, additional projects are planned in nearly all the sub-catchments as well as along the main river. Possible irrigation projects have been investigated over a number of years (e.g. Lahmeyer, 1962; USBR, 1964; JICA, 1977; WAPCOS, 1990; BCEOM, 1998) and the total potential irrigated area is estimated to be 815,581 ha, comprising 45,856 ha of small-scale, 130,395 ha of medium-scale and 639,330 ha of large-scale schemes. Of this, 461,000 ha are envisaged to be developed in the long term (BCEOM, 1998).

In the Ethiopian Blue Nile, more than 120 potential hydropower sites have been identified (WAPCOS, 1990). Of these, 26 were investigated in detail during the preparation of the Abay River Basin Master Plan (BCEOM, 1998). The major hydropower projects currently being contemplated in Ethiopia have a combined installed capacity of between 3634 and 7629 MW (cf. the Aswan High Dam, which has an installed capacity of 2100 MW). The exact value depends on the final design of the dams and the consequent head that is produced at each. The four largest schemes being considered are dams on the main stem of the Blue Nile River. Of these schemes the furthest advanced is the Karadobi project for which the pre-feasibility study was conducted in 2006 (Norconsult, 2006).

In addition to the single-purpose hydropower schemes, electricity generation is expected to be added to several of the proposed irrigation projects where dams are being built. This is estimated to provide an additional 216 MW of capacity (BCEOM, 1998). The total energy produced by all the hydropower schemes being considered is in the range 16,000–33,000 GWh yr^{-1}. This represents 20 to 40 per cent of the technical potential in the Ethiopian Blue Nile (i.e. 70,000 GWh yr^{-1}) estimated by the Ministry of Water Resources (Beyene and Abebe, 2006). Currently, it is anticipated that much of the electricity generated by these power stations will be sold to Sudan and possibly to other countries in the Nile Basin.

Sudan is also planning to increase the area irrigated in the BNB. Additional new projects and extension of existing schemes are anticipated to add an additional 889,340 ha by 2025. An additional 4000 million m^3 of storage will be created by raising the height of the existing Roseries dam by 10 m (Omer, 2010). However, currently there are no plans for additional dams to be constructed on the Blue Nile.

Technically feasible hydropower energy production in the Nile Basin of Sudan is estimated to be 24,137 GWh yr^{-1} (Omer, 2009), most of which is on the main Nile downstream of the White and Blue Nile confluence. Currently, the Merowe dam, with an installed capacity of 2500 MW, is being commissioned on the Nile downstream of Khartoum. Several other major hydropower dams are being planned, none of which are to be located on the Blue Nile River.

Application of the Water Evaluation And Planning model

Model description

Developed by the Stockholm Environment Institute (SEI), the WEAP model is intended to be used to evaluate planning and management issues associated with water resources development. The WEAP model essentially calculates a mass balance of flow sequentially down a river system, making allowance for abstractions and inflows. The elements that comprise the water demand–supply system and their spatial relationship are characterized within the model. The system is represented in terms of its various water sources (e.g. surface water, groundwater and water reuse elements); withdrawal, transmission, reservoirs, wastewater treatment facilities; and water demands (i.e. user-defined sectors, but typically comprising industry, mines, irrigation and domestic supply; Yates *et al.*, 2005; SEI, 2007).

Typically, the model is configured to simulate a 'baseline' year, for which the water availability and demands can be confidently determined. It is then used to simulate alternative scenarios to assess the impact of different development and management options. For each scenario, the model optimizes water use in the catchment using an iterative linear programming algorithm, the objective of which is to maximize the water delivered to demand sites, according to a set of user-defined priorities. All demand sites are assigned a priority between 1 and 99, where 1 is the highest priority and 99 the lowest. When water is limited, the algorithm is formulated to progressively restrict water allocation to those demand sites given the lowest priority. In this study, the model was configured to simulate the 16 major sub-catchments of the basin (Figure 14.4a). It was assumed that because hydropower generates greater income it would be considered more important than irrigation by both governments. Consequently, within WEAP it was given a higher priority than irrigation. However, all schemes in Ethiopia and Sudan were given the same priority (i.e. no attempt was made to reflect differences between upstream and downstream locations).

Description of the scenarios

Scenarios are commonly used to investigate complex systems that are inherently unpredictable or insufficiently understood to enable precise predictions. In this instance, although there is reasonable (but not total) knowledge of current (i.e. 2008) water demand, there is considerable uncertainty about how future water resources development will proceed. Consequently, a scenario approach was adopted.

The model was set up to simulate four scenarios, each of which provides a coherent, internally consistent and plausible description of water demand within the catchment (Table 14.3; Figure 14.4a–d).

Table 14.3 Water resources development scenarios simulated using the Water Evaluation And Planning model

Scenario	*Description*
Natural	No human-made storage and no abstractions so that flows are assumed to be natural. This scenario provides a 'baseline' against which all the other scenarios can be assessed.
Current	The current water resources development situation (around 2008) including all major irrigation and hydropower schemes.
Medium-term future	Water resources development (irrigation and hydropower) in the medium-term future (around 2010–2025) including all schemes for which feasibility studies have been conducted.
Long-term future	Water resources development (irrigation and hydropower) in the long-term future (around 2025–2050) including schemes that are included in basin master plans, but have not yet reached the feasibility study stage of planning.

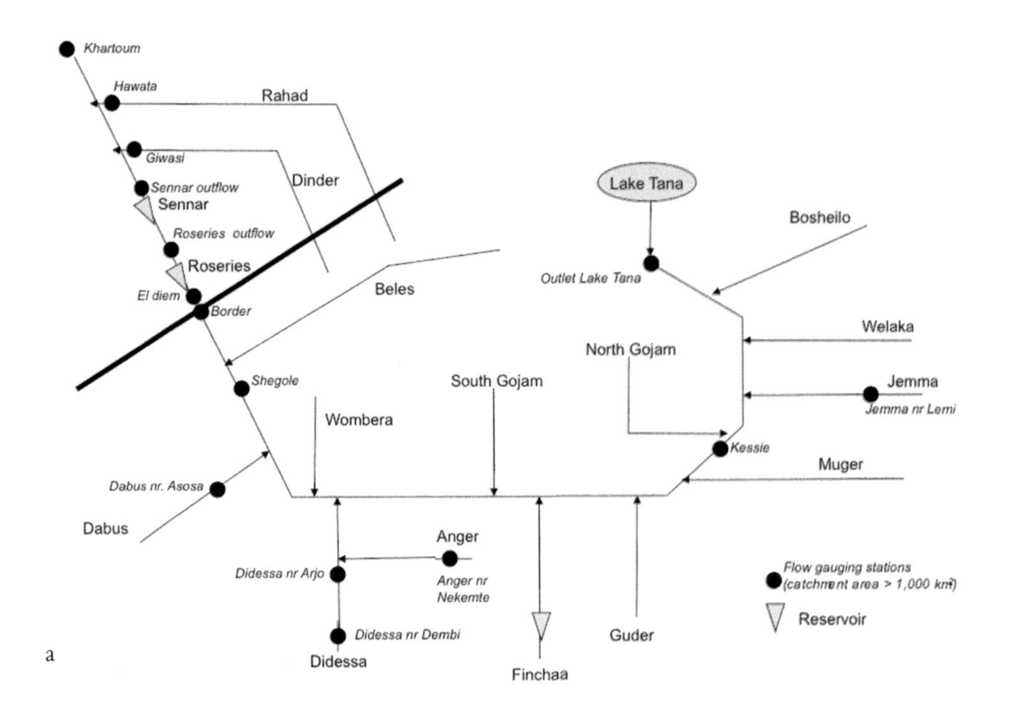
Khartoum
Hawata
Rahad
Giwasi
Dinder
Sennar outflow
Sennar
Roseries outflow
Roseries
El diem
Border
Beles
Lake Tana
Bosheilo
Outlet Lake Tana
Welaka
North Gojam
Jemma
Jemma nr Lemi
Shegole
South Gojam
Wombera
Kessie
Muger
Dabus nr. Asosa
Dabus
Anger
Didessa nr Arjo
Anger nr Nekemte
Flow gauging stations (catchment area > 1,000 km²)
Reservoir
Didessa nr Dembi
Guder
Didessa
Finchaa
a

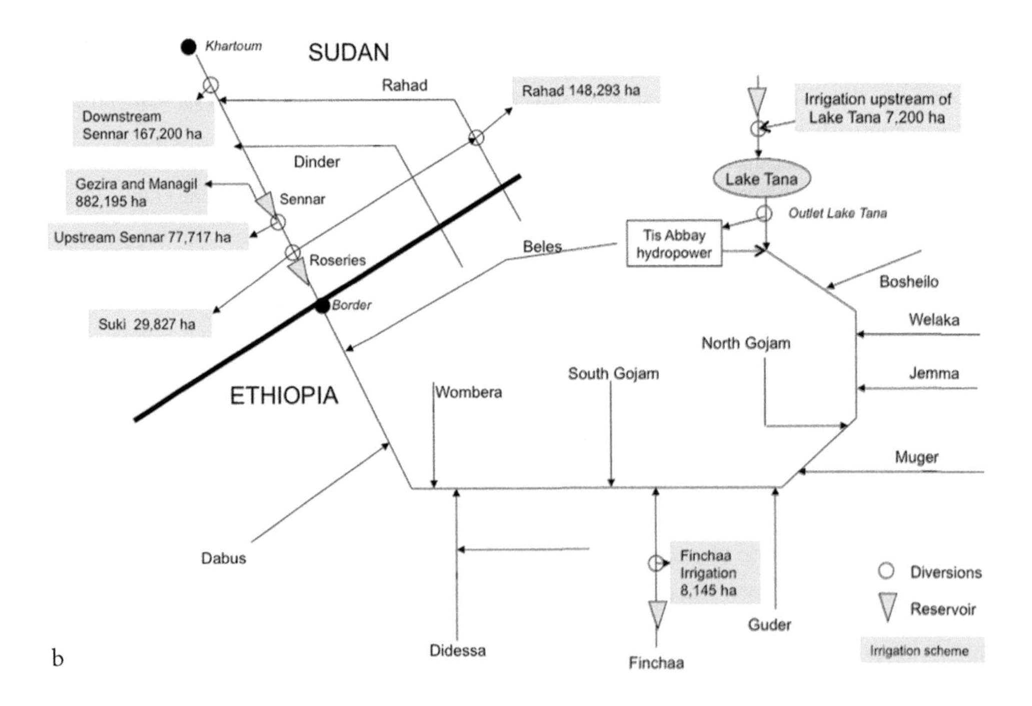
Khartoum
SUDAN
Rahad
Rahad 148,293 ha
Downstream Sennar 167,200 ha
Irrigation upstream of Lake Tana 7,200 ha
Dinder
Gezira and Managil 882,195 ha
Sennar
Lake Tana
Upstream Sennar 77,717 ha
Outlet Lake Tana
Tis Abbay hydropower
Beles
Roseries
Bosheilo
Border
Suki 29,827 ha
Welaka
North Gojam
South Gojam
Jemma
ETHIOPIA
Wombera
Muger
Dabus
Finchaa Irrigation 8,145 ha
Diversions
Reservoir
Guder
Didessa
Irrigation scheme
Finchaa
b

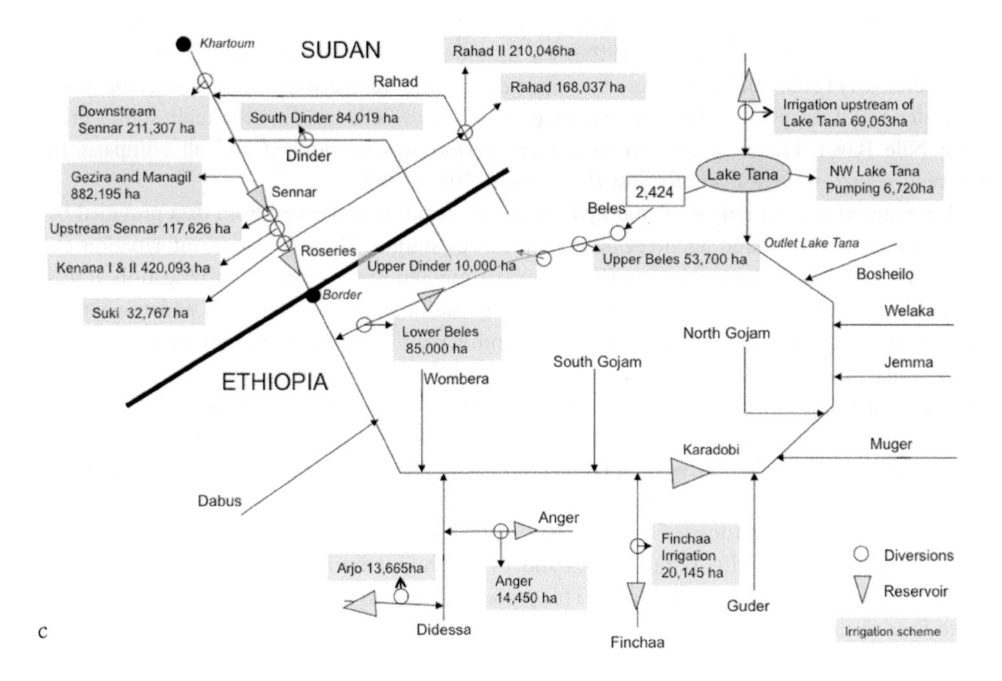

c

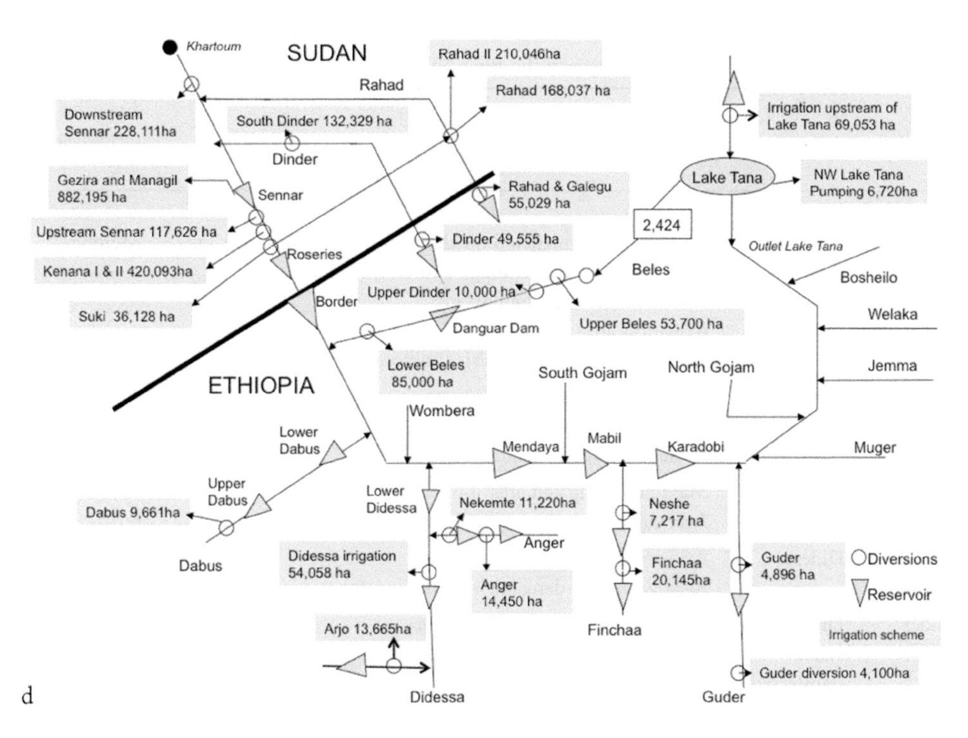

d

Figure 14.4 Schematic of the model configuration for different scenarios: (a) the natural situation, (b) the current (2008) situation, (c) the medium-term (2010–2025) future, and (d) the long-term (2025–2050) future

Time series of monthly naturalized flow data for the period 1960–1992, obtained from the Abay Basin Master Plan (BCEOM, 1998), and modified slightly based on more recent feasibility studies (ENTRO, 2007), were used as input data. In the future scenarios, considerable time is needed to fill the planned large reservoirs, particularly those located on the main stem of the Blue Nile River. Hence, a 20-year 'warm-up' period was introduced and all comparisons between scenarios were made for just the 13 years 1980–1992.

Estimates of current irrigation and hydropower demand were derived from data provided by government ministries and agencies or from previous studies. These included information on water passing through the turbines of the power stations and water diverted for irrigation. It was necessary to make several assumptions, particularly about irrigation demands and the return flows from irrigation schemes. Net evaporation from Lake Tana and the reservoirs was estimated from rainfall and potential evaporation data obtained from the meteorological station located closest to each dam. These data were obtained from the FAO LocClim database (FAO, 2002).

For the medium-term and long-term scenarios, the sizes of planned hydropower and irrigation development schemes were derived from the basin master plan for the Ethiopian Blue Nile and through discussion with academics and water resource planners in Sudan. New schemes, proposed extension of existing irrigation schemes as well as planned hydropower developments were identified (Tables 14.4 and 14.5). The medium-term scenario includes the Tana–Beles transfer scheme in Ethiopia. This project, which involves the transfer of water from Lake Tana to the Beles River to generate hydropower, actually came on line in 2010, but after the modelling had been undertaken (McCartney *et al.*, 2010).

Table 14.4 Proposed irrigation development in the Blue Nile River Basin

Scheme	*Sub-basin*	*Description*	*Estimated completion date*
Ethiopia			
Lake Tana	Lake Tana	Dams to be constructed on the major inflows to Lake Tana (i.e. Megech, Ribb, Gumara and Gilgel Abay) Total storage: 1028 million m^3 Irrigation area: 61,853 ha Average annual demand: 516 million m^3	Medium term
Beles	Beles	Upper Beles scheme: 53,700 ha Lower Beles scheme: 85,000 ha Average annual demand: 1554 million m^3	Medium term
Anger	Anger	Maximum irrigated area: 14,450 ha Average annual demand: 202 million m^3	Medium term
Arjo	Didessa	Arjo scheme: 13,665 ha Average annual demand: 92.1 million m^3	Medium term
Dinder	Dinder (but water transferred from Beles)	Upper Dinder scheme: 10,000 ha Average annual demand: 98.2 million m^3	Medium term
Finchaa	Finchaa	Extension of existing scheme Additional area: 12,000 ha Average annual demand: 456.6 million m^3	Medium term

Table 14.4 Continued

Scheme	*Sub-basin*	*Description*	*Estimated completion date*
Rahad and Galegu	Rahad	Rahad and Galegu scheme: 15,029 ha Average annual demand: 607 million m^3	Long term
Dinder	Dinder	Dinder scheme: 49,555 ha Average annual demand: 556 million m^3	Long term
Guder	Guder	Guder diversion: 4100 ha Guder: 4896 ha Average annual demand: 54.4 million m^3	Long term
Nekemte	Anger	Nekemte scheme: 11,220 ha Average annual demand: 71.5 million m^3	Long term
Didessa	Didessa	Didessa irrigation scheme: 54,058 ha Average annual demand: 769.4 million m^3	Long term
Sudan			
Raising Roseries Dam	Blue Nile main stem	Roseries dam raised by 10 m to provide total (gross) storage of 7400 million m^3.	Medium term
Extension of Rahad irrigation scheme	Rahad	Additional irrigation area: 19,740 ha Rahad II irrigation scheme: 210,000 ha Total average annual demand: 2433 million m^3	Medium term
Extension of Suki irrigation scheme	Blue Nile main stem	Additional irrigation area: 2940 ha/3361 ha Total average annual demand: 201 million m^3/221 million m^3	Medium/ long term
Extension of Upstream Sennar	Blue Nile main stem	Additional irrigation area: 39,910 ha Total average annual demand: 745 million m^3	Medium term
Extension of Downstream Sennar	Blue Nile main stem	Additional irrigation area: 44,110 ha/6804 ha Total average annual demand: 1414 million m^3/ 1526 million m^3	Medium/ long term[a]
Kenana II and III	Blue Nile main stem	Additional irrigation area: 420,093 ha Average annual demand: 2352 million m^3	Medium term
South Dinder	Dinder	Additional irrigation area: 84,019 ha/48,318 ha[b] Average annual demand: 541 million m^3/851 million m^{3b}	Medium/ long term

Notes: [a] Schemes are extended partially in the medium-term future and partially in the long-term future
[b] The slash in the third column demarcates values between the medium-term future and the long-term future

For many potential schemes there is currently considerable uncertainty about the dates when they will be completed. In the current study it was assumed that, for Ethiopian schemes, if prefeasibility studies have been undertaken then the scheme will be completed in the medium term. For all other planned schemes it was assumed that they will be completed in the long term. For the Sudanese schemes, information on likely completion dates was obtained

Table 14.5 Proposed hydropower development in the Blue Nile River Basin

Scheme	*Sub-basin*	*Description*	*Estimated completion date*
Ethiopia			
Tana–Beles	Tana and Beles	Transfer of water from Lake Tana to Beles catchment for hydropower production and irrigation Hydropower capacity: 460 MW Average annual transfer: 2424 million m^3	Medium term
Anger	Anger	Linked to the Anger irrigation scheme Hydropower capacity: 1.8–9.6 MW	Medium term
Arjo	Didessa	Linked to the Arjo irrigation scheme Hydropower capacity: 33 MW	Medium term
Karadobi	Blue Nile main stem	Height of dam: 250 m Total storage: 40,220 million m^3 Hydropower capacity: 1600 MW	Medium term
Mendaya	Blue Nile main stem	Height of dam: 164 Total storage: 15,900 million m^3 Hydropower capacity: 1620 MW	Medium term
Border	Blue Nile main stem	Height of dam: 90 m Total storage: 11,100 million m^3 Hydropower capacity: 1400 MW	Long term
Mabil	Blue Nile main stem	Height of dam: 170 m Total storage: 17,200 million m^3 Hydropower capacity: 1200 MW	Long term
Lower	Didessa	Didessa Height of dam: 110 m Total storage: 5510 million m^3 Hydropower capacity: 190 MW	Long term
Dabus	Dabus	Linked to the Dabus irrigation scheme. Hydropower capacity: 152 MW	Long term
Danguar	Beles	Height of dam: 120 m Total storage: 4640 million m^3 Hydropower capacity: 33 MW	Long term
Lower	Dabus	Dabus Height of dam: 50 m Total storage: 1290 million m^3 Hydropower capacity: 164 MW	Long term

from discussions with water resources experts within the country. However, clearly the two scenarios reflect only an approximate timeline for water resources development in the basin. In reality, development is dependent on many external factors and so it is impossible to predict exactly when many planned schemes will actually be implemented, or indeed the exact sequencing of schemes. As they stand the medium-term and long-term future scenarios represent a plausible development trajectory, but it is unlikely that it will actually come to pass in exactly the way envisaged.

The water withdrawals for irrigation schemes were derived from a variety of sources, including the Ethiopia Basin master plan and, where available, feasibility studies. For Sudan, useful information on irrigation water use was obtained from a study of the Roseries irrigation scheme (Ibrahim *et al.*, 2009). In schemes for which there were no data, it was assumed that withdrawals per hectare would be similar to those at the nearest scheme where data were available, with some allowances for differences in rainfall where this differed significantly between locations. Irrigation return flows were estimated from existing feasibility studies, and averaged approximately 20 per cent of withdrawals in Ethiopia and 15 per cent in Sudan. Where it is planned to extend irrigation schemes the future withdrawals and return flows were estimated based on current values but weighted by the new area. Thus, no allowance was made for possible future improvements in irrigation efficiency. Furthermore, no allowance was made for inter-annual variations in rainfall, which might affect irrigation demand between years.

Results

Model validation

Figure 14.5 shows the simulated and observed flows at the Ethiopia–Sudan border and at Khartoum for the current situation. At Khartoum, observed data (obtained from the Global Data Runoff Centre) were only available for the period 1960–1982. Over this period the error in the simulated mean annual flow was 1.9 per cent. As a result of current abstractions, primarily for irrigation in Sudan, the flow at Khartoum is estimated to be approximately 7.8 billion $m^3\ yr^{-1}$ less than would have occurred naturally over this period (i.e. 42.4 billion $m^3\ yr^{-1}$ rather than 50.2 billion $m^3\ yr^{-1}$). At the border there are two flow gauging stations. One is operated by the Government of Ethiopia, and just a few kilometres downstream another is operated by the Government of Sudan. Possibly because of differences in periods of missing data, observed flows at these two stations differ, and there is a 10 per cent difference in mean annual flow over the period 1960–1992: 50.6 billion m^3 measured by Ethiopia and 45.5 billion m^3 measured by Sudan. Without a detailed analysis, which was beyond the scope of the present study, it is not possible to know which of the two flow series is the more accurate. The WEAP model simulation falls between the two with a mean annual discharge of 46.2 billion m^3.

Figure 14.6 compares the simulated and observed water levels of Lake Tana, also over the period 1960–1992. Although the average simulated water level (1786.3 masl) is close to the observed average (1786.0 masl), it is clear that the variability in the simulated water levels does not quite match that of the observed levels. Nevertheless, these results in conjunction with the flow results indicate that the WEAP simulation of the current situation is reasonably accurate and provides credibility for the results of the simulated future scenarios.

Comparison of scenarios

Currently, irrigation water withdrawals in Sudan greatly exceed those in Ethiopia because of the differences in irrigated area. The total irrigation demand in Sudan is estimated to average 8.45 billion $m^3\ yr^{-1}$. This compares with an average of just 0.26 billion $m^3\ yr^{-1}$ in Ethiopia. With the planned irrigation development, demand is estimated to increase to 13.39 billion and 2.38 billion $m^3\ yr^{-1}$ in the medium-term scenarios, and to 13.83 billion and 3.81 billion $m^3\ yr^{-1}$ in the long-term scenarios in Sudan and Ethiopia, respectively (Table 14.6). If all planned dams are constructed the total reservoir storage in Ethiopia is estimated to increase to 70 billion m^3 (i.e. 1.5 times the mean annual flow at the border) in the mid-term and to 167 billion m^3 (i.e.

3.6 times the mean annual flow at the border) in the long term. Hydropower generated in Ethiopia, from the Tis Abay and Finchaa power stations, is currently estimated to be 1383 GWh yr^{-1}. With the construction of the Tana Beles transfer, the Karadobi dam and other smaller schemes, this is estimated to increase to 12,908 GWh yr^{-1} in the medium term. With Border, Mendaya and Mabil hydropower stations, as well as with additional smaller schemes, electricity production in the long term could increase to 31,297 GWh yr^{-1}.

Hydropower generated on the Blue Nile in Sudan is currently estimated to be just over 1000 GWh yr^{-1}, but there are no publicly available data to confirm this estimate. Because of the additional head and increased storage, the raising of the Roseries dam will result in a very small increase to 1134 GWh yr^{-1} in the medium term and to 1205 GWh yr^{-1} in the long term. The increase in the long term is due entirely to greater dry season flows, resulting from increased regulation upstream in Ethiopia.

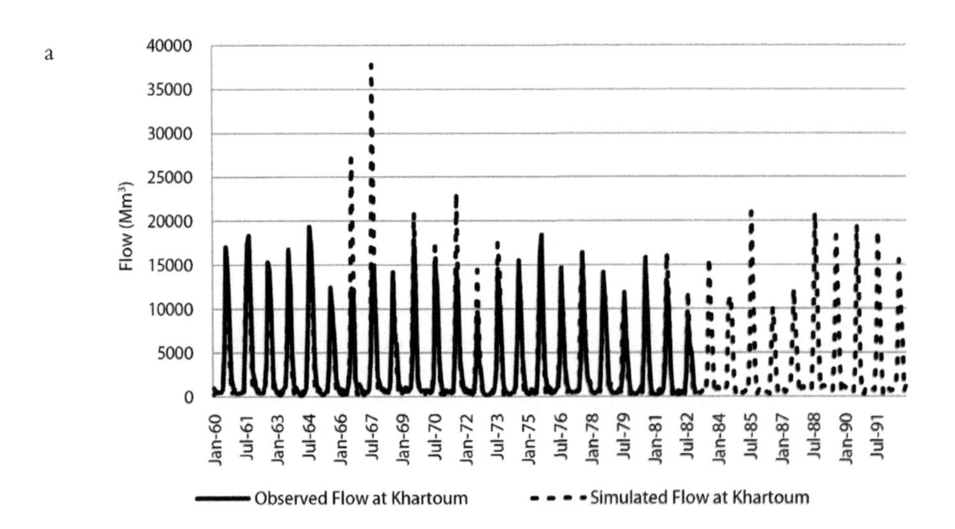

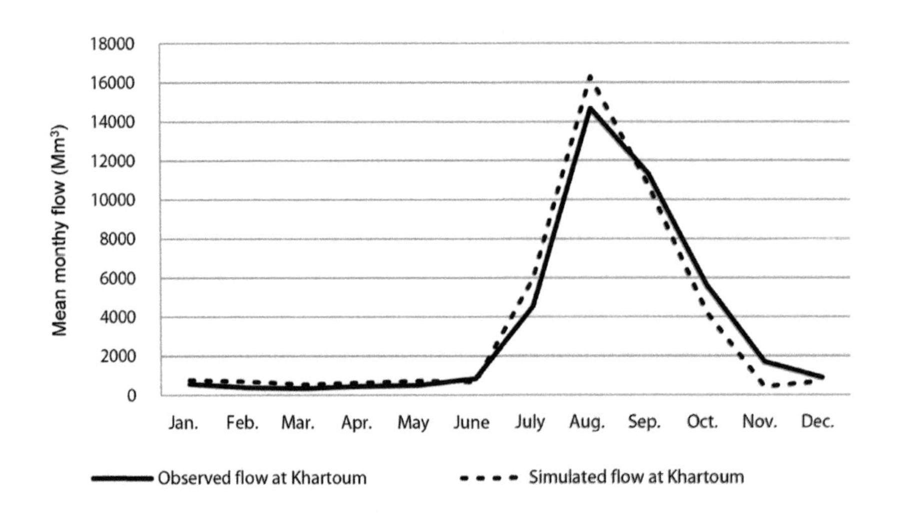

b

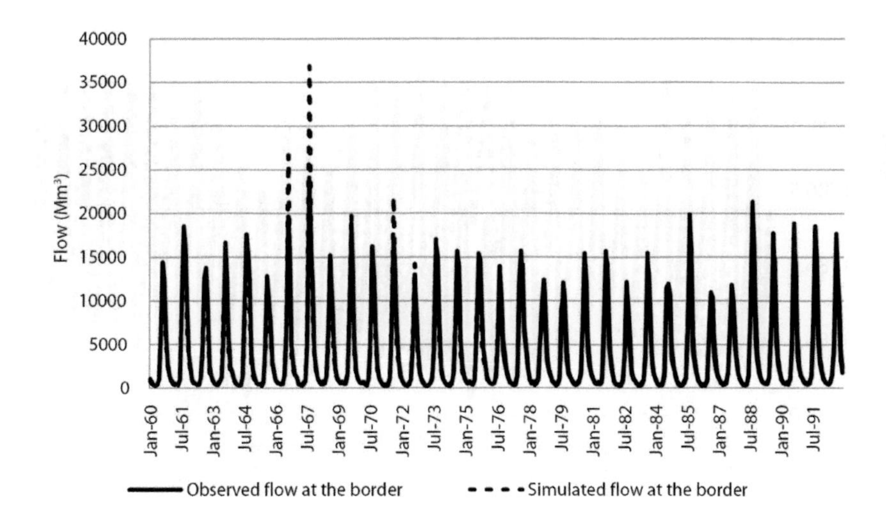

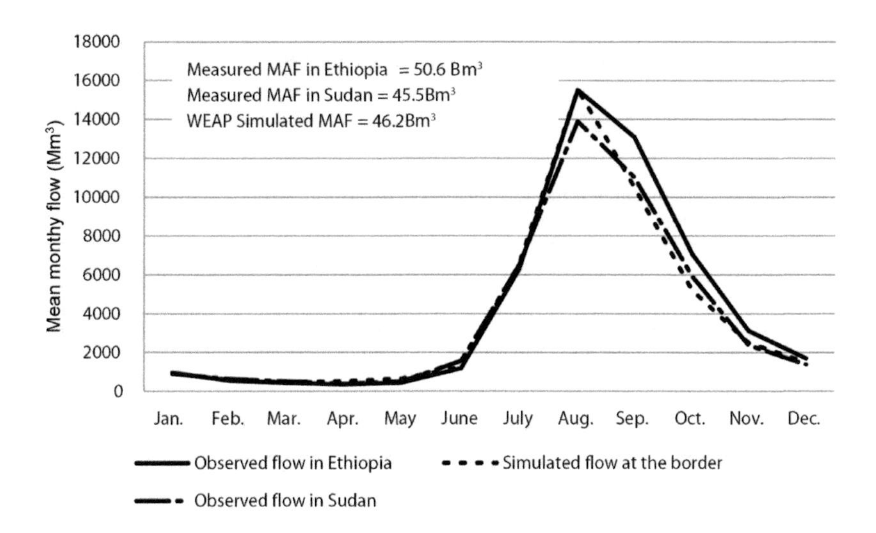

Figure 14.5 Simulated and observed flow series and mean monthly flows (1960–1992) for the Blue Nile (current situation) at (a) Khartoum and (b) the Ethiopia–Sudan border

Comparison of the mean monthly flows at Khartoum for the simulated natural condition, current situation and the medium- and long-term scenarios, for the 13 years 1980–1992, indicates how the mean annual flow is progressively reduced as a consequence of greater upstream abstractions (Table 14.7). Wet season flows are reduced significantly, but dry season flows are increased as a consequence of flow regulation (Figure 14.7; Table 14.7). Under natural conditions, 84 per cent of the river flow occurs in the wet season months (July–October). In the medium-term and long-term scenarios this is reduced to 61 and 37 per cent, respectively.

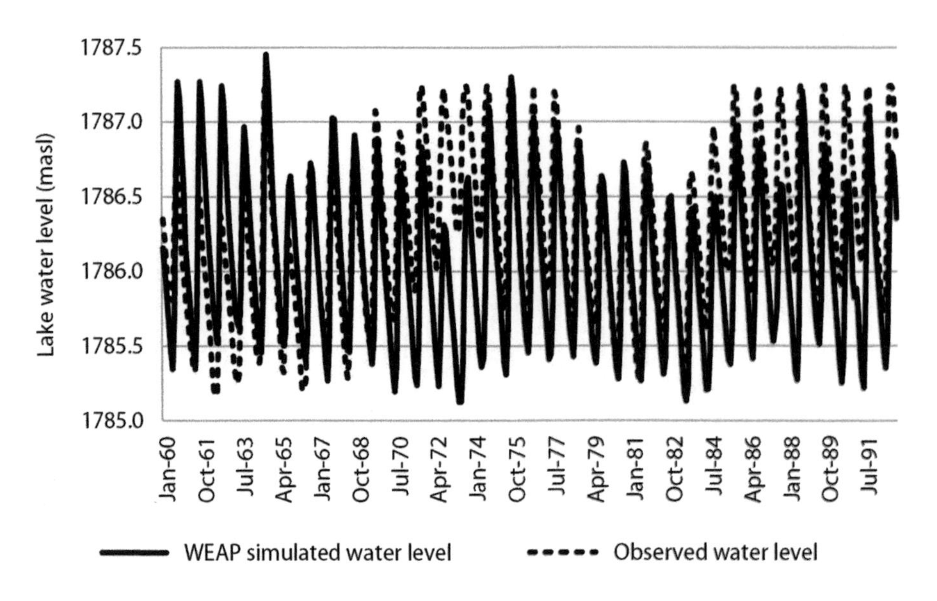

Figure 14.6 Simulated and observed water levels in Lake Tana (1960–1992)

Table 14.6 Comparison of current and future irrigation demand and hydropower production in the Ethiopian and Sudanese parts of the Blue Nile

	Current		*Medium-term future*		*Long-term future*	
	Ethiopia	*Sudan*	*Ethiopia*	*Sudan*	*Ethiopia*	*Sudan*
Total storage (million m³)	11,578	3370[a]	70,244	10,770	167,079	10,770
Formal irrigation						
Area (ha)	<10,000	1,305,000	210,000	2,126,000	461,000	2,190,000
Water withdrawals per year (million m³ yr⁻¹)	0.26	8.45	2.38	13.39	3.81	13.83
Hydropower						
Installed capacity (MW)	218	295	2194	295	6426	295
Production (GWh yr⁻¹)	1383	1029	12,908	1134	31,297	1205

Note: [a] Allowance made for sedimentation of both the Roseries and Sennar reservoirs

At the Ethiopia–Sudan border the current situation is almost identical to the natural condition, so this is not shown (Figure 14.7). Mean annual flow is reduced from 45.2 to 43.2 and 42.7 billion m³ in the medium-term and long-term future scenarios, respectively. As in Khartoum, there is a significant reduction in wet season flows, but there are significant increases in dry season flows as a consequence of flow regulation (Figure 14.7; Table 14.7). Under natural conditions 81 per cent of the river flow occurs in the wet season, but this decreases to 59 and 43 per cent in the medium-term and long-term scenarios, respectively. The total decrease in border flow in the long-term scenario is less than might be expected given the increased

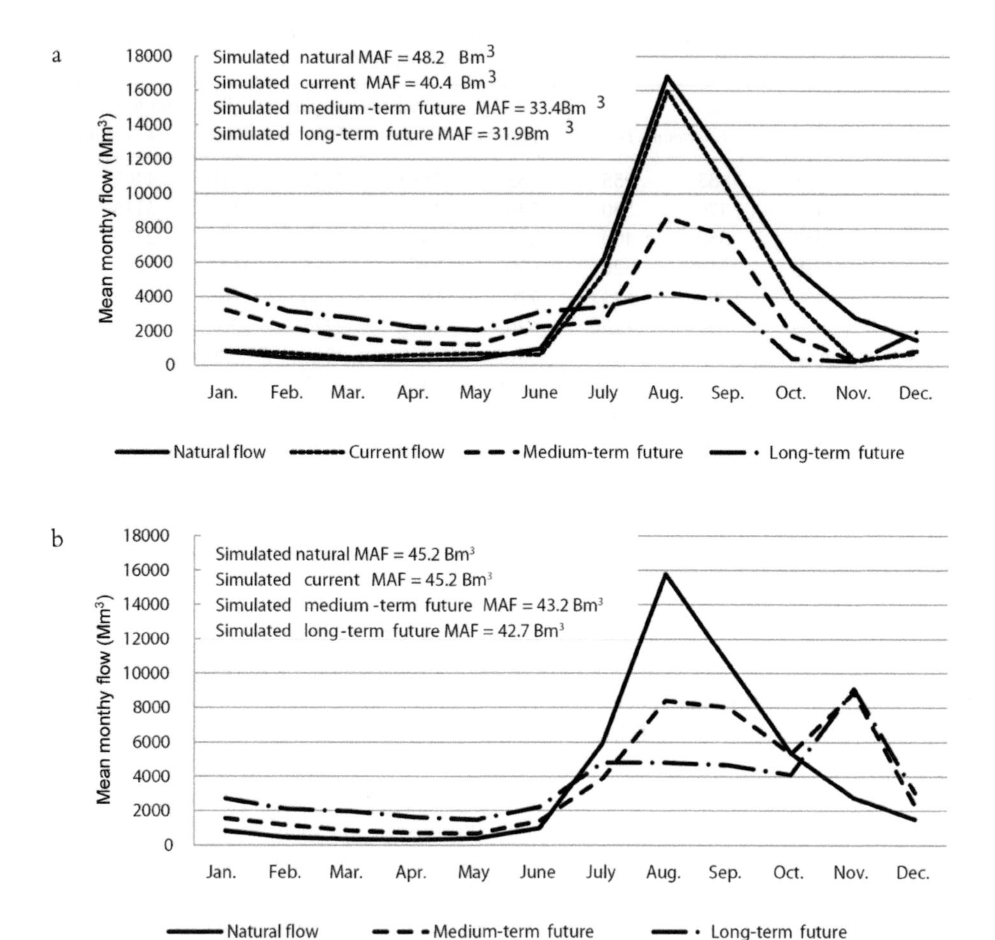

Figure 14.7 Comparison of simulated mean monthly flow derived for natural, current, medium-term and long-term future scenarios at (a) Khartoum and (b) the Ethiopia–Sudan border

irrigation demand in Ethiopia. The reason is partly that less water is diverted from the Tana to the Beles catchment and more flow is routed down the main stem of Blue Nile.

Currently, shortfalls (i.e. failure in any given month to supply the full amount of water needed for irrigation withdrawals or hydropower needs) in Ethiopia are negligible. However, in the medium-term scenario shortfalls increase to 0.8 and 5.0 billion m^3 yr^{-1} for irrigation and hydropower, respectively. In the long term, the increased storage means that shortfalls will average 0.4 billion m^3 yr^{-1} for irrigation and 0.7 billion m^3 yr^{-1} for hydropower. In comparison, under current conditions, there is an average shortfall of 0.8 billion m^3 yr^{-1} in water for the Sudanese irrigation schemes. However, because of the improved flow regulation there are no shortfalls in irrigation or hydropower in Sudan in either the medium- or the long-term scenarios. These results reflect the fact that, in each scenario, the Sudanese schemes were given the same priority as those in Ethiopia. Hence, although in the medium term and long term more water is stored in Ethiopia, in these scenarios, no preference was given to the schemes in Ethiopia.

Table 14.7 Simulated mean monthly flow (million m^3) at the Ethiopia–Sudan border and Khartoum for natural, current, medium- and long-term future scenarios (1980–1992)

Month	*Natural*		*Current*		*Medium-term future*		*Long-term future*	
	Border	*Khartoum*	*Border*	*Khartoum*	*Border*	*Khartoum*	*Border*	*Khartoum*
January	835	835	955	855	1565	3220	2710	4405
February	470	470	580	740	1180	2220	2110	3175
March	350	350	400	475	845	1615	1980	2770
April	310	310	520	620	710	1315	1635	2250
May	390	390	645	710	680	1235	1485	2055
June	980	990	1230	640	1390	2275	2205	3125
July	5930	6235	6105	5365	3870	2560	4820	3420
August	15,770	16,830	15,430	15,950	8400	8615	4820	4245
September	10,590	11,680	10,130	10,165	8020	7490	4665	3760
October	5360	5825	4970	3865	5315	1740	4105	420
November	2750	2795	2615	310	8870	260	9095	250
December	1510	1510	1575	740	2305	850	3055	1990
Total	45,245	48,220	45,155	40,435	43,150	33,395	42,685	31,865

Net evaporation (i.e. the difference between evaporation from a reservoir and the rainfall directly onto its surface) from the Ethiopian reservoirs currently averages 0.8 billion m^3 yr^{-1}. However, by far the bulk of this is from Lake Tana which is a natural geographic feature and would be evaporating even without regulation. By comparison, net evaporation from the Sudanese reservoirs is 0.4 billion m^3 yr^{-1}. In the medium term this increases to 1.2 billion m^3 yr^{-1} in Ethiopia (0.3 billion m^3 yr^{-1} excluding Lake Tana) and to 1.4 billion m^3 yr^{-1} in Sudan. The increase in Sudan is due to the increased area of the Roseries reservoir arising from raising the Roseries Dam and the fact that water levels in both Roseries and Sennar reservoirs are maintained at higher levels because of the higher dry season inflows. In the long term, total net reservoir evaporation increases to 1.7 billion m^3 yr^{-1} in Ethiopia (0.8 billion m^3 yr^{-1} excluding Lake Tana) and remains at 1.4 billion m^3 yr^{-1} in Sudan. However, evaporation losses per cubic metre of water stored are considerably lower in Ethiopia than in Sudan in all the scenarios. In fact, as a result of the locations of the planned reservoirs, as storage increases in Ethiopia, losses per cubic metre of water stored decrease significantly over time (Table 14.8).

Table 14.8 Simulated average annual net evaporation from reservoirs in Ethiopia and Sudan for each of the scenarios

Scenario	*Ethiopia*[a]			*Sudan*		
	Total storage (million m^3)	*Total evaporation (million m^3)*	*Evaporation from storage (m^3 m^{-3})*	*Total storage (million m^3)*	*Total evaporation (million m^3)*	*Evaporation from storage (m^3 m^{-3})*
Current	11,578	846	0.07	3370	443	0.13
Medium term	70,244	1158	0.02	10,770	1363	0.13
Long term	167,079	1732	0.01	10,770	1387	0.13

Notes: [a] Including from Lake Tana, which is a natural lake, though regulated by the Chara Chara weir

Discussion

The results presented in this chapter are based on many assumptions. Lack of data on flow and water demand and use, particularly in Sudan, makes it very difficult to validate the model for the current situation. However, where it has been possible to verify them the model results appear to be reasonably accurate. For example, in the current scenario, simulated flows closely match the observed flows at key locations on the main stem of the river and the simulated water levels in Lake Tana were reasonably accurate (Figures 14.5 and 14.6). Consequently, though the results should be treated with caution, they are believed to be broadly indicative of the likely impacts arising from the development currently being considered in both Ethiopia and Sudan. By illustrating what may occur the scenarios provide information that is useful for resource planning, and the results provide a basis for discussion.

Climate, and hence hydrological variability, possibly increased by future climate change, will remain key factors in the economic development of both Ethiopia and Sudan in the future. As in the past, future water resources development in the Blue Nile will be driven predominantly by the need for water for agriculture and hydropower, and hence the need for large volumes of stored water. Irrigation will remain by far the largest user of water and the future scenarios indicate significantly increased water withdrawals as a consequence of increasing irrigation, predominantly in Sudan, and also increasingly in Ethiopia. The construction of the dams, particularly the very large hydropower dams proposed by Ethiopia, though not consuming large amounts of water, will significantly alter the flow regime of the river, resulting in lower wet season flows and much greater dry season flows. The results of this are likely to be beneficial for Sudan. The frequency of flooding, which occurs every few years in the flat areas of the country and is particularly devastating in and around Khartoum, may be reduced. Higher dry season flows mean greater availability of water at a time when it is naturally scarce and hence increased opportunities for withdrawals for irrigation and other uses. Thus increased water storage in Ethiopia has the potential to provide benefits for Sudan too.

As a result of higher rainfall and lower evaporative demand, net evaporation loss per cubic metre of water stored in the Ethiopian reservoirs (including Lake Tana, which is a natural lake) is currently approximately 50 per cent of that in Sudan. As more water is stored in Ethiopia, this ratio decreases, so that in the long term it could be as low as 8 per cent of that in Sudan (Table 14.8). This confirms that one of the most significant benefits of storing water in the Ethiopian Highlands, rather than in the lower, more arid, regions of Sudan (or indeed in Egypt) is significantly reduced evaporation losses.

For all scenarios, the model was run as a single system, making no allowance for the fact that Ethiopia and Sudan are separate countries. Water demands in Sudan were given the same priority as those in Ethiopia and water was released from reservoirs in Ethiopia to meet downstream demands in Sudan. This assumes a much higher level of cooperation between the two states, in relation to both the planning and management of water resources, than at present.

Future research is needed to refine the model. Key to improving the simulations are:

- improved estimates of irrigation water demand;
- improved estimates of the dates on which schemes will become operational;
- more realistic dam operating rules;
- detailed economic, livelihood and environmental assessments of the cumulative impacts of all the proposed schemes; and
- evaluation of the possible hydrological implications of climate change.

An important issue not considered in the current model simulations is the transient stages of reservoir filling. Given the large cumulative volume of the planned reservoirs in Ethiopia, it is essential that reservoir filling is planned and managed in such a way that adverse downstream impacts, including potentially negative environmental and social impacts, are minimized. The need to give due consideration to dam operation that provides for environmental flows, to avoid degradation of riverine ecosystems, has recently been emphasized (Reitburger and McCartney, 2011).

Conclusion

The WEAP model has been configured to simulate the impacts of water resources development in the BNB. Currently, Ethiopia utilizes very little water for irrigation, but does regulate some flow for hydropower production. In contrast, Sudan uses some water for hydropower production and also abstracts large volumes for irrigation. Both countries plan to develop water resources substantially in the near future. The extent to which actual water resources development will match the plans of both countries in the long term is unclear and will depend a lot on unpredictable social and economic factors. However, in both Ethiopia and Sudan, hydropower and irrigation are widely perceived as critical to national development and, in both countries, current investment in water infrastructure is substantial and increasing. Consequently, pressures on water resources are rising and will increase substantially in the near future.

The results of this study have confirmed that, if the states cooperate effectively, mutually beneficial scenarios are possible; upstream regulation in Ethiopia reduces evaporation losses, probably reduces the frequency of flooding and provides opportunities for greatly increased water development in Sudan. However, maximizing benefits and minimizing potential adverse impacts (e.g. to the environment), especially when the large reservoirs are being filled, require much greater cooperation than currently exists between the riparian states. The key to success is the establishment of pragmatic institutional arrangements that enable the water resources of the basin to be planned and managed as a single entity (i.e. without consideration of national borders), in the most effective and equitable manner possible. It is to be hoped that such arrangements will be devised through the protracted negotiations currently under way.

References

Awulachew, S. B., McCartney, M. P., Steenhuis, T. S., Ahmed, A. A. (2008) *A Review of Hydrology, Sediment and Water Resource Use in the Blue Nile Basin*, International Water Management Institute, Colombo, Sri Lanka.

BCEOM (1998) *Abbay River Basin Integrated Development Master Plan, Section II, Volume V, Water Resources Development, Part 1: Irrigation and Drainage*, Ministry of Water Resources, Addis Ababa, Ethiopia.

Beyene, T. and Abebe, M. (2006) Potential and development plan in Ethiopia, *Hydropower and Dams*, 13, 61–64.

Block, P., Strzepek, K. and Rajagopalan, B. (2007) *Integrated Management of the Blue Nile Basin in Ethiopia: Hydropower and Irrigation Modeling*, IFPRI Discussion Paper 00700, International Food Policy Research Institute, Washington, DC, p22.

Cascão, A. E. (2009) Changing power relations in the Nile River Basin: Unilateralism vs. cooperation? *Water Alternatives*, 2, 2, 245–268.

Elala, G. (2008) Study of Mainstem Dams on the Blue Nile, MSc thesis, Arba Minch University, Ethiopia, pp94.

ENTRO (Eastern Nile Technical Regional Office) (2007) *Pre-feasibility study of Border Hydropower Project, Ethiopia*, ENTRO, Addis Ababa, Ethiopia.

FAO (Food and Agriculture Organization of the United Nations) (2002) *Announcing LocClim, the FAO Local Climate Estimator CD-ROM*, www.fao.org/sd/2002/EN1203a_en.htm, accessed June 2009.

Georgakakos, A. P. (2003) Nile Decision Support Tool (Nile DST) Executive Summary, Report submitted to FAO and the Nile riparian states, Rome, Italy, June.

Guariso, G. and Whittington, D. (1987) Implications of Ethiopian water development for Egypt and Sudan, *International Journal of Water Resources Development*, 3, 2, 105–114.

Ibrahim, Y. A., Ahmed, A. A. and Ramdan, M. S. (2009) Improving water management practices in the Rahad Scheme (Sudan), in *Improved Water and Land Management in the Ethiopian Highlands: Its Impact on Downstream Stakeholders Dependent on the Blue Nile*, Intermediate Results Dissemination Workshop held at the International Livestock Research Institute (ILRI), 5–6 February, S. B. Awulachew, T. Erkossa, V. Smakhtin, and F. Ashra (eds), Addis Ababa, Ethiopia.

Jägerskog, A., Granit, J., Risberg, A. and Yu, W. (2007) Transboundary water management as a regional public good, in *Financing Development: An Example from the Nile Basin*, Report no 20, SIWI, Stockholm, Sweden.

JICA (Japan International Cooperation Agency) (1997) *Feasibility Report on Power Development at Lake Tana Region*, Japan International Cooperation Agency, Tokyo, Japan.

Lahmeyer Consulting Engineers (1962) *Gilgel Abay Scheme*, Imperial Ethiopian Government, Ministry of Public Works, Addis Ababa, Ethiopia.

McCartney, M. P., Alemayehu, T., Shiferaw, A. and Awulachew, S. B. (2010) Evaluation of current and future water resources development in the Lake Tana Basin, Ethiopia, IWMI Research Report 134, International Water Management Institute, Colombo, Sri Lanka.

Metawie, A. F. (2004) History of co-operation in the Nile Basin, *Water Resources Development*, 20, 1, 47–63.

Monteith, J. L. (1981) Evaporation and surface temperature, *Quarterly Journal of the Royal Meteorological Society*, 107, 1–27.

Norconsult (2006) *Karadobi Multipurpose Project: Pre-feasibility Study*, Report to the Ministry of Water Resources, The Federal Republic of Ethiopia, Addis Ababa, Ethiopia.

Omer, A. M. (2009) *Hydropower Potential and Priority for Dams Development in Sudan*, www.scitopics.com/Hydropower_potential_and_priority_for_dams_development_in_Sudan.html#, accessed November 2010.

Omer, A. M. (2010) *Sudanese development*, 20 September, www.waterpowermagazine.com/story.asp?sc=2057614, accessed 29 April 2012.

Reitburger, B. and McCartney, M. P. (2011) Concepts of environmental flow assessment and challenges in the Blue Nile Basin, Ethiopia, in *Nile River Basin: Hydrology, Climate and Water Use*, A. M. Melesse (ed.), pp337–358, Springer, Heidelberg, Germany.

SEI (Stockholm Environment Institute) (2007) *WEAP: Water Evaluation And Planning System – User Guide*, Stockholm Environment Institute, Boston, MA.

Sutcliffe, J. V. and Parks, Y. P. (1999) *The Hydrology of the Nile*, IAHS, Wallingford, UK.

USBR (United States Bureau of Reclamation) (1964) *Land and Water Resources of the Blue Nile Basin, Ethiopia*, United States Bureau of Reclamation, Main report, United States Bureau of Reclamation, Washington, DC.

WAPCOS (1990) *Preliminary Water Resources Development Master Plan for Ethiopia*, Final Report, prepared for EVDSA, Addis Ababa, Ethiopia.

Whittington, D., Wu, X. and Sadoff, C. (2005) Water resources management in the Nile Basin: the economic value of cooperation, *Water Policy*, 7, 227–252.

Yates, D., Sieber, J., Purkey, D. and Huber-Lee, A. (2005) WEAP 21: a demand, priority and preference driven water planning model, part 1: model characteristics, *Water International*, 30, 487–500.

Yilma, D. A. and Awulachew, S. B. (2009) Characterization and atlas of the Blue Nile Basin and its sub basins, in *Improved Water and Land Management in the Ethiopian Highlands: Its Impact on Downstream Stakeholders Dependent on the Blue Nile*, Intermediate Results Dissemination Workshop held at the International Livestock Research Institute (ILRI), 5–6 February, S. B. Awulachew, T. Erkossa, V. Smakhtin, and F. Ashra (eds), Addis Ababa, Ethiopia.

15

Water management intervention analysis in the Nile Basin

Seleshi B. Awulachew, Solomon S. Demissie, Fitsum Hagos, Teklu Erkossa and Don Peden

Key messages

- Agricultural water management (AWM) interventions in the Nile Basin are a key to improve agricultural production and productivity. AWM interventions can be categorized based on spatial scales, sources of water and type of technologies for water management in control, lifting, conveyance and application. Various combinations of these interventions are available in the Nile Basin. Successful application of AWM interventions should consider the full continuum of technologies in water control, conveyance and field applications.
- AWM technology intervention combined with soil fertility and seed improvement may increase productivity up to threefold. Similarly, data sets used from a representative sample of 1517 households in Ethiopia shows that the average treatment effect of using AWM technologies is significant and has led to an income increase of US$82 per household per year, on average. The findings indicated that there are significantly low poverty levels among users compared to non-users of AWM technologies, with about 22 per cent less poverty incidence among users compared to non-users of *ex situ* AWM technologies.
- The Nile basin has 10 major man-made water control structures that are used for various purposes including irrigation, hydropower, flood and drought control, and navigation. The Water Evaluation And Planning (WEAP) model is applied to the Nile Basin, considering existing infrastructure, and scenarios of water use under current, medium term and long term. The major water use interventions that affect water availability in rivers are related to irrigation development. Accordingly, the irrigation areas of the current, medium-term and long-term scenarios in the Nile Basin are, respectively, about 5.5, 8 and 11 million ha, with water demands of 65,982 million m^3, 94,541 million m^3 and 127,661 million m^3, respectively. The total irrigation water demand for the current scenario is lower than the Nile mean annual flow. The total irrigation water demand for the medium-term scenario exceeds the Nile mean annual flow marginally. The irrigation demands for the long-term scenario are considerably greater than the mean annual flow of the Nile basin, assuming the existing management practice and irrigation water requirement estimation of the countries. The river water would therefore not satisfy irrigation water demands in the long term unless the irrigation efficiency is improved, water saving measures are implemented and other sources of water and economic options are explored.

Introduction

The major objective of AWM interventions is to enhance growth of agricultural productivity, poverty reduction and livelihood improvement. This can be achieved through increasing the positive role of water and reducing the negative impacts of water. The purpose of this chapter is to identify the major types of water management intervention that exist in the Nile Basin, analyse options that may be considered for further development and management, and evaluate their impacts, particularly focusing on interventions already implemented and planned for the future to improve access to water. If the interventions are carefully planned and implemented, they contribute to national and regional economic transformations and development. The methods used here include inventorying and characterization of existing interventions in various parts of the basin and production systems, review of performance of existing interventions, trade-off analysis, ranking, scenario analysis and modelling to select and evaluate the high-impact interventions and implementation strategy.

Interventions may be categorized as:

- interventions based on water availability, access and management;
- agricultural and non-agricultural water use interventions;
- water interventions based on the production system, livelihood and hydro-economic modelling; and
- small- and large-scale interventions.

In this chapter we will use the last type of categorization. The next section deals with detailed identification, listing and characterization of smallholder water interventions, shortlisting of interventions as they fit the various agro-ecologies, and associated impacts on productivity and poverty with considerations of typical case studies. Subsequent sections deal with the large-scale interventions, modelling, scenario analysis and implications on access to water and availability in the basin.

Small-scale water interventions in the Nile Basin

The water management interventions for agriculture

The small-scale interventions here are primarily those of AWM (Molden, 2007) that range from field conservation practices to irrigation and drainage associated with crop production. However, the broader definition of AWM may include water not only for crops but also for animals, agro-forestry and a combination with multiple uses such as drinking water, environment, and so on. Rain-fed agriculture, supported to some extent by small-scale irrigation (SSI) and water harvesting systems, is the dominant form of agriculture in the upstream countries of the Nile such as Ethiopia, Rwanda and Uganda, whereas the downstream countries – Egypt and Sudan – are dominated by irrigated agriculture in large-scale irrigation (LSI) schemes. In the transition, the system is dominated by pastoral and agro-pastoral systems. Rainfall management strategies through (i) on-farm water management, (ii) maximizing transpiration and reducing soil evaporation, (iii) collecting excess run-off from farm fields and using it during dry spells and as supplementary irrigation, (iv) draining of waterlogged farm areas, and (v) enhancing livestock productivity are crucial for transformation of rain-fed agriculture to higher productivity. In addition, stream diversions and groundwater management with appropriate technology for control, conveyance and application in supplementary and full irrigation are the interventions that may enhance smallholder agricultural productivity.

AWM interventions include water control, water lifting, conveyance, field application and drainage/reuse technologies. Figure 15.1 provides an illustration of the major categories of small-scale water management interventions, with emphasis on crop production (see also Molden *et al.*, 2010). Most of the categories related to water control and management are also applicable to the livestock sector and some for fishery and aquaculture, with certain modifications on the part of conveyance and application/use.

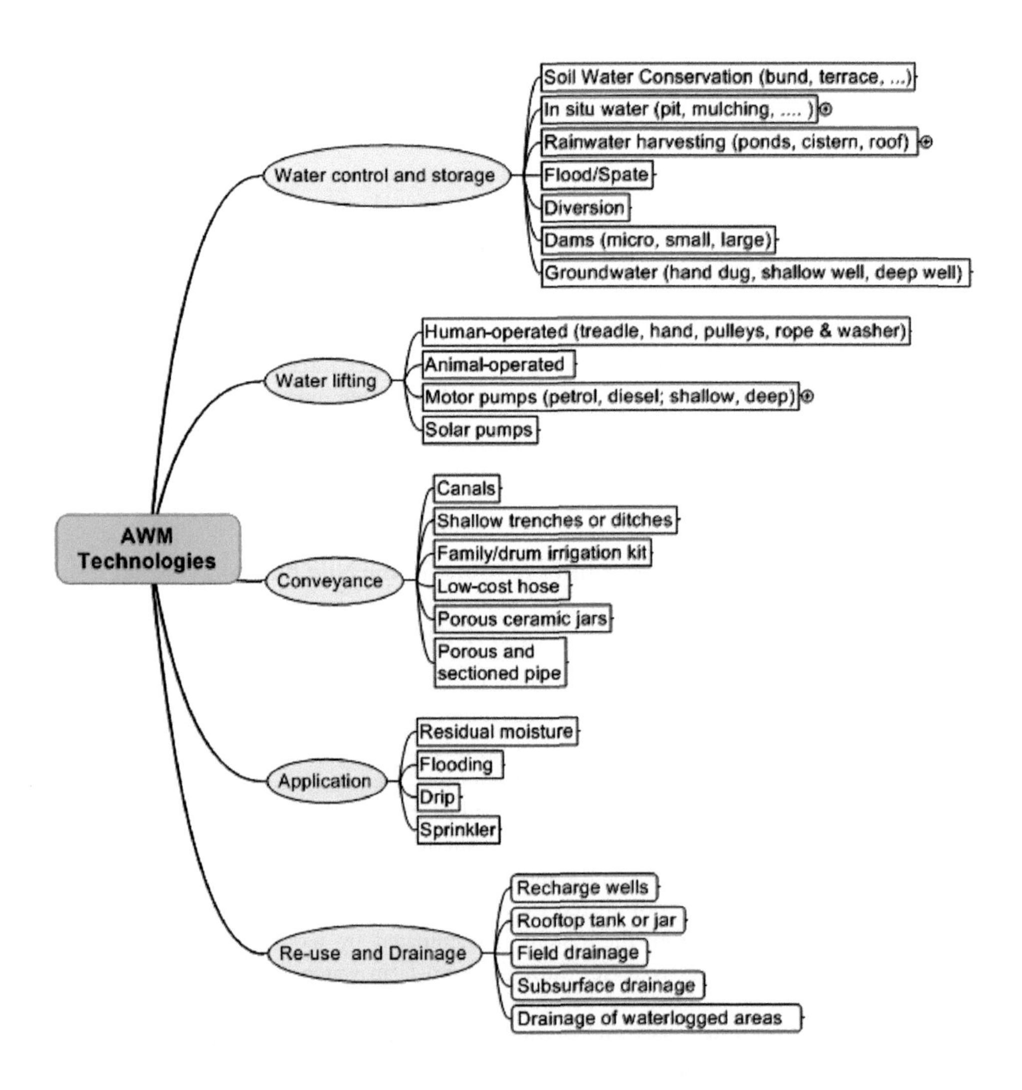

Figure 15.1 Agricultural water management continuum for control, lifting, conveyance and application

Furthermore, numerous combinations of this continuum are possible, creating what is termed here as 'AWM technology suites' that can be applicable at the household or farm level, community or small catchment/watershed level, sub-basin, basin or regional level. Table 15.1 lists these suites categorized by the scale of application and source of water. An inventory of SSI practised in the Nile Basin countries is given in Anderson and Burton, 2009.

Impacts of interventions on productivity

The impacts of AWM interventions on productivity and poverty alleviation may be evaluated using simple and complex techniques ranging from simple mean separation tests, estimation of average treatment effects using propensity score matching, poverty analysis and modelling. Here, impacts related to productivity and poverty reduction are evaluated by taking Ethiopian Highlands as an example.

The rampant rain-fed mixed crop-livestock farming system in the Ethiopian Highlands is characterized not only by growing one crop per year but also by poor land and water productivity, which perpetuates poverty and vulnerability to shocks caused by climate variability, among others. Low productivity is reinforced by continued decline in landholding per household due to rapid population growth and severe land degradation. In order to overcome these constraints, technological interventions are essential. The possibilities to (i) improve productivity of maize under the prevailing climatic conditions and a range of soil fertility management and (ii) enhance the productive use of water are examined here as an example. Maize is one of the dominant crops in crop livestock system of Ethiopian Highlands. It is typical for areas with high rainfall and relatively productive soils.

The Food and Agriculture Organization of the United Nations (FAO) AquaCrop model was used after validation with data from agricultural research stations in and around the basin in Ethiopian Highlands. The attempt was made to simulate the productivity of maize under varying soil fertility levels (poor, near-optimal and non-limiting) using hybrid seed under the prevailing climatic conditions, and to examine potential gains of productivity that can be achieved. Results suggest that improving soil fertility can tremendously enhance grain and biomass productivity (Anchala *et al.*, 2001; Erkossa *et al.*, 2009). Grain yield increased from 2.5 t ha^{-1} under poor to 6.4 and 9.2 tha^{-1} with near-optimal and non-limiting soil fertility conditions, respectively. Correspondingly, soil evaporation decreased from 446 to 285 and 204 mm, while transpiration increased from 146 to 268 and 355 mm. Consequently, grain water productivity increased by 48 and 54 per cent, respectively, due to the near-optimal and non-limiting soil fertility conditions. The model predicts that about 593 mm of the seasonal rainfall are lost as run-off. If harvested, this can be used to grow a second crop on a fraction or the whole area depending on the type of crop, irrigation efficiency and availability of labour. Part of the excess water can also fulfil domestic needs or livestock consumption. Both productivity gain during the main season and the secondary production constitute evidence of significant untapped potential in the area and similar agro-ecosystems in sub-Saharan Africa. This result also clearly shows that the lack of integration of measures such as fertilizers, seeds and management of rainfall is limiting productivity potential.

Impacts of interventions on poverty and food security

In the past, a lack of clear understanding of the issues that link AWM to poverty reduction and agricultural productivity has been one of the reasons for underdevelopment of agriculture (Anderson and Burton, 2009). AWM technologies are expected to have significant impact on

Table 15.1 Agricultural water management technology suites and scale of application

Scale	*Water source*	*Water control*	*Water lifting*	*Conveyance*	*Application*	*Drainage and reuse*
Smallholder farmlevel	Rainwater	*In situ* water Farm ponds Cistern and underground ponds Harvesting roof water Recession agriculture	Treadle pumps Water cans	Drum Channels Pipes	Flooding Direct application Drip	Drainage of waterlogging Surface drainage channels Recharge wells
Smallholder farmlevel	Surface water	Spate and flooding Diversion Pumping	Micro pumps (petrol, diesel) Motorized pumps	Channels Canals Pipes (rigid, flexible)	Flood and furrow Drip Sprinkler	Surface drainage channels Drainage of waterlogging
Smallholder farmlevel	Groundwater	Spring protection Hand-dug wells Shallow wells	Gravity Treadle pumps Micro pumps (petrol, diesel) Hand pumps	Channels Canals Pipes (rigid, flexible)	Flood and furrow Drip Sprinkler	Surface drainage channels Drainage of waterlogging Recharge wells
Community or catchment	Rainwater	Soil Water Conservation Communal ponds Recession agriculture Sub-surface dams	Treadle pumps Water cans	Drum Channels Pipes	Flooding Direct application Drip	Drainage of waterlogging Surface drainage channels
Community or catchment	Surface water	Spate and flooding Wetland Diversion Pumping Micro dams	Micro pumps (petrol, diesel) Motorized pumps Gravity	Channels Canals Pipes (rigid, flexible)	Flood and furrow Drip Sprinkler	Surface drainage channels
Community or catchment	Groundwater	Spring protection Hand-dug wells Shallow wells Deep wells	Gravity Treadle pumps Micro pumps (petrol, diesel) Hand pumps Motorized pumps	Channels Canals Pipes (rigid, flexible)	Flood and furrow Drip Sprinkler	Surface drainage channels Recharge wells and galleries
Sub-basin, Basin	Surface water	Large dams	Gravity Large-scale motorized pumps	Channels Canals Pipes (rigid, flexible)	Flood and furrow Drip Sprinkler	Surface drainage channels Drainage reuse

household well-being through increasing food production and income (Namara *et al.*, 2007). The Comprehensive Assessment of Water Management in Agriculture (Molden, 2007) states that:

> Improving access to water and productivity in its use can contribute to greater food security, nutrition, health status, income and resilience in income and consumption patterns. In turn this can contribute to other improvements in financial, human, physical and social capital simultaneously alleviating multiple dimensions of poverty.

An attempt is made to explore whether adoption of AWM technologies has led to such improvements and, if so, to identify which technologies have a higher impact. The study quantified the average treatment effect of using AWM technologies. Analysis on the state of poverty among sample farm households with and without access to AWM technologies can reveal the impact of these technologies on poverty. In this study welfare indicators such as per capita income and expenditure per adult equivalent are used in matching econometrics and in poverty analysis, respectively. The inflation adjusted poverty lines equivalent to US$200 and US$120 were adopted to show overall poverty and food poverty/insecurity, respectively (MoFED, 2006). Data sets from a representative sample of 1517 households from 30 kebeles in four regions of Ethiopia have been used. The interventions include rainwater harvesting, groundwater, surface water using ponds, wells, diversions and small dams. The results indicate that the average effect of using AWM technologies is significant and has led to an income increase of, on average, US$82 per household. It also shows that there is about 22 per cent less poverty incidence among users compared to non-users of ex-situ AWM technologies. Furthermore, from the poverty analysis (severity indices), it is found that AWM technologies are not only effectively poverty-reducing but also equity-enhancing interventions.

The magnitude of poverty reduction is found to be technology-specific. Accordingly, deep wells, river diversions and micro dams are associated with reductions in poverty incidence of 50, 32 and 25 per cent, respectively, compared with the purely rain-fed systems. The use of modern water withdrawal technologies (treadle pumps and motorized pumps) was also found to be strongly related to lower poverty. The use of motorized pumps was associated with a reduction in poverty incidence of more than 50 per cent. Similarly, households using gravity irrigation had significantly lower poverty levels than those using manual (cans) applications because of scale benefits. While access to AWM technologies seems to unambiguously reduce poverty, the study also indicates that there is a plethora of factors that can enhance this impact. Figures 15.2 and 15.3 show sample results of poverty and food security status of reduction of users and non-users of technologies and the relative impacts of poverty reduction with respect to technology.

It was also found that the most important determinants of poverty include asset holdings, educational attainment, family labour and access to services and markets. To enhance the contribution of AWM technologies to poverty reduction, therefore, there is a need to (i) build assets, (ii) develop human resources and (iii) improve the functioning of labour markets and access to markets (input or output markets).

In summary:

- Various AWM technologies for water control, lifting, conveyance and field applications exist. It is essential to identify the best suites of AWM technologies.
- Based on the sample survey data, access to AWM in water control and management help farmers to decrease poverty incidence by about 22 per cent. Some technologies, such as deep wells, reduced poverty by 50 per cent.

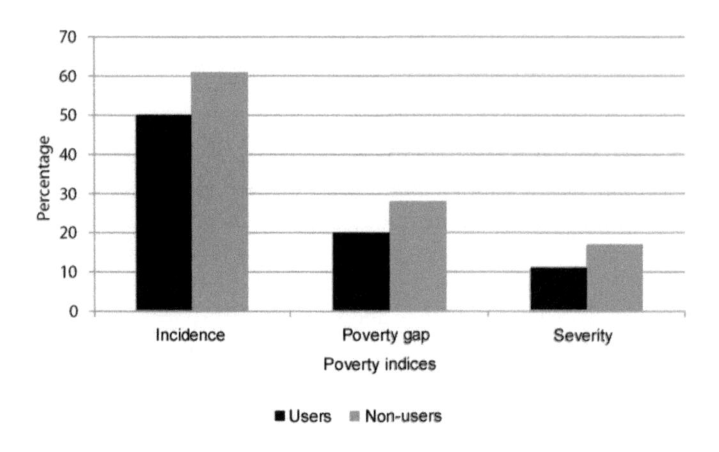

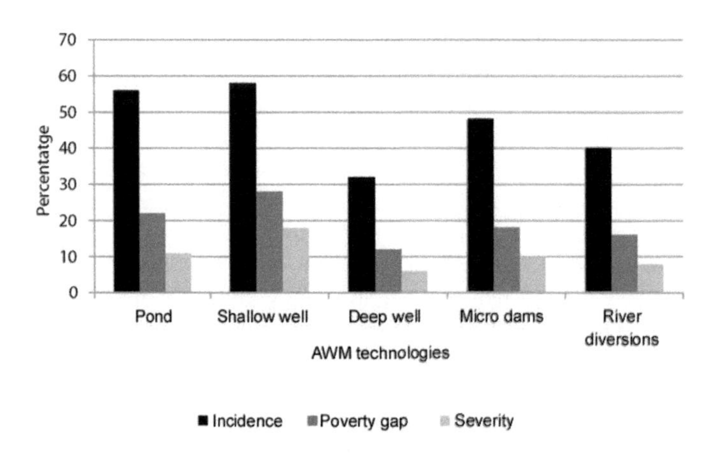

Figure 15.2 Poverty profiles and agricultural water management technologies

- Rainwater harvesting technologies are generally successful in areas with high variability and low rainfall to increase household agricultural production for food, cash crops and livestock production.
- The impact on productivity gain can be tripled if access to AWM technology can be increased and combined with access to improved soil fertility (fertilizer use) management and seeds.

There is therefore significant scope for managing rain-fed and small-scale irrigation systems in the Nile Basin to increase productivity, reduce poverty and enhance food availability. The combined interventions for more gains should be exploited.

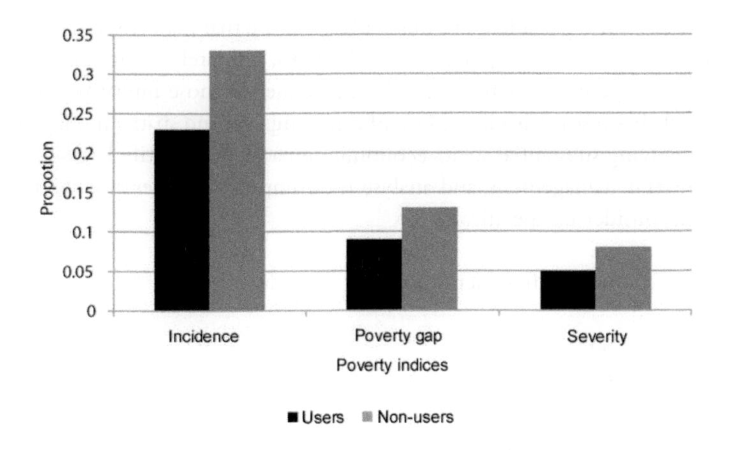

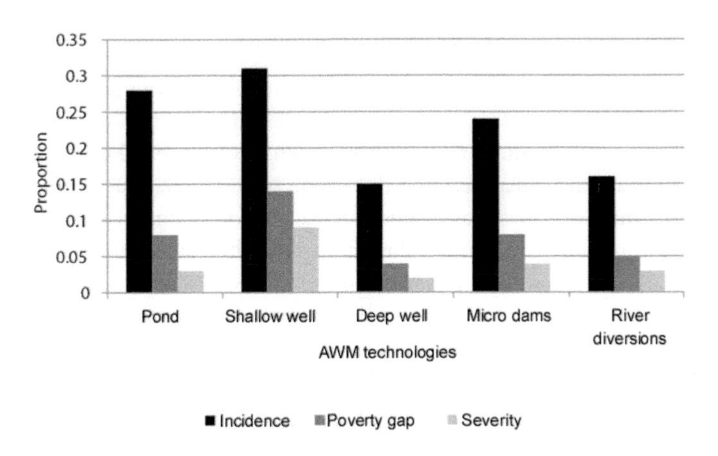

Figure 15.3 Food poverty profiles and agricultural water management technologies

Large-scale interventions in the Nile

The Nile River Basin is characterized by complex topography, high climate variability, low specific discharge and high system losses through conveyance and evaporation (see also previous chapters). Most of the Nile flow is generated from the Ethiopian Highlands plateau and the equatorial lakes regions that cover only 20 per cent of the basin, and only 25 per cent of the basin receives rainfall exceeding 1000 mm (see also Chapters 3–5). The remainder of the basin is in arid and semi-arid regions where the demand for water is comparatively large due to high evaporation and seepage losses. In order to provide a buffer for climatic and hydrological variability, large storage infrastructures were built along the Nile River in Egypt and Sudan. More large-scale infrastructures are planned for meeting the food and energy demands of the fast-growing population of the Nile Basin.

The large infrastructures considered in this study are those that mainly serve district (provincial), national and trans-national (regional) spatial domains, and rarely community or household levels directly. These large infrastructures can also be defined as those interventions undertaken at river basin or sub-basin scales leading to significant temporal and spatial modifications of the natural flow or implying substantial socio-economic impacts. We identify large-scale interventions relevant to water management, and analyse their impact on water availability and access in the Nile Basin, considering specifically:

- water control and storage infrastructures (single or multi-purpose);
- irrigation schemes;
- hydropower plants; and
- environment and wetlands.

The Nile water and its infrastructure

Operational systems

Water control infrastructures have been used for a long time in the Nile Basin to regulate and utilize the seasonally varying river flow for irrigation, hydropower and flood-control purposes. They are located either at the outlet of natural lakes, such as Owen Fall Dam at Lake Victoria and CharaChara weir at Lake Tana, or along the major river courses. The High Aswan Dam provides storage over the year. The storage dams in Sudan are losing significant amounts of storage volume through time due to sediment flow from the Ethiopian Highlands. For example, the capacity of the Roseires reservoir was reduced from about 3.4 billion m^3 in 1966 to 1.9 billion m^3 in 2007 (Bashar and Mustafa, 2009). The details of control and storage infrastructures listed in Table 15.2 were compiled from published literature (Yao and Georgakakos, 2003), national master plan documents (TAMS Consulting, 1997; BCEOM, 1998; NEDECO, 1998) and from personal communication with experts in the basin.

Table 15.2 Existing water control structures in the Nile Basin

Dam	*Country (million m^3)*	*Live storage*	*Year built*	*Purpose*
Abobo	Ethiopia	57	1992	Irrigation; not yet used
Finchaa	Ethiopia	1050	1971	Irrigation, hydropower
Aswan	Egypt	105,900	1970	Irrigation, hydropower
Jebel El Aulia	Sudan	3350	1937	Irrigation, hydropower
KhashmEl Gibra	Sudan	835	1964	Irrigation, hydropower
Koga	Ethiopia	80	2008	Irrigation
Chara Chara	Ethiopia	9126	2000	Hydropower
Owen Falls	Uganda	215,586	1954	Irrigation, hydropower
Roseires	Sudan	2322	1966	Irrigation, hydropower
Sennar	Sudan	753	1925	Irrigation, hydropower

Emerging developments

The Nile Basin countries are facing challenges of meeting food and energy demands for their rapidly growing populations. Therefore, a number of water resource developments have been planned by the riparian countries. Some of the planned projects are already being implemented. The Merowe Dam in Sudan and the Tekeze Dam in Ethiopia were recently constructed for hydropower generation, and these dams would have become fully operational in 2010. The construction of the Bujagali hydropower plant in Uganda is under progress. Sudan will raise the height of the Roseires Dam by 10 m to further increase its storage capacity. Ethiopia is currently undertaking the Tana-Beles hydropower project through intra-basin diversion of 77 $m^3 s^{-1}$ of water from Lake Tana to the Beles River (tributary of the Abbay River) and is planning to build other storage infrastructures mainly for hydropower.

Apart from these emerging water resources developments, the riparian countries are unilaterally planning to expand their irrigated agriculture and hydropower generation. Most countries have developed integrated master plans for parts of the Nile Basin within their territories. Under the subsidiary action programmes of the Nile Basin Initiative (NBI), the regional offices, Nile Equatorial Lakes Subsidiary Action Program (NELSAP) and Eastern Nile Technical Regional Organization (ENTRO), are also planning joint multi-purpose projects that benefit the riparian countries.

The Nile Basin modelling framework

The WEAP System model was applied to the entire Nile Basin for simulating the water supply and demands of the large-scale intervention scenarios. WEAP has the capability of integrating the demand and supply sides of water accounting with policy and management strategies (SEI, 2007). The model (Figure 15.4) was set up for the Nile Basin at monthly time intervals. For better illustration, the basin-wide topology (framework) of the WEAP model is independently displayed for the major regions of the basin in Figures 15.5–15.8.

The release rules from natural lakes are defined as flow requirements downstream of the lakes. The flow rate at these nodes of the release rules is defined in terms of the water level of the lakes. The ecological water needs of wetlands are represented as flow requirement nodes that take up the predefined percentage of the incoming flow into the wetland system. The contribution of wetlands to the dry season river flow is schematized in the WEAP model as streams, such as Ghazal Swamps and Machar Return (Figure 15.6).

The details of the WEAP schematization depend upon availability of climatic, hydrological and infrastructural information. The tributaries in the equatorial lakes region are aggregated into a number of streams since the datasets obtained for that region are very minimal. However, the WEAP modelling schematics is well detailed for the Ethiopian and, to some extent, for the Sudanese parts of the Nile Basin as the required datasets are obtained from master plans and project reports.

Wubet *et al.* (2009), Ibrahim *et al.* (2009) and McCartney *et al.* (2009) successfully applied Mike Basin, HEC-Res and WEAP models for Ethiopia, Sudan and Blue Nile, respectively, to evaluate the impacts of consumptive water use on water availability and implications on the water balance. The current WEAP schematization of the Nile Basin has attempted to incorporate their modelling features. However, the Nile Basin WEAP modelling was conducted using mean values of monthly flow and net evapotranspiration.

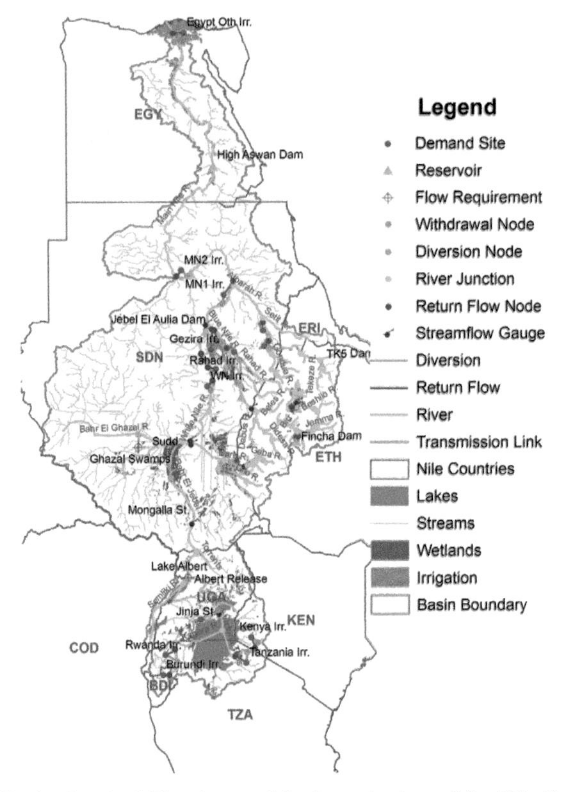

Figure 15.4 Water Evaluation And Planning model schematization of the Nile Basin for the current situation

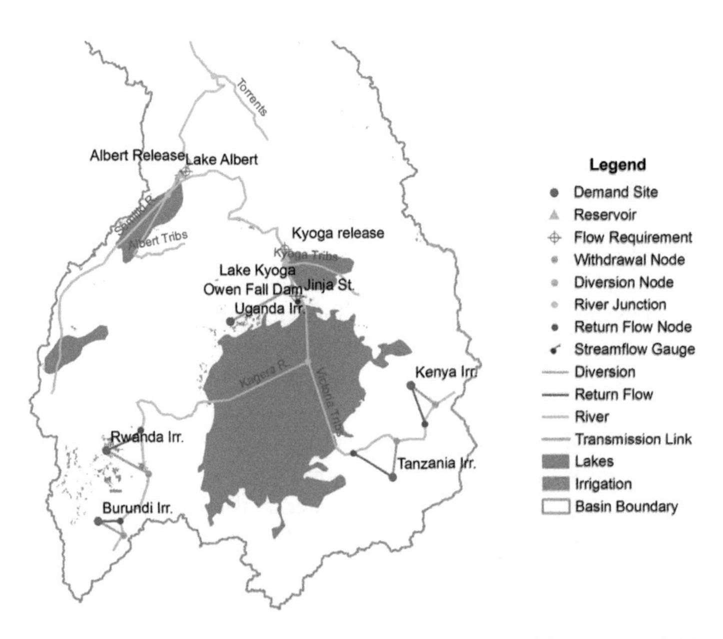

Figure 15.5 Water Evaluation And Planning model schematization of the equatorial lakes part of the Nile Basin

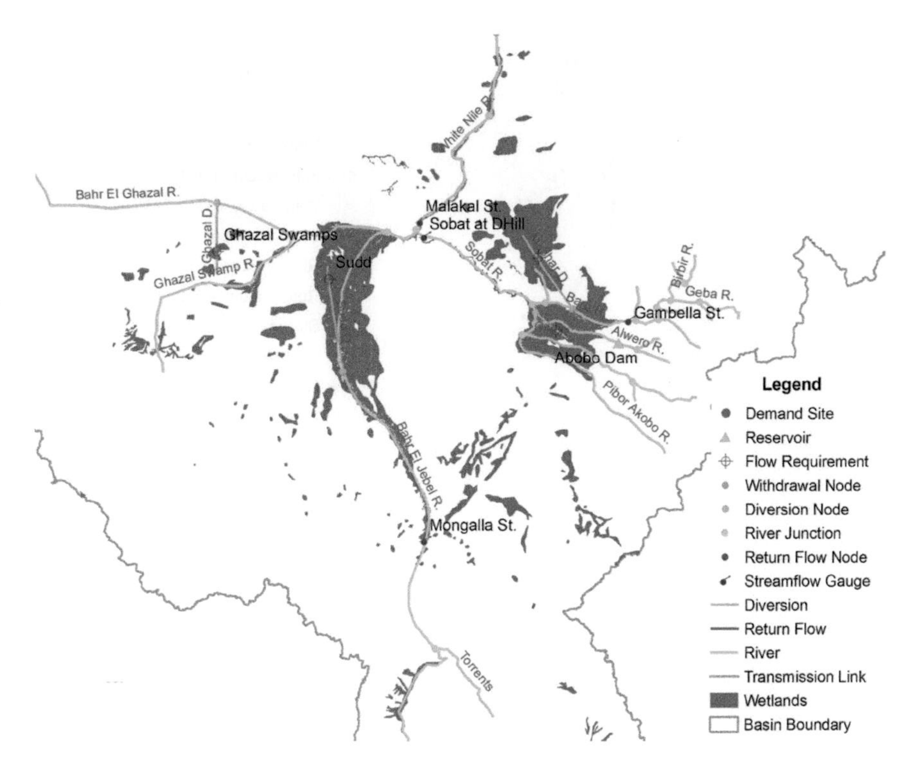

Figure 15.6 Water Evaluation And Planning model schematization of the wetlands and Sobat-Baro parts of the Nile Basin for the current situation

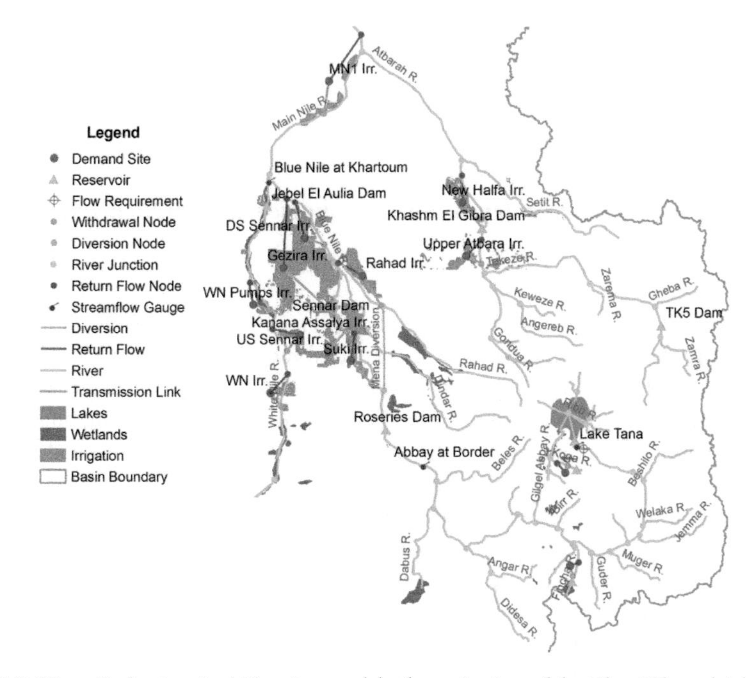

Figure 15.7 Water Evaluation And Planning model schematization of the Blue Nile and Atbara-Tekeze parts of the Nile Basin for the current situation

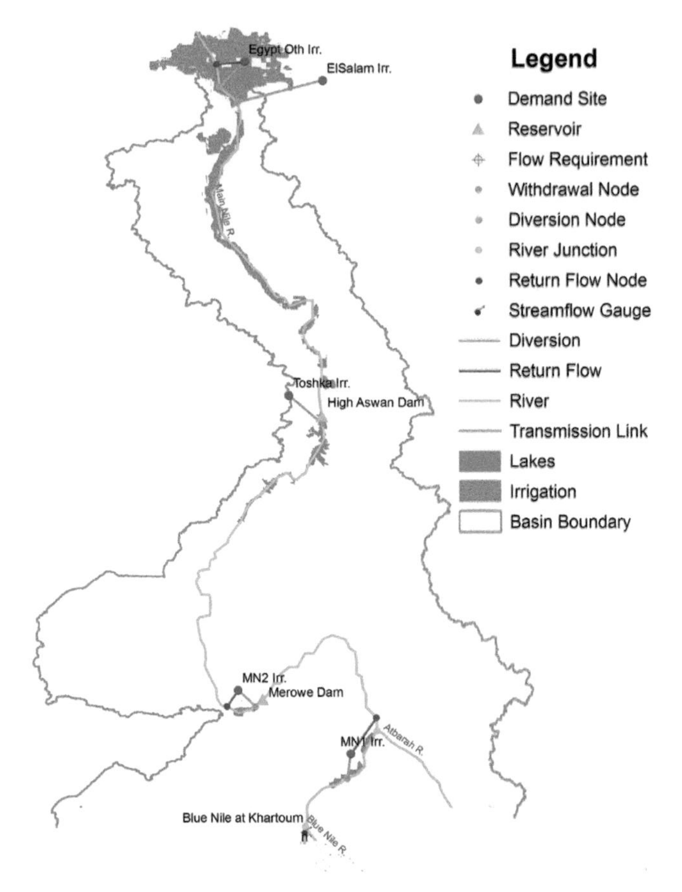

Figure 15.8 Water Evaluation And Planning model schematization of the main Nile part of the Nile Basin

The water resources development scenarios and implications

Scenarios

The large-scale water development and management interventions that are operational, emerging and planned in the entire Nile Basin are categorized into three different scenarios for the purpose of analysing their plausible impacts on the availability of, and access to, water. While the operational interventions form the *current (baseline) scenario*, the emerging and fast-track (planned) interventions are considered as the *medium-term scenario*. Other planned large-scale interventions that approach towards utilizing the potential land and water resources are categorized under *long-term scenarios*. It may not be possible to assign a strict timeline between these development scenarios since the riparian countries have different planning horizons. Some countries, for example Sudan, have clearly identified their development plans for the medium and long term. When such information is not available, about one-third of potential developments of countries is assumed to be implemented during the medium-term scenario period, and the remaining near-potential developments are also assumed to be realized during the long-term scenario period.

The existing and planned irrigation areas of the riparian countries and regions in the Nile Basin for the three development scenarios (Table 15.3) are determined from country-specific feasibility studies and master plans (TAMS Consulting, 1997; BCEOM, 1998; NEDECO, 1998), published literature (FAO, 2000) and project documents (ENTRO, 2007). Accordingly, the irrigation areas of the current, medium-term and long-term scenarios in the Nile Basin are about 5.5, 8 and 11 million ha, respectively.

The water requirements of the irrigation scenarios are (i) determined from literature on the annual rate of irrigation and (ii) compiled from project documents, feasibility studies and relevant master plans cited above. The monthly distributions of the irrigation water requirement are either compiled from the above sources whenever available or determined from rainfall and evapotranspiration data. The percentage of water returning from the irrigation system to the river network is assumed to be based on the topography of the irrigation field. In flat irrigation fields no return flow is considered. As shown in Table 15.4, the annual irrigation water requirement for the Blue Nile part of Sudan is less than that for the Ethiopian part. Even though this conflicts with prevailing climatic conditions, the figures are retained in this study in order to value both sources of data.

The environmental water requirements are expressed in terms of the percentage of incoming flow to the wetland in the previous month. The one month lag is adopted due to model restrictions in accessing the incoming flow of the current month. However, the lag helped to account for the routing effect of the wetlands.

Table 15.3 The irrigation areas (ha) for the current, medium- and long-term scenarios

Country/sub-basin	*Current*	*Medium term*	*Long term*
Burundi	0	18,160	80,000
Egypt	3,324,300	3,764,733	4,205,166
Nile valley	*3,324,300*	*3,521,133*	*3,717,966*
El-Salam	*0*	*130,200*	*260,400*
Toshka	*0*	*113,400*	*226,800*
Ethiopia	15,900	343,503	1,216,130
Blue Nile	*15,900*	*217,023*	*489,726*
Baro-Akobo-Sobat	*0*	*71,954*	*536,904*
Tekeze-Atbara	*0*	*54,526*	*189,500*
Kenya	5600	70,000	200,000
Rwanda	5000	50,000	155,000
Sudan	2,175,600	3,574,620	4,503,240
Tekeze-Atbara	*391,440*	*412,440*	*731,640*
Blue Nile	*1,304,940*	*2,125,620*	*2,194,080*
Main Nile	*130,620*	*449,820*	*781,200*
White Nile	*348,600*	*586,740*	*796,320*
Tanzania	475	10,000	30,000
Uganda	9120	80,000	247,000
Total	5,535,995	7,911,016	10,636,536

The total irrigation water demand for the current scenario is lower than the Nile mean annual flow. The total irrigation water demand for the medium-term scenario exceeds the Nile mean annual flow marginally. However, the irrigation demand for the long-term scenario is considerably greater than the mean annual flow of the Nile Basin. This shows that the river water would not suffice for future irrigation water demands unless irrigation efficiency is improved, measures of water saving and loss are implemented and other sources of water and economic options are explored.

Table 15.4 The annual irrigation requirement rate ($m^3.ha^{-1}$) and total irrigation water demands (million.m^3) for the current, medium- and long-term scenarios

Country/Sub-basin	*Rate*	*Current*	*Medium term*	*Long term*
Burundi	11,000	0	200	880
Egypt		43,216	48,942	54,668
Nile valley	*13,000*	*43,216*	*45,775*	*48,334*
El-Salam	*13,000*	*0*	*1693*	*3385*
Toshka	*13,000*	*0*	*1474*	*2948*
Ethiopia		152	4190	15,178
Blue Nile	*10,196*	*152*	*2497*	*5523*
Baro-Akobo-Sobat	*13,140*	*0*	*945*	*7055*
Tekeze-Atbara	*13,566*	*0*	*748*	*2600*
Kenya	8500	48	595	1700
Rwanda	12,500	63	625	1937
Sudan		22,425	39,239	50,992
Tekeze-Atbara	*13,776*	*5392*	*5682*	*10,079*
Blue Nile	*9861*	*11,565*	*21,266*	*21,949*
Main Nile	*13,250*	*1720*	*5879*	*10,203*
White Nile	*13,000*	*3749*	*6413*	*8761*
Tanzania	11,000	5	110	330
Uganda	8000	73	640	1976
Total		65,982	94,541	127,661

Implications

The water availability in the Nile River system was found to decrease for the medium-term and long-term scenarios than in the current scenario. The impact of the development interventions on water availability increases along the river course following the direction of flow (Table 15.5). For both medium- and long-term scenarios, the inflows to Lake Victoria and Lake Nasser are expected to decrease. During the future scenarios, the river flow from Lake Victoria to the Sudd wetland are not significantly affected since more water is released from the equatorial lakes to satisfy the downstream irrigation demands.

The spatial distribution of the mean annual river flow for the long-term scenario is portrayed in Figure 15.9. Other development scenarios have similar patterns of river flow volume.

Table 15.5 Mean annual flow (km^3) at major nodes in the Nile Basin for current, medium- and long-term scenarios

River junction	*Current*	*Medium term*	*Long term*
Main Nile after Egypt irrigation	28.56	11.83	2.42
Main Nile at Aswan outlet	69.61	53.95	51.70
Main Nile at Aswan inlet	80.62	64.93	54.04
Main Nile after Atbara	82.44	71.35	65.29
Main Nile after Blue Nile	74.46	63.22	58.37
Atbarah at Kilo3	8.57	8.94	8.22
Atbarah after Tekeze inflow	9.21	8.66	8.25
Tekeze at Sudan border	6.56	6.13	5.81
Blue Nile at Khartoum	40.49	31.54	30.82
Blue Nile at Sudan Border	48.20	46.11	46.27
White Nile at Khartoum	33.97	31.68	27.55
White Nile at Malakal	38.76	37.64	35.03
Sobat at outlet	13.66	13.36	11.14
Baro at outlet	9.42	8.98	7.49
Baro before Machar	12.73	12.00	9.61
Bahr El Ghazal at oulet	0.30	0.60	0.31
Bahr El Ghazal before swamp	11.33	11.33	11.33
Bahr El Jebel after Sudd	24.80	23.68	23.58
Bahr El Jebel before Sudd	47.61	44.33	46.95
Kyoga Nile at lake outlet	41.02	39.05	41.35
Victoria Nile at lake outlet	40.23	38.84	41.26
Inflow to Lake Victoria	22.87	21.97	19.89

All irrigation water demands are satisfied for the current (baseline) scenarios as expected. However, the irrigation demands for the medium-term and long-term scenarios are not fully met. Most of the unmet irrigation demands could be satisfied by improving irrigation efficiency, saving water through alternative storage strategies and implementing carryover storages on seasonal tributaries and sub-basins.

In summary, an integrated basin-wide simulation of the large-scale water development and management interventions in the Nile Basin revealed that the Nile flow would not meet the irrigation water demands for long-term development scenarios, and somewhat short for medium-term scenario, taking 84.5 billion m^3 as the benchmark for average water availability. The parts of the basin that have pronounced seasonal flows, the Blue Nile and the Tekeze-Atbara sub-basins, are the most affected regions in terms of meeting irrigation demands.

The water availability in the Nile River system was found to significantly decrease for the medium- and long-term development scenarios. The impact of large-scale interventions on the river flow increases along the river course in the direction of flow. This pattern of future water availability could be explained by higher water demands in the downstream part of the basin.

The impact of the large-scale water management interventions on the water availability and irrigation schemes could be mitigated by adopting interventions in water-saving and water-demand management. The current irrigation water requirement is very high. In order to meet future challenges, the following recommendations can be made:

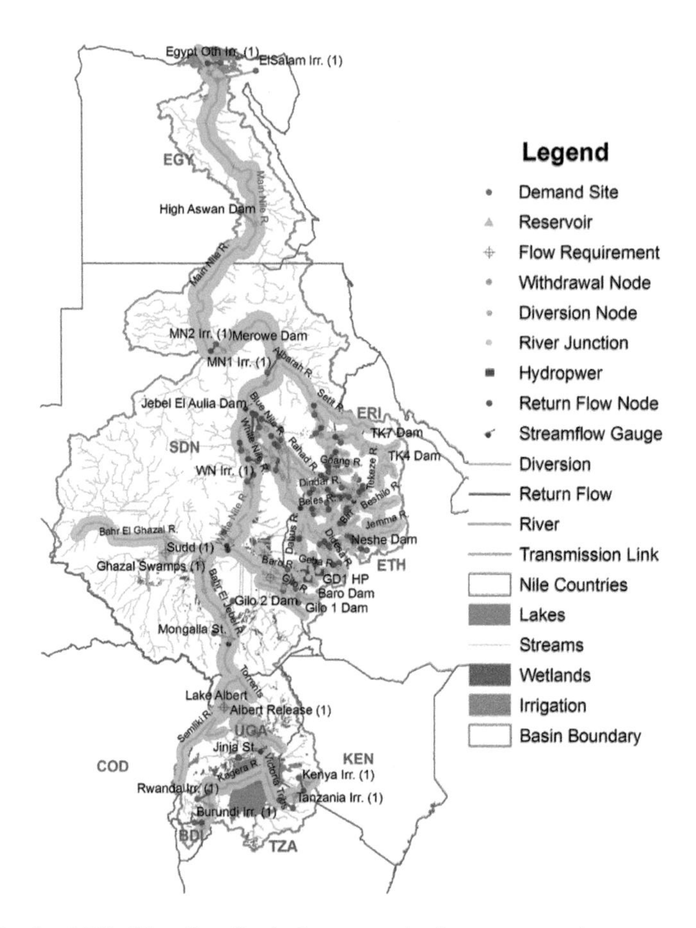

Figure 15.9 Simulated Nile River flow for the long-term development scenario

- Reservoirs developed for hydropower and irrigation with carryover storage capacity could provide more reliable water for the planned irrigation schemes. This demands integrated management of reservoirs as one unit and placing new storage schemes in the highland areas, where higher storage per surface area and less evaporation are attained.
- Irrigation demand could be substantially reduced by improving the efficiency of irrigation systems. Most of the current irrigation efficiency is assumed to be about 50 per cent, and if the efficiency could be increased to 80 per cent, over 40 billion m^3 of water can be saved in the long-term scenario, which can nearly offset the deficit even in the long-term scenario.
- Water productivity should be improved by shifting water from the economic sector that uses more water per unit production to that which uses less water (more value per unit of water). For example, the water used for cooling thermal energy plants could be used for other productive systems by importing hydropower energy from other riparian states, even at more competitive costs.
- Reduce non-consumptive water losses through efficient reservoir operation and irrigation water management; this could also improve water availability in the basin.

- Manage occurrences of high system losses due to evaporation and seepage, and implement water storage in less-evaporative areas.
- Explore alternative sources of water such as groundwater, which may be lost in the system, without contributing to river flows and/or irrigation demands.
- Manage flooding regime in the wetlands.

The above are recommendations, which are amenable to further research on their implications and impact. On the other hand, it was shown in the previous section of this chapter that, upgrading rain-fed systems with the scope of enhancing beneficial use of rainfall can also contribute significantly to meet the food production and demand in the basin.

Conclusion

Water management interventions are complex in river system, and these range from what we undertake at household or micro watershed level to the national, regional and basin scales to improve water access for productive, consumptive and environmental purposes.

We have analysed options of small-scale agricultural interventions focusing on water control in rain-fed systems, small-scale irrigation technologies and suites, their productivity and poverty reduction impacts, and large-scale interventions with respect to meeting future water needs and water availability.

Other types of interventions such as those related to policy, institutions and benefit-sharing are discussed in other relevant chapters of this book. Future intervention analysis work can link hydronomic zones covered in Chapter 4 with interventions, so that proposed interventions take into consideration the various biophysical factors and resources availability. The hydronomic zoning combined with production system zoning provides numerous options that have potential within the Nile, but need to be tailored to site-specific needs in terms of technologies choices and scales.

The poverty analysis pointed to the widespread rural poverty. It also showed that access to water, productivity gains and actions to reduce vulnerability would help reduce poverty. This shows the clear role of water management interventions. The sections on water availability (related to Chapters 4 and 5) and the above WEAP modelling results demonstrate that there is a certain scope for large-scale irrigation development, but that there is ample water (as rainfall) in rain-fed systems that can be managed. Where poverty is high, water productivity is low. Basically, the main message in poverty reduction is clear and simple – there is ample work that needs to be done to improve water access and water productivity to reduce poverty. In a sense, nearly all rural water actions within the basin have poverty implications (except in Egypt where other actions outside agriculture probably have more impact on poverty reduction than in agriculture). The real work is identifying where and how to make these interventions.

Our key recommendation is to transform rain-fed systems by focusing on water access for agriculture, and good agricultural practices. In the small-scale and smallholder interventions, we have developed generic and comprehensive lists of AWM interventions that are most common in the basin, which can enhance agricultural water access in rain-fed, small-scale irrigated and livestock production systems. The generic tabular matrix of Table 15.1 can help with identification of AWM interventions for water control, lifting, conveyance and applications customized per sources of water as rainfall, surface water and groundwater (including reuse and drainage). In addition to AWM technologies, other factors related to economic, policy, institutions, social factor, environment and health factors as well as operation and maintenance influence the success of use of AWM technologies. Furthermore a combination of interventions beyond

AWM techniques creates the expected optimal impact on productivity. Supported by experimental evidence and modelling, it was shown that productivity can be gained up to threefold from a single harvest by integration of AWM, soil fertility and improved seed.

In relation to large-scale interventions, the whole Nile Basin was modelled as one integrated system, and current, medium- and long-term scenarios were analysed considering irrigation, hydropower, environment and wetlands. While the irrigation, environment and wetland requirements are sensitive, the hydropower demand, which is a non-consumptive use, was taken as unimportant in affecting the water availability in the basin. A thorough study of the plans of the countries reveals that planned irrigation in various countries is 10.6 million ha, compared with the current total of 5.5 million ha. With the current level of water application, absence of reservoir management and irrigation efficiency the total water withdrawal requirement in the long term would be 127 billion m^3, far beyond the 84.5 billion or 88.4 billion m^3 of available water (see Chapter 5). While there is scope for some irrigation expansion, in order to come close to future plans, mitigation measures are required that include improvements in water productivity, increase in the storage capacity upstream to reduce evaporation in the downstream storage, enhanced carryover storage and implementation of demand management and water saving practices. Countries should also consider which priority areas of investment should be taken on board and work together to achieve optimal benefits from the available common resource.

All the above are first-time baseline results that point to areas of further research and analysis. Research detailing specific interventions per hydronomic zone, further refinement of the small-scale interventions and analysis per agro-ecological and spatial area, more in-depth analysis of impacts of interventions on poverty alleviation, analysis of suggested options to balance future demand and water balance – all these deserve further investigation. While there is scope for such strategic research, there is an even more pressing need for immediate implementation of already identified efficient interventions.

References

Anchala, C., Deressa, A., Dejene, S., Beyene, F., Efa, N., Gebru, B., Teshome, A. and Tesfaye, M. (2001) Research center based maize technology transfer efforts and achievements, in *Enhancing the Contribution of Maize to Food Security in Ethiopia. Proceedings of the Second National Maize Workshop conducted in Ethiopia, 12–16 November 2001*, M. Nigussie, D. Tanner, and S. Twumasi-Afriyie (eds), Addis Ababa, Ethiopia, http://apps.cimmyt.org/english/docs/proceedings/fse/covertoOpening.pdf, accessed 28 April 2012.

Anderson, I. M. and Burton, M. (2009) *Best Practices and Guidelines for Water Harvesting and Community Based (Small Scale) Irrigation in the Nile Basin, Part I to III*, unpublished report, Nile Basin Initiative, Entebbe, Uganda.

Bashar, K. E. and Mustafa, M. O. (2009) Water balance assessment of the Roseires Reservoir, in *Improved Water and Land Management in the Ethiopian Highlands: Its Impact on Downstream Stakeholders Dependent on the Blue Nile, Intermediate Results Dissemination Workshop Held at the International Livestock Research Institute (ILRI), Addis Ababa, Ethiopia, 5–6 February 2009*, Awulachew, S. B., Erkossa, T., Smakhtin, V. and Fernando, A. (eds), International Water Management Institute (IWMI), Colombo, Sri Lanka, pp38–49.

BCEOM (Bureau Central d'Etudes pour les Equipements d'Outre-Mer) (1998) *Abbay River Basin Integrated Development Master Plan Project, Phase 2, Section II, Volume III – Water Resources, Part 1&2 – Climatology and Hydrology*, Ministry of Water Resources, Addis Ababa, Ethiopia.

ENTRO (Eastern Nile Technical Regional Office) (2007) *One System Inventory (OSI) for the Joint Multi-Purpose Projects*, Eastern Nile Technical Regional Office, Addis Ababa, Ethiopia.

Erkossa, T., Bekele, S., Hagos, F. (2009) Characterization and productivity assessment of the farming systems in the upper part of the Nile Basin, *Ethiopian Journal of Natural Resources*, 11, 2, 149–167.

FAO (Food and Agriculture Organization of the United Nations) (2000) *Irrigation potential in Africa: A Basin Approach*, FAO, Rome, Italy.

Ibrahim, Y. A., Elnil, M. S. R. and Ahmed, A. A. (2009) Improving water management practices in the Rahad Scheme, in *Improved Water and Land Management in the Ethiopian Highlands: Its Impact on Downstream Stakeholders Dependent on the Blue Nile*, Intermediate Results Dissemination Workshop Held at the International Livestock Research Institute (ILRI), Addis Ababa, Ethiopia, 5–6 February, Awulachew, S. B., Erkossa, T., Smakhtin, V. and Fernando, A. (eds), International Water Management Institute (IWMI), Colombo, Sri Lanka, pp50–69.

McCartney, M. P., Alemayehu, T., Awulachew, S. B. and Seleshi, Y. (2009) *Evaluation of Current and Future Water Resource Development in the Blue Nile*, IWMI Research Report 134, International Water Management Institute (IWMI), Colombo, Sri Lanka.

MoFED (Ministry of Finance and Economic Development) (2006) *Ethiopia: Building on progress, A Plan for Accelerated and Sustained Development to End Poverty (PASDEP)*, (2005/06-2009/10), Volume I, September, MoFED, Addis Ababa, Ethiopia.

Molden, D. (ed.) (2007) *Water for Food, Water for Life: A Comprehensive Assessment of Water Management in Agriculture*, Earthscan, London, pp279–310.

Molden, D., Awulachew, S. B., Conniff, K., Rebelo, L.-M., Mohamed, Y., Peden, D., Kinyangi, J., van Breugel, P., Mukherji, A., Cascão, A., Notenbaert, A., Demissie, S. S., Neguid, M.A. and El Naggar, G. (2010) *Nile Basin Focal Project: Synthesis Report*, CGIAR Challenge Program on Water and Food, Colombo, Sri Lanka.

Namara, R. E., Hussain, I., Bossio, D. and Verma, S. (2007) Innovative land and water management approaches in Asia: productivity impacts, adoption prospects and poverty outreach, *Irrigation and Drainage Journal*, 6, 335–348.

NEDECO (Netherlands Engineering Consultants) (1998) *Tekeze River Basin Integrated Development Master Plan Project, Sectorial Reports – Water Resources, Volumes VI, VII and X – Climatology, Hydrology, and Dams, Reservoirs, Hydropower and Irrigation Development*, Ministry of Water Resources, Addis Ababa, Ethiopia.

SEI (Stockholm Environment Institute) (2007) *WEAP: Water Evaluation and Planning System – User Guide*, Stockholm Environment Institute, Boston, MA.

TAMS Consulting (1997) *Baro-Akobo River Basin Integrated Development Master Plan Study, Volume V – Prefeasibility Studies, Annex 1 – Water Resources, Part 1 and 2 – Climatology and Hydrology*, Ministry of Water Resources, Addis Ababa, Ethiopia.

Wubet, F., Awulachew, S. B. and Moges, S. A. (2009) Analysis of water use on a large river basin using MIKE BASIN Model: a case study of the Abbay River Basin, Ethiopia, in *Improved Water and Land Management in the Ethiopian Highlands: Its Impact on Downstream Stakeholders Dependent on the Blue Nile*, Intermediate Results Dissemination Workshop Held at the International Livestock Research Institute (ILRI), Addis Ababa, Ethiopia, 5–6 February 2009, Awulachew, S. B., Erkossa, T., Smakhtin, V. and Fernando, A. (eds), International Water Management Institute (IWMI), Colombo, Sri Lanka, pp70–77..

Yao, H. and Georgakakos, A. (2003) *Nile Decision Support Tool (Nile DST): River Simulation and Management*, Georgia Water Resources institute, Atlanta, GA.

INDEX

9781032921501